C0-AYS-033

# Formation and Differentiation of Early Embryonic Mesoderm

# NATO ASI Series

**Advanced Science Institutes Series**

*A series presenting the results of activities sponsored by the NATO Science Committee, which aims at the dissemination of advanced scientific and technological knowledge, with a view to strengthening links between scientific communities.*

The series is published by an international board of publishers in conjunction with the NATO Scientific Affairs Division

| | | |
|---|---|---|
| A | **Life Sciences** | Plenum Publishing Corporation |
| B | **Physics** | New York and London |
| C | **Mathematical and Physical Sciences** | Kluwer Academic Publishers |
| D | **Behavioral and Social Sciences** | Dordrecht, Boston, and London |
| E | **Applied Sciences** | |
| F | **Computer and Systems Sciences** | Springer-Verlag |
| G | **Ecological Sciences** | Berlin, Heidelberg, New York, London, |
| H | **Cell Biology** | Paris, Tokyo, Hong Kong, and Barcelona |
| I | **Global Environmental Change** | |

***Recent Volumes in this Series***

*Volume 225*—Computational Aspects of the Study of Biological Macromolecules by Nuclear Magnetic Resonance Spectroscopy
edited by Jeffrey C. Hoch, Flemming M. Poulsen, and Christina Redfield

*Volume 226*—Regulation of Chloroplast Biogenesis
edited by Joan H. Argyroudi-Akoyunoglou

*Volume 227*—Angiogenesis in Health and Disease
edited by Michael E. Maragoudakis, Pietro Gullino, and Peter I. Lelkes

*Volume 228*—Playback and Studies of Animal Communication
edited by Peter K. McGregor

*Volume 229*—Asthma Treatment—A Multidisciplinary Approach
edited by D. Olivieri, P. J. Barnes, S. S. Hurd, and G. C. Folco

*Volume 230*—Biological Control of Plant Diseases: Progress and Challenges for the Future
edited by E. C. Tjamos, G. C. Papavizas, and R. J. Cook

*Volume 231*—Formation and Differentiation of Early Embryonic Mesoderm
edited by Ruth Bellairs, Esmond J. Sanders, and James W. Lash

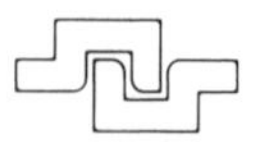

*Series A: Life Sciences*

# Formation and Differentiation of Early Embryonic Mesoderm

Edited by

Ruth Bellairs

University College London
London, United Kingdom

Esmond J. Sanders

University of Alberta
Edmonton, Alberta, Canada

and

James W. Lash

University of Pennsylvania
Philadelphia, Pennsylvania

Plenum Press
New York and London
Published in cooperation with NATO Scientific Affairs Division

Proceedings of a NATO Advanced Research Workshop
on Formation and Differentiation of Early Embryonic Mesoderm,
held October 25-27, 1991,
in Banff, Alberta, Canada

**NATO-PCO-DATA BASE**

The electronic index to the NATO ASI Series provides full bibliographical references (with keywords and/or abstracts) to more than 30,000 contributions from international scientists published in all sections of the NATO ASI Series. Access to the NATO-PCO-DATA BASE is possible in two ways:

—via online FILE 128 (NATO-PCO-DATA BASE) hosted by ESRIN, Via Galileo Galilei, I-00044 Frascati, Italy.

—via CD-ROM "NATO-PCO-DATA BASE" with user-friendly retrieval software in English, French, and German (© WTV GmbH and DATAWARE Technologies, Inc. 1989)

The CD-ROM can be ordered through any member of the Board of Publishers or through NATO-PCO, Overijse, Belgium.

**Library of Congress Cataloging-in-Publication Data**

Formation and differentiation of early embryonic mesoderm / edited by Ruth Bellairs, Esmond J. Sanders, and James W. Lash.
p. cm. -- (NATO ASI series. Series A, Life sciences ; v. 231)
"Proceedings of a NATO Advanced Research Workshop on Formation and Differentiation of Early Embryonic Mesoderm, held October 25-27, 1991, in Banff, Alberta, Canada"--T.p. verso.
"Published in cooperation with NATO Scientific Affairs Division."
Includes bibliographical references and index.
ISBN 0-306-44236-1
1. Mesoderm--Congresses. 2. Cell differentiation--Congresses. I. Bellairs, Ruth. II. Sanders, Esmond J. III. Lash, James W., 1929- . IV. North Atlantic Treaty Organization. Scientific Affairs Division. V. NATO Advanced Research Workshop on Formation and Differentiation of Early Embryonic Mesoderm (1991 : Banff, Alta.) VI. Series.
[DNLM: 1. Cell Differentiation--physiology--congresses. 2. Chick Embryo--growth & development--congresses. 3. Embryo--growth & development--congresses. 4. Mesoderm--physiology--congresses. WQ 205 F724 1991]
QL951.F67 1992
596'.0333--dc20
DNLM/DLC
for Library of Congress 92-19796
CIP

ISBN 0-306-44236-1




A Division of Plenum Publishing Corporation
233 Spring Street, New York, N.Y. 10013



Printed in the United States of America

# PREFACE

Mesoderm is a key tissue in early development. It is involved in the differentiation of almost every organ in the body, not merely as a structural component, but as an active participant in the establishment of diverse cell types. All mesoderm is derived from ectoderm. Its appearance signals the start of a significant new phase in the development of the embryo. At this time all three germ layers are now present and myriad sequences of cell and tissue interactions begin to occur which will eventually give rise to the entire embryo.

The control of the growth and differentiation of the mesoderm is critical for the production of a normal individual. Indeed, disturbance of the patterning of the mesoderm or of its interaction with other tissues plays a critical part in the formation of most congenital anomalies. The main focus of this book is therefore on the establishment, divergence and specialisation of mesodermal derivatives.

The central role of the mesoderm in development has long been appreciated and a wide literature exists on its activity in certain specialised situations. Recently, however, an impetus to its study has been provided by new approaches opened up through biotechnological advances. Many of these advances are reflected in the reports in this volume. Scientists from various disciplines have become drawn to mesodermal tissues, and this volume may help them find a framework within which their work will fit.

The formation and differentiation of early embryonic mesoderm was the theme for a NATO Advanced Research Workshop which was held in Banff, Alberta, Canada, 25-29 October 1991. The Canadian Rockies provided an ideal setting for stimulating discussions and friendly exchange of views. This book is the product of that meeting and we are particularly grateful to NATO for providing us with this opportunity.

Ruth Bellairs
Esmond J. Sanders
James W. Lash

CONTENTS

# MESOBLAST ANLAGE FIELDS IN THE UPPER LAYER OF THE CHICKEN BLASTODERM AT STAGE 5V

Hilde Bortier[1] and Lucien C.A. Vakaet[2]

[1] Senior Research Assistant to the National Fund for Scientific Research (Belgium)
[1,2] Laboratory of Embryology, University of Gent
Godshuizenlaan, 4
B-9000 Gent, Belgium

## INTRODUCTION

The intraembryonic mesoblast of the chicken embryo originates from the upper layer by ingression through the primitive streak during gastrulation. There is no agreement about the disposition of the Anlage fields of the mesoblast, as appears from the reviews by Rudnick (1944) and Waddington (1952) and the maps of Rosenquist (1966) and Vakaet (1984).

One of the reasons for this inconsistency is the difference in developmental stage of the blastoderms that were used for investigation. The most used table, by Hamburger and Hamilton (1951) is adequate for older stages, but between stages 3HH and 4HH there is a gap of about twelve hours. In that period Vakaet (1970) distinguishes stages 4V, 5V, and 6V, stage 7V corresponding to stage 4HH.

Between stages 3HH and 4HH, avian gastrulation turns from polyingression of individual cells to ingression through the primitive streak.
- At stage 4V the streak is rod-like and built up by an upper layer underlain by prospective endoblast cells (Fig. 1A). These cells accumulated by polyingression and convergence in the middle of the so-called sickle of Koller.
- The beginning of stage 5V is marked by the appearance of a groove, rostrally in the upper layer of the primitive streak. This grooving coincides with the start of the insertion of the definitive endoblast cells into the layer of primary hypoblast (Fig. 1B). Concomitantly, the mesoblast cells in the upper layer overlying the streak begin to ingress. Ingression is a combined process in which upper layer cells deepithelialize and subsequently migrate away from the streak, between the upper and the deep layers.

*Formation and Differentiation of Early Embryonic Mesoderm*
Edited by R. Bellairs *et al.*, Plenum Press, New York, 1992

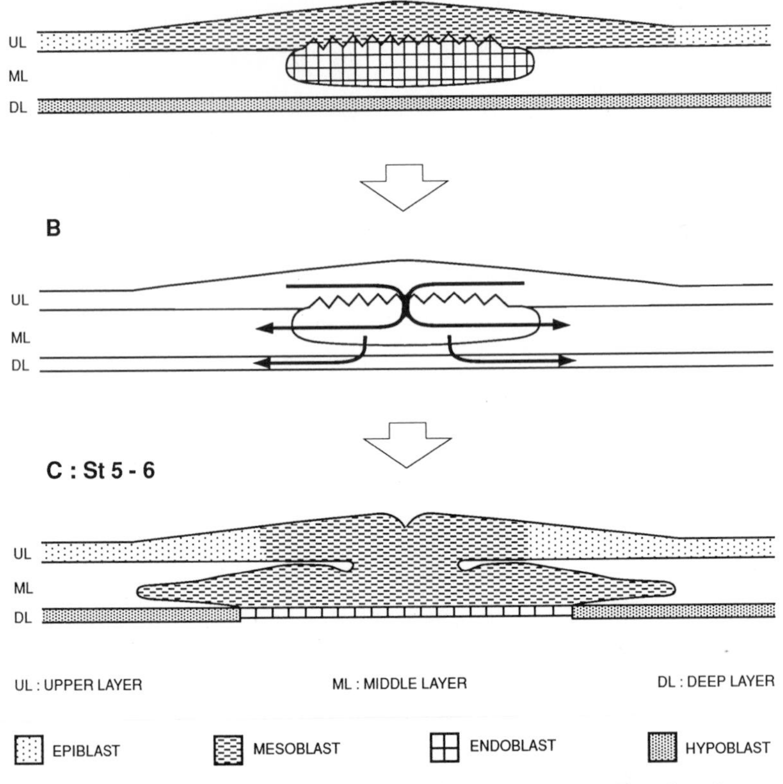


Fig. 1. Transition from stage 4V to stage 5V-6V (open arrows).
The definitive endoblast is first inserted into the hypoblast. This is followed by the ingression of the mesoblast, close to the midline (arrows).

- At the beginning of stage 6V, Hensen's node and the primitive groove are fully developed. During this stage, ingression culminates (Fig. 1C).

We have studied the Anlage fields of the intraembryonic mesoblast in the upper layer in blastoderms at the beginning of stage 5V, when ingression has not yet started. We used time-lapse videography to follow chick-quail xenografts (Le Douarin, 1973). This has allowed us to construct a fate map of the boundary of the intraembryonic mesoblast and the disposition of its axial, paraxial and lateral parts.

MATERIALS AND METHODS

Chicken eggs (White Rock, from the Rijksstation voor

Pluimveeteelt, Merelbeke, B) and quail eggs (from our laboratory stock) were incubated at 38°C for 15 hrs to obtain stage 5V blastoderms. They were cultured by New's (1955) technique, except that the substrate used was a mixture of 25 ml thin egg white and a gel made of 150 mg Bacto-agar Difco (Detroit, Michigan) in 25 ml Ringer's solution, instead of pure thin egg white. This semisolid medium allowed microsurgery and further culturing on the same substratum.

For microsurgery we used a Pasteur pipette with a diameter of 0.20 mm to 0.25 mm at its tip. The experiments consisted in punching out and discarding a circular fragment of the upper layer. It was replaced with an isotopic and isopolar piece of quail upper layer that was punched out with the same pipette. Care was taken not to graft middle layer or deep layer cells (Bortier and Vakaet, 1987a; Schoenwolf et al., 1989).

The grafts were followed with Polaroid photography and time-lapse videography (Bortier and Vakaet, 1987b). As the quail cells contain less yolk than the chicken cells, some grafts could be followed in vivo until they ingressed through the primitive streak. We reincubated the chimeras for 24 hours, which is long enough for the grafted cells to become incorporated in the tissues of the chimeras.

All chimeras were studied histologically. The culturing was interrupted by fixing the blastoderms in a mixture of absolute alcohol, formaldehyde 4%, acetic acid (75:20:5, v:v:v). After paraffin embedding they were stained after Feulgen and Rossenbeck (1924).

## RESULTS

In the results we describe the boundary and the internal disposition of the intraembryonic mesoblast Anlage fields lateral to the rostral half of the primitive streak. They are based on the observation of 43 chimeras that developed normally by in ovo standards. The results are summarized in Fig. 2.

With videography we observed that the grafts migrated towards the streak from rostrolateral to mediocaudal. The grafts formed clearly delineated patches after their integration in the upper layer. The patches elongated in the direction of their migration towards the primitive streak. On reaching the streak, they extended along its length. This strip narrowed and faded away.

Histological examination confirmed that the graft cells kept together and did not intermingle with host cells as long as the graft was in the upper layer. They deepithelialized close to the midline of the streak. In sections where graft cells were found in the upper layer, rows of single graft cells were found orientated ventrolaterally. These rows might consist of 6-7 cells. They were not visible in sections rostral to the site of ingression. There the graft cells were found scattered

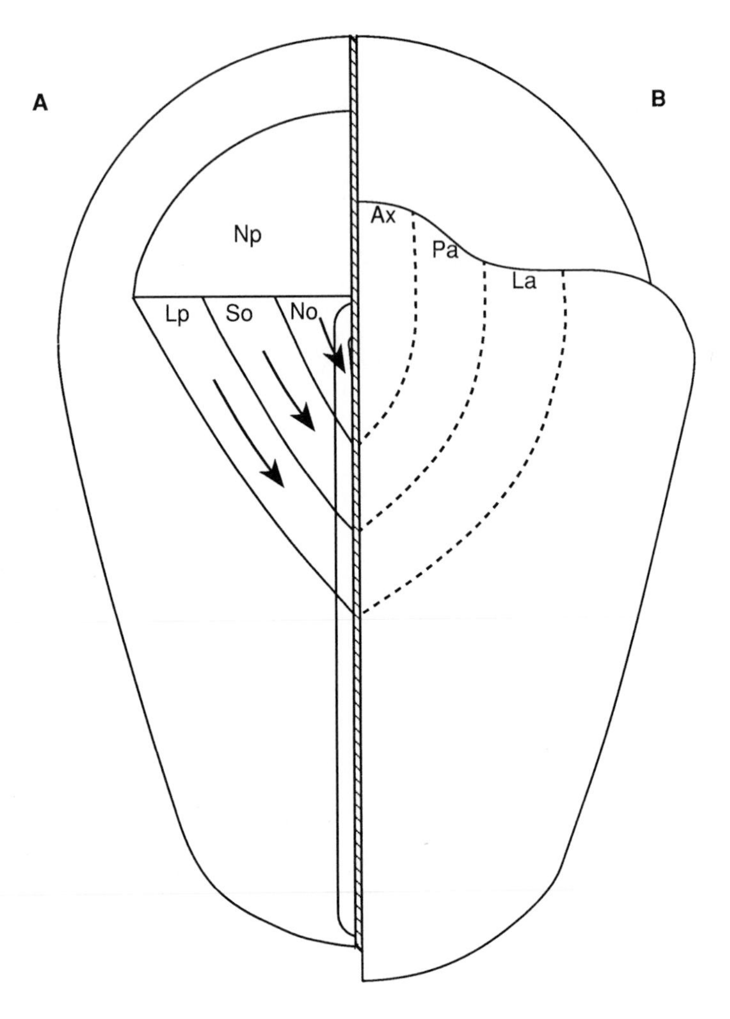


Fig. 2. Dorsal views :
A : Stage 5V : disposition of the Anlage fields of the intraembryonic mesoblast. A transverse border of these fields with the neural plate (Np) is drawn. The arrows show the movements within the fields of future notochord (No), somites (So) and lateral plate (Lp)
B : Stage 6V : disposition of the axial (Ax), paraxial (Pa) and lateral (La) mesoblast after ingression. Note the rostral bending of these fields.

between host cells on both sides of the median structures. In the trunk region we found graft cells in bilateral paraxial and lateral tissues issued from unilateral grafts. The graft side often contained more graft cells than the contralateral side.

Combining videography and histology, we found a border between mesoblast and neural plate along a transverse line just rostral to the node (Fig. 2A). In grafts rostral to this border only neural tissue was found; caudal to this border only mesoblast cells were found. Grafts on the border yielded neural tissue as well as mesoblast cells.

The disposition of the mesoblast Anlage fields caudal to this border was similar in all chimeras :

- In a medio-rostral triangle we found future notochord. The rostral side of this triangle was about one third of the border wide; its medial side extended over about one third of the rostral half of the primitive streak.
- Bilaterally flanking the notochord area we found a strip of future somitic tissue. This ribbon-like area occupied rostrally also a third of the border while its length on the midline occupied the second third of the rostral half of the primitive streak.
- A third strip flanked the somite area on both sides. It yielded nephrotome and lateral plate cells. Its periphery was bordered with intraembryonic epiblast cells.

The upper layer along the caudal half of the primitive streak formed extraembryonic mesoblast of the area vasculosa: blood vessels and blood cells.

## DISCUSSION

The Anlage field of the intraembryonic mesoblast at stage 5V has the shape of an inverted equilateral triangle (Fig. 2A). Its base lies transversely, just rostral to the node at the border of the mesoblast field and the neural plate. Every subdivision of the trunk mesoblast contributes to the mesoblast side of this border.

From our results we deduced that during gastrulation the border between the neural plate and the mesoblast fields describes a movement similar to that of scissor blades that would pivot in Hensen's node. This movement is accompanied by a narrowing of the mesoblast area and by ingression of the future mesoblast cells through the primitive streak.

The deepithelialization close to the midline of the ingressing mesoblast cells explains how cells may cross the midline to join contralateral tissues. The presence of bilateral mesoblast is a histological proof of positioning through gastrular ingression.

We have observed that rows of up to seven cells are visible in transverse sections at the level of the fossa rhomboidalis where graft cells are still at the surface of the groove. They are cells with a similar destination migrating along the same pathway. Graft cells are not seen in rows rostrally to the level of ingression. This is probably due to bending of the rows in a more rostral direction. The bending of the axial, paraxial and lateral mesoblast is brought about by the regression of the anterior half of the primitive streak from St 6V on (Vakaet, 1960). It imposes on the mesoblast fields the shape represented in Fig. 2B.

The destination of grafts in the mesoblast Anlage fields is correlated with topographical features :
- Grafts implanted farther away from the primitive streak will be found in more lateral structures after ingression. In Fig. 2A a graft inserted in the lateral plate Anlage field will be found in a more lateral structure in Fig. 2B than a graft inserted in the somite Anlage field which will be found in a paraxial position.
- Within the Anlage fields, caudal grafts will ingress first. Due to the concomitant regression of the primitive streak, cells of more caudal grafts will be found in a more rostral position after ingression than more rostral grafts that ingress later and therefore settle in a more caudal position.

ACKNOWLEDGEMENTS

We gratefully acknowledge the support of the N.F.S.R. (Belgium) grant Nr 3.9001.87 during this work.

REFERENCES

Bortier, H., and Vakaet, L., 1987a, Wound healing in the upper layer of the chicken blastoderm. Cell Diff., 20/suppl., p. 114S.

Bortier, H., and Vakaet, L., 1987b, Videomicrography in the study of morphogenetic movements in the early chick blastoderm. Cell Diff., 20/suppl., p. 114S.

Feulgen, R., and Rossenbeck, H., 1924, Mikroskopisch-chemischer Nachweis einer Nucleinsäure vom Typus der Thymonucleinsäure und die darauf beruhende elektive Färbung von Zellkernen in mikroskopischen Präparaten. Hoppe Seylers Z. Physiol. Chem., 135:203-252.

Hamburger, V., and Hamilton, H.L., 1951, A series of normal stages in the development of the chick embryo. J. Morphol., 88:49-92.

Le Douarin, N.M., 1973, A Feulgen positive nucleolus. Exp. Cell Res., 77:459-468.

New, D.A.T., 1955, A new technique for the cultivation of the chick embryo in vitro. J. Embryol. exp. Morphol., 3/4:326-331.

Rosenquist, G.C., 1966, A radioautographic study of labeled grafts in the chick blastoderm. Development from primitive-streak stages to stage 12. Contrib. Embryol., 38:71-110.

Rudnick, D., 1944, Early history and mechanics of the chick blastoderm. Quart. Rev. Biol., 19:187-212.

Schoenwolf, G.G., Bortier, H., and Vakaet, L., 1989, Fate mapping the avian neural plate with quail/chick chimeras : origin of prospective median wedge cells. J. Exp. Zool., 249:272-278.

Vakaet, L., 1960, Quelques précisions sur la cinématique de la ligne primitive chez le poulet. J. Embryol. Exp. Morphol., 8/3:321-326.

Vakaet, L., 1970, Cinephotomicrographic investigations of gastrulation in the chick blastoderm. Arch. Biol., 81/3:387-426.

Vakaet, L., 1984, Early development of birds. In : "Chimeras in Developmental Biology," N. Le Douarin & A Mc. Laren, eds., Academic Press, London,:71-88.

Waddington, C.H., 1952, The epigenetics of birds, Univ. Press, Cambridge.

# THE AVIAN MARGINAL ZONE AND ITS ROLE IN EARLY DEVELOPMENT

Hefzibah Eyal-Giladi

Department of Zoology, Hebrew University of Jerusalem

## WHAT IS THE AVIAN MARGINAL ZONE?

The term avian marginal zone is descriptive but in a way misleading, as it automatically reminds us of the same term used for a nonhomologous structure in amphibian development. In the amphibian blastula the marginal zone is a belt whose cells are already determined to invaginate during gastrulation and to form the different mesodermal components. In the chick the marginal zone (MZ) is also a belt-like region at the periphery of the area pellucida, but its contours start to show up at the posterior end at a stage earlier than the blastula (stage X E.G&K: Eyal-Giladi and Kochav, 1976), and become completely delineated at stage XIII E.G&K which is comparable to the amphibian blastula (Waddington, 1933; Eyal-Giladi, 1984, 1991). A stage XIII blastoderm can be divided according to morphological criteria into two components:

1) The central, blastular part senso stricto, which is double layered. The upper layer - the epiblast - can be compared to the amphibian animal hemisphere, while the lower layer - the hypoblast - can be compared to the amphibian vegetal hemisphere. The narrow space separating the two layers is comparable to the blastocoelic cavity.

2) The peripheral ring surrounding the blastular part, includes the marginal belt adjacent to the central part, which in turn is encircled by the area opaca (Fig. 1).

## THE DEVELOPMENTAL POTENCIES OF THE DIFFERENT BLASTODERMAL REGIONS

When a stage XIII central part is separated from the peripheral ring, it will continue its development and form a primitive streak (PS) followed by the formation of an embryo (Azar and Eyal-Giladi, 1979). The double layered central disc should therefore contain the anlagen of all three germ layers. The mesodermal anlage (whatever it is) must be therefore included in this central disc and would be homologous to the amphibian marginal zone belt. However in contrast to amphibia a mesodermal anlage cannot be isolated at stage XIII from the central disc. When the epiblastic and the hypoblastic components of the central blastular part are separated from each other, neither of them is capable of forming axial mesodermal structures. This means that at stage XIII the axial mesodermal anlage which is about to materialize in the epiblast, in the form of a PS, is not yet definitely determined (induced). In the avian blastoderm the determination of axial mesoderm seems therefore to take place within the blastular component relatively later in development and during a much shorter time-span than in amphibians.

Another important conclusion from the above analysis is that in the avians, the stage XIII marginal zone belt is dispensable (can be ablated), does not seem to directly contribute substantially to mesoderm formation and cannot therefore be compared to the amphibian marginal zone. This leaves us with the problem of what the avian marginal zone actually is and whether it has at all any impact on embryonic development.

THE DEVELOPMENTAL POTENCIES OF THE AVIAN MARGINAL ZONE

Azar and Eyal-Giladi (1979) have demonstrated, as mentioned above, that 1) A central double layered stage XIII disc is self-sufficient and can form a PS. 2) When the hypoblastic layer of such a central disc is removed and the isolated epiblastic fragment is grown in culture, a new lower layer is regenerated but no PS is formed in the regenerative double layered blastula-like structure. In addition it was shown that 3) if the

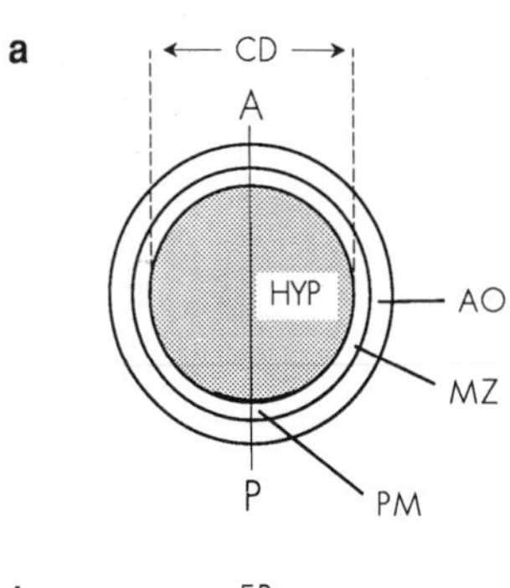


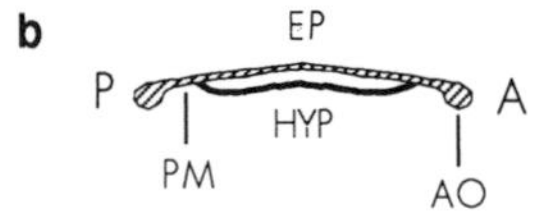


Fig. 1. Scheme of a stage XIII E.G&K blastoderm
a - view of its lower surface facing the yolk.
b - sagittal section of the same blastoderm.

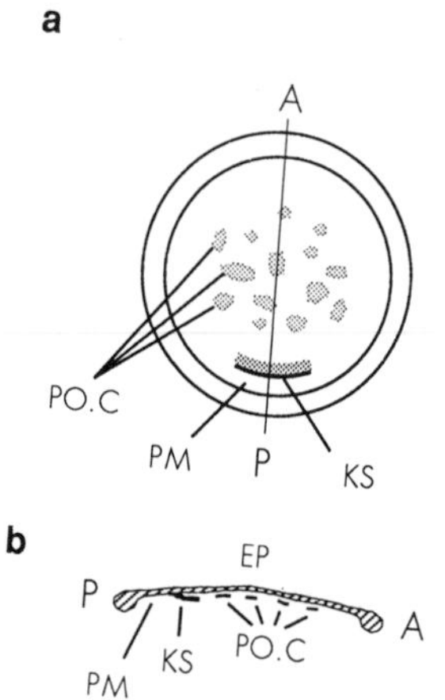


Fig. 2. Scheme of a stage X E.G&K blastoderm.
a - view of lower surface
b - sagittal section

Abbreviations:
A - anterior, AO - area opaca, CD - central disc (including EP & HYP), EP - epiblast, HYP - hypoblast, KS - Koller's sickle, MZ - marginal zone, P - posterior, PM - posterior MZ, PO.C - polyinvaginating cells.

hypoblast is removed but the marginal zone is retained, a new lower layer is regenerated but contrary to (2) a PS is formed in the epiblast of the regenerated blastula. The difference between (2) and (3) is the presence of the marginal zone in (3), which made it clear that the MZ is important for the normal regeneration process and its importance must be by a cellular contribution which it makes to the hypoblast. In (2) a defective hypoblast regenerated in the absence of the MZ, in which the cellular contribution of the latter was lacking, and this resulted in the absence of a PS. If the assumption that such a cellular contribution is required also for normal PS development, one should look for it between stages X-XIII, during the normal growth of the hypoblast.

Waddington (1933) and Azar and Eyal-Giladi (1981) rotated the hypoblast of a stage XIII (Azar and Eyal-Giladi) and older blastoderms (Waddington) by 90° and 180° and were able thereby to change the orientation and position of the developing PS in the epiblast. The above authors therefore concluded that it is the hypoblast which induces the formation of a PS in the epiblast. If we combine all the above information the inevitable conclusion would therefore be, that a cellular contribution of the marginal zone to the primary hypoblast is instrumental in the latter's capacity to induce a PS.

The morphological data (Eyal-Giladi and Kochav, 1976; Kochav et al., 1980) as well as earlier descriptive information (Vakaet, 1962) supported the idea that the primary hypoblast of stage XIII is composed of two different cell populations which are indistinguishable from each other in the final product (a stage XIII hypoblast). One population is comprised of polyinvaginated cells, which descend directly from the epiblast into the hypoblast at many different loci. The other population has a much more orderly appearance and seems to grow anteriorly from a sickle shaped border line (Koller's sickle - KS) separating the posterior section of the MZ (PM) from the central disc of the epiblast (Fig. 2). The two cell populations gradually merge with each other to form the primary hypoblast.

EXPERIMENTAL TRIALS TO ESTABLISH THE ROLE OF THE MARGINAL ZONE IN PS INDUCTION

Our working hypothesis based on the experimental results of Azar and Eyal-Giladi (1979, 1981) included the following assumptions:

1) The polyinvaginated cells are incapable of PS induction, while the cells growing anteriorly from Koller's sickle (KS) are the ones that induce the PS in the epiblast. 2) Taking into account the indications that the MZ is probably indispensable for the inductivity of the hypoblast we assumed that the inductive cells growing anteriorly from Koller's sickle are derived from the marginal zone. 3) In order to ensure the formation of a single PS the inductive cells must enter into a defined area in the hypoblast which would have initially a shape more or less similar to that of the PS. 4) An anteriorly directed growth of cells from the posterior section of the marginal zone (PM) can be ensured only if a circular developmental gradient exists in the MZ with a maximum at the PM region. Such a gradient has already been predicted by Spratt (1963)

Our first experimental approach, was to check more extensively the impact of the marginal zone on PS formation in normal development as related to parameters of time and space. The stages chosen for the following studies were X-XIII, during which the primary hypoblast is newly formed, as opposed to the above experiments of Azar and Eyal-Giladi (1979, 1981) which were performed on stage XIII blastoderms in which normally the impact of the MZ is probably already minimal.

Khaner and Eyal-Giladi (1986) used blastoderms younger than stage XIII, and performed several experimental series involving the rotation of the central disc at 90° to the peripheral belt (MZ + AO). In addition, also transplantation experiments of MZ fragments (PM's and LM's) into positions other than their original ones within the MZ, were performed (Khaner and Eyal-Giladi, 1989; Eyal-Giladi and Khaner, 1989). The aim of the above series was to establish whether the formation of the PS could be influenced by manipulations involving a change in the position of the PM. I have chosen some examples to demonstrate the results on which I would like to base our conclusions. Fig. 3 summarizes three homoplastic experimental series in which PM and LM sections of the MZ were either ablated, transplanted or exchanged. The experiments were done on three different developmental stages, X, XI and XII. In series A an LM fragment was ablated and replaced by a PM fragment from the same blastoderm, which

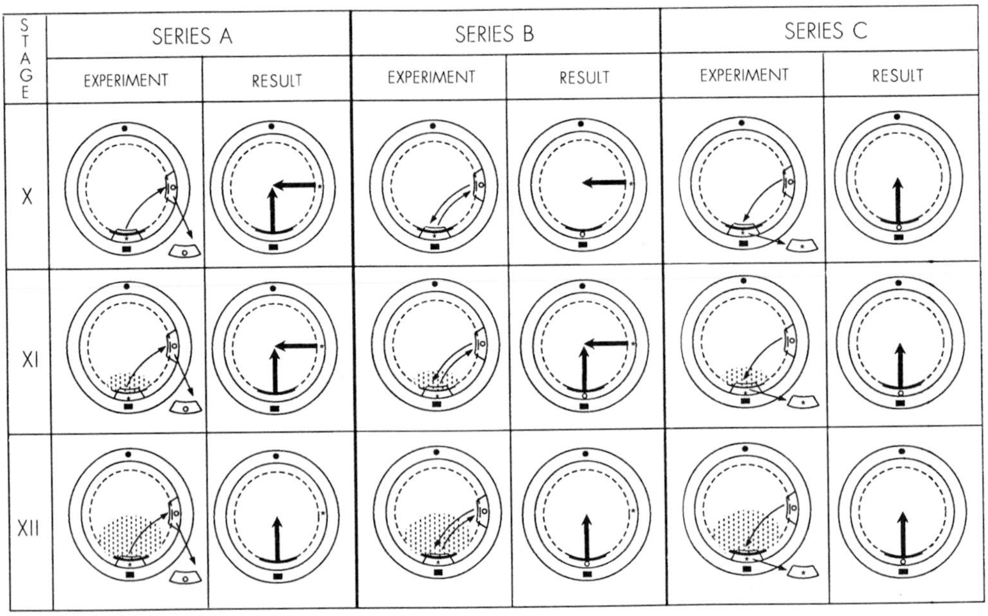

| STAGE | SERIES A | | SERIES B | | SERIES C | |
|---|---|---|---|---|---|---|
| | EXPERIMENT | RESULT | EXPERIMENT | RESULT | EXPERIMENT | RESULT |
| X | | | | | | |
| XI | | | | | | |
| XII | | | | | | |

Fig. 3. Homoplastic transplantations
Series A: The LM was removed and a PM from the same blastoderm was inserted into the lateral gap.
Series B: LM and PM fragments were mutually exchanged.
Series C: The PM was removed and replaced by an LM from the same blastoderm.

was of a similar size. The gap at the posterior side was allowed to heal without further intervention. In series B there was no ablation, but the positions of the PM and LM fragments were interchanged. In series C the PM was ablated and its place was taken by an LM from the same blastoderm, while the gap at the lateral side was allowed to heal. All the fragments in the above series were cut so as to include in addition to the MZ fragment per se, also a small strip central to it, which is part of the CD. The outcome of the manipulations in stage X blastoderms, in all three series, indicates that a stage X PM when transplanted to a lateral position (series A and B) promotes the formation of a lateral PS. When the empty gap at the posterior side (series A) is allowed to heal without further interference an additional PS will develop posteriorly probably as a result of the inductive activity of the MZ cells, at both sides of the closing hole, which are next to the PM in the hierarchy within the MZ belt. When the gap is filled up with an LM (series B), which is a fragment with a low inductive potential, no PS will form at the posterior side. In series A the results of stage XI are similar to those of stage X, while in series B, at stage XI, in addition to the lateral PS also a posterior one develops. At stage XII the outcome of the experiments in both series A and B is similar to that of all three stages of series C, namely only a single PS developed from the posterior side of the blastoderm. The conclusions from the above series were that the posterior region of the marginal zone (PM) is involved in the induction of the PS also in normal development. The effect of the PM was assumed by us to be the consequence of the migration of PS inductive cells from the PM into the primary hypoblast during its progressive growth anteriorly. Series B supports this idea quite well, since when the PM was transplanted laterally at stage X (prior to the beginning of hypoblast growth) and its place was occupied by an LM, no remarkable anteriorly directed migration of inductive cells could have taken place yet into the very rudimentary primordium of the hypoblast, which explains why no posterior PS has been formed. However, at stage XI the number of PM cells that have already migrated anteriorly is probably sufficient to already allow for a PS induction. At stage XII the amount of PM cells that have already migrated anteriorly into the hypoblast is probably larger and quite adequate for the induction of a PS. From the same series (B) one can also conclude that while the inductive potential of the hypoblast is increasing from stage X onwards, the inductive potential of the PM is decreasing. This could be explained either by the depletion of inductive cells from the MZ, or by the gradual loss of inductive potential of the MZ cells, or by a combination of the two.

## QUANTIFICATION OF THE REGIONAL INDUCTIVE POTENTIAL OF THE MARGINAL ZONE AND ITS STAGE DEPENDENCE

Our working hypothesis concerning the inductive potential of the MZ, was therefore the following: The MZ belt is a gradient field of PS-inductive potential with a maximum at the posterior side (PM). The potential at any point of the MZ belt is maximal at stage X and decreases as development progresses. In normal development only a single PS can be formed, anterior to the MZ section which has the maximal inductivity. The other regions of the MZ, whose inductivity is not expressed, are probably suppressed or inhibited by the peak of the field.

The experiments planned to test the above hypotheses were done at two different levels (Eyal-Giladi and Khaner, 1989). To check the gradient within the MZ belt, stage X blastoderms were used. The PM fragment of the host blastoderm was cut out, but left in place to heal as if it was a transplant. An LM fragment was also cut out of the same blastoderm and removed and in its place a PM transplant from another stage X blastoderm was inserted (Fig. 4). The idea was to let the two PMs compete with each other in their ability to initiate PS formation while being located at theoretically unequal positions in the gradient belt, which should affect their performance. The sizes of all the PM fragments, either posteriorly or laterally situated, were variable. Immediately after the operation the blastoderms were photographed to allow an exact measurement of the size of the PM fragments at the time of the operation. After 24 h of incubation the blastoderms were checked for PS formation. The outcome of the above experiments (Eyal-Giladi and Khaner, 1989) very clearly supported the idea of a gradient in the marginal belt. A laterally inserted PM was shown to be able to materialize its potential inductivity, but only if it was 1.3 times larger than the posteriorly inserted PM. In such a case not only the lateral but also the posterior PM expressed their potential inductivity, which resulted in the formation of two PS's in the same blastoderm, at 90° to each other. Only in those cases in which the laterally inserted PM was at least 1.4 times larger than the posterior one, a single lateral PS developed and the posterior PM did not materialize its potencies which indicates again that it was inhibited by a stronger axial center. As far as the gradient is concerned, we concluded that the potency of the field in a lateral section of the belt is about 1.3 times less than that of the posterior section and that is why a compensation is needed in the form of a 1.3 times larger PM.

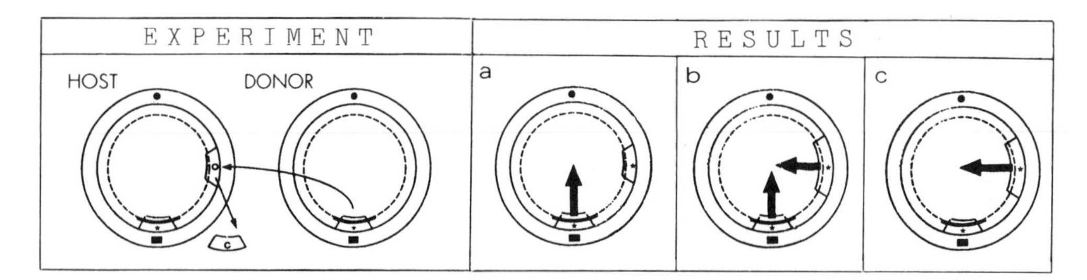


Fig. 4. Experiment: two stage X blastoderms were used. LM fragments (of variable sizes) were ablated from the host blastoderms and replaced by PM fragments from the donor blastoderm. An original PM section of the host was separated by cutting from its surrounding tissue but left in situ.

Results: a - The laterally implanted PM was less than 1.3 times larger than the posterior PM. A single PS developed from the posterior PM.

b - The laterally implanted PM was about 1.3 times larger than the posterior PM. Two PS's developed from each of the PM fragments.

c - The laterally implanted PM was at least 1.4 times larger than the posterior PM. A single PS developed in each blastoderm from the lateral PM.

To quantitatively check the decrease of the potential inductivity of the MZ along the time axis, another set of experiments was designed in which the potential inductivity of PM fragments of different sizes and of different developmental stages was tested. In order to demonstrate the results, I have chosen to compare the potential inductivity of PM fragments, of stages X and XI, transplanted laterally into the MZ of a stage X blastoderm (Fig. 5). In the host blastoderm, the gap resulting from the extirpation of the PM, was filled up with the LM fragment which was removed to make place for the PM transplant.

When the stage X, laterally inserted PM transplant, contained less than 1000 cells, no PS was initiated at the side, while a single posterior PS developed from the implanted LM. When the laterally inserted PM transplant was taken from a stage XI blastoderm, and put into an equal system as above, at least 3000 cells were needed for the initiation of a lateral PS. This meant to us that a PM loses about 2/3 of its inductive potential between stages X and XI.

In the quantitative results of the above experimental series there was also a hint that not only the potential inductivity of the MS is rapidly lost from stage X onwards, but also the ability to inhibit the formation of supernumerary PS's. This conclusion is based on the following observations: when a stage X PM is large enough (1400 cells or more) to inititate PS development in a lateral location, it will inhibit the formation in the same blastoderm of a PS from the posterior site into which the LM fragment was transplanted. However, when the laterally implanted PM was of stage XI, and large enough for that stage to initiate a lateral PS (3000-5000 cells), it never inhibited the formation of a PS from the posteriorly implanted stage X LM and in all the studied cases two PS's were found in the blastoderm, one lateral and one posterior.

THE INHIBITORY EFFECT OF A "STRONG" MARGINAL ZONE SECTION ON THE ABILITY OF "WEAKER" MARGINAL ZONE SECTIONS TO PROMOTE THE FORMATION OF AN AXIS

In the analysis of several of the above described experiments we

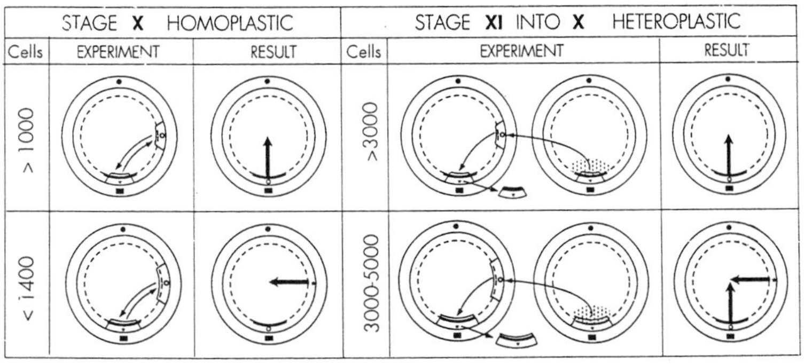

Fig. 5. A quantitative assessment of decrease in the Posterior margin's potential inductivity between stages X and XI.

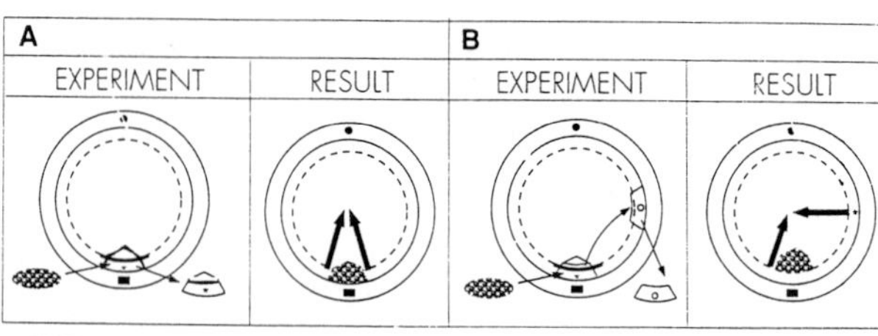


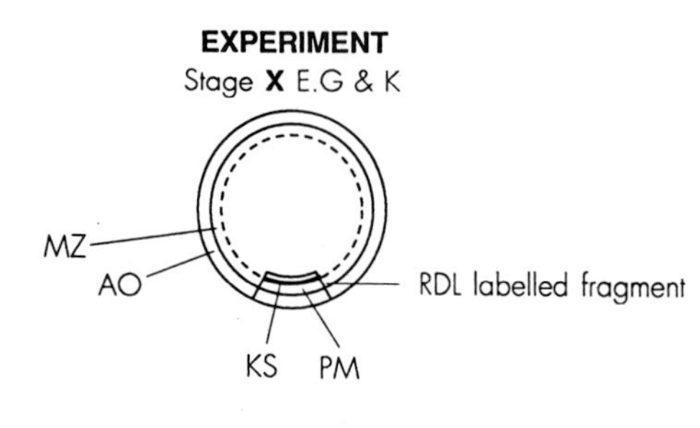


**RESULTS**

| | A<br>Stage **XII** E.G & K | B<br>Stage **2** H & H | C<br>Stage **3** H & H | D<br>Stage **4** H & H |
|---|---|---|---|---|
| **1** Hypoblast | | | | |
| **2** Epiblast | | | | |
| **3** Mesoblast | ———— | ———— | ———— | |

RDL labelled cells   Non-labelled hypoblastic cells
Non-labelled mesoblastic cells

Fig. 6. The circular inhibitory affect of a developing axis on other potential axis forming areas (stage X).
A - following the removal of a stage X PM and mechanically blocking the closure of the gap, a PS forms on each side of the gap.
B - Same experiment as in A, but the PM is implanted laterally and causes the formation of a lateral PS. The expected PS on the side of the gap close to the lateral PS, does not develop.

Fig. 7. The contribution of a stage X PM and a narrow epiblastic strip anterior to it to the blastodermic layers of stages XII E.G&K to 4 H&H.
Experiment - The posterior fragment was excised, labelled with RDL and returned to its original location. The blastoderm was then incubated until it reached one of the desired stages.
A - Development was stopped when the blastoderm reached stage XII. A1 shows the lower side of the blastoderm. A median strip of labelled cells occupies a median position in the not yet fully developed hypoblast. A2 - There is no invasion of labelled cells into the epiblast, beyond the anterior border of the implant.
B - Development was stopped at stage 2 H&H. Many labelled cells occupy a large section of the hypoblast. Only the peripheral anterior and lateral sections of the hypoblast are not labelled (B1). In the epiblast (B2) a median protrusion of labelled cells forms the initial PS.
C - Development was stopped at stage 3 H&H. The labelled area in the hypoblast (C1) is similar to stage 2 H&H. In the epiblast (C2) the labelled strip extends more anteriorly and is again congruent with the PS.
D - Development was stopped at stage 4 H&H. Here already three germ layers are found. In the hypoblast (D1) the labelled area is relatively more expanded than in earlier stages. In the epiblast (D2) there is a full PS but only its most anterior and posterior edges contain labelled cells. Most of the PS is not labelled. In the mesoblast (D3) labelled cells are found in the peripheral areas, while the central part underneath the PS is not labelled.

indicated that the results could be explained only if we assume that a "strong" MZ section is capable of inhibiting the potential inductivity of the rest of the MZ belt. If the above assumption is correct the inhibitory effect should be circular, within the belt. The following experiment was therefore designed to check the above assumption (Khaner and Eyal-Giladi 1989). A PM fragment together with a small section of the CD attached to it was removed from stage X blastoderms. The gap at the posterior side was filled with biocompatible polymeric beads to delay its closure and form a mechanical barrier. With the PM missing, both MZ sections flanking the gap, which now became the "strongest" in the field, took over as initiators of axis formation, and a PS developed on each side of the bead barrier.

The second series was again done with stage X blastoderms which were treated exactly like the first, but in additon the excised PM fragment was implanted laterally, into the MZ of the same blastoderm. In these blastoderms a PS developed from the direction of the laterally implanted PM, while posteriorly a change occurred,as compared to the first experiment. On the side of the bead barrier nearer to the PM implant no PS developed, whereas on the far side of the barrier a PS did develop (Fig. 6). The conclusion was that an inhibitory signal originating from the laterally implanted PM spreads circularly within the MZ belt and does not normally allow for supernumerary PS formation. However, when such a signal is stopped by the bead barrier, the MZ on the far side of the barrier is able to materialize its potential inductivity.

DOES CELL MIGRATION FROM THE MARGINAL ZONE INTO THE HYPOBLAST REALLY TAKE PLACE?

After having all the above indications about the involvement of the MZ in PS formation, we had also to positively clarify whether cell migration really occurs. For this purpose we had to use a reliable labelling method, which would allow for all the cells of the tested fragment to be intensively labelled, so as to be detected in the daughter cells after several cell divisions (24 h of incubation). The label should also be bound to the cells and not diffuse into neighbouring cells and not impair normal development. The label that was found to be most adequate was Rhodamine-Dextran-Lysine (RDL). In an extensive series of experiments (Eyal-Giladi and Debby, in preparation) PM fragments either with or without KS, from blastoderms of stages X to XII, were cut out and incubated for 30 min at 37°C in a solution of RDL in Ringer's solution (7.2 mg/µl). In the meantime the donor blastoderm was kept in the cold to avoid healing and further development, so that it would be suitable to also serve as a host for the same explant. After careful rinsing, the labelled PM was inserted into the gap in the original blastoderm in the right orientation and the blastoderm was then incubated at 37°C. Only blastoderms in which the ex-implant was properly healed were used for the study. The blastoderms were fixed, serially sectioned and observed under a fluorescence microscope.

Series 1

Stage X blastoderms whose development was interrupted at different stages (XII E.G&K; 2,3 and 4 H&H Hamburger and Hamilton, 1951) were used.

The labelled section included a section of the AO, the PM anterior to it, the KS and a small strip of the CD anterior to it (Fig. 7 Experiment). This series supplied us with information about the behavior of the labelled cells during both hypoblast and PS formation.

Formation of the hypoblast

By stage XII the hypoblast had grown anteriorly to about half its final length. It was found to contain centrally a tongue-like strip of fluorescent cells which stretched from the labelled implant to the anterior border of the growing hypoblast. On both sides of the labelled cells, non-labelled cells were seen to occupy most of the hypoblastic area (Fig. 7 A). By stages 2, 3 and 4 H&H progressively more and more labelled cells were found to occupy an increasingly larger area of the hypoblast. The expansion of the labelled cells was in both an anterior and lateral direction, but even at stage 4 they did not reach either the anterior or the lateral borders of the hypoblast, and were surrounded on those sides by non-labelled cells. Only posteriorly, the labelled hypoblastic area was continuous with the labelled fragment.

THE CELLULAR CONTRIBUTION TO THE PRIMITIVE STREAK

An unexpected phenomenon was observed in the epiblasts of the above series (Fig. 7 B2, C2, D2). While after incubation from the initial stage X (when the implantation was done) to stage XII, the labelled cells seemed to move only into the hypoblast, at stage 2 H&H labelled cells were also found in a typical pattern in the epiblast. At stage 2 the PS is already forming as a thickening of the epiblast and is about 1/3 of its final size (stage 4). The entire PS thickening in our stage 2 experimental group was composed of labelled cells. The same phenomenon was seen at stage 3 when the PS has extended anteriorly. At stage 3 the central disc was still composed of only two germ layers, the epiblast and the hypoblast, without any signs of a forming mesodermal layer. At stage 4 however, a dramatic change occurred. Gastrulation was underway and labelled cells disappeared almost entirely from the PS except for a small anterior section containing Hensen's node. More labelled cells were found in a posterior section of the blastoderm probably identical to the original implant and in lateral and anterior cells of the newly formed mesoblastic layer (Fig. 7 D3). The most obvious interpretation of the above results is that the initial PS (until the onset of gastrulation) is entirely formed of cells derived from the posterior section of the epiblast, which shoots anteriorly and stretches out within the epiblastic framework to constitute the PS. Our assumption was that the above cells originate from the epiblastic strip anterior to KS, which was usually included in the labelled section. When gastrulation starts (between stages 3 and 4) the cells of the initial PS invaginate and move laterally and anteriorly, their place in the PS is being taken by laterally situated epiblastic cells that move medially. Some of those non-labelled cells have already ingressed by stage 4 and are found in a median position within the middle layer. The epiblastic replacement cells of the PS do not migrate from the posterior part of the blastoderm anteriorly like the founder cell of the streak, as they are not labelled, while a large section of labelled cells is still to be found posteriorly.

Experiments done on older stages (XI, XII) (Eyal-Giladi and Debby, in preparation) together with the above results of stage X substantiate the following scenario: From stage X onwards a massive migration and expansion of PM cells (KS included) into the growing hypoblast takes place. At the same time or probably slightly later, epiblastic cells situated anteriorly to KS start to shift in a medio-anterior direction and are later found to form the PS. The expansion of the MZ cells within the hypoblast is much more extensive and precedes the cellular rearrangement leading to the formation of the PS in the epiblast. It is probably the interaction between the cell population of MZ origin that entered the hypoblast and the overlying epiblastic cells that result in the determination of a PS and the following cellular rearrangements within the epiblast.

## WHEN DO THE MARGINAL ZONE CELLS BECOME INDUCTIVE?

Following the conclusion that within the hypoblast it is the MZ derived cell population that is the inductive one, and without which there would be no PS formation, we decided to check when do the MZ cells start to be inductive. For the above purpose a possible inductive effect of PMs was compared to the known inductive effect of hypoblasts. Central epiblastic discs from stage XIII blastoderms were used as the reacting competent system (Eyal-Giladi and Levin, in preparation). The lower surface of the epiblastic CDs was completely covered in the experimental series, with stage X PM fragments either with or without KS. In the control series similar CDs were covered with strips cut from stage XIII hypoblasts. Many variations of the above experiments have been done concerning the orientation and the integrity of the tested tissues. The unequivocal result was that while the PM fragments (either with or without KS) were incapable of inducing a PS in the CD, the hypoblastic strips were capable of inducing a single PS in the CD in the original posterio-anterior orientation. This means that the MZ cells while still in the marginal belt are only potentially inductive. They become effectively inductive only after their migration via Koller's sickle into the hypoblast.

## REFERENCES


Azar, Y., and Eyal-Giladi, H., 1979, Marginal zone cells - the primitive streak-inducing component of the primary hypoblast in the chick, J.Embryol.Exp.Morph., 52:79-88.

Azar, Y., and Eyal-Giladi, H., 1981, Interaction of epiblast and hypoblast in the formation of the primitive streak and the embryonic axis in chick, as revealed by hypoblast-rotation experiments, J.Embryol.Exp.Morph., 61:133-144.

Eyal-Giladi, H., 1984, The gradual establishment of cell commitments during the early stages of chick development, Cell Differ., 14:245-255. Review.

Eyal-Giladi, H., 1991, The early embryonic development of the chick as an epigenetic process, Crit. Rev. Poultry Biol., 3:143-166.

Eyal-Giladi, H., and Debby, A., The contribution of the marginal zone and Koller's sickle, to axis formation in the chick, in preparation.

Eyal-Giladi, H., and Khaner, O., 1989, The chick's marginal zone and primitive streak formation. II. Quantification of the marginal zone's potencies - temporal and spatial aspects, Develop. Biol., 134:215-221.

Eyal-Giladi, H., and Kochav, S., 1976, From cleavage to primitive streak formation: A complementary normal table and a new look a the first stages of the development of the chick. I. General morphology, Develop. Biol., 49:321-337.

Eyal-Giladi, H., and Levin, T., Marginal zone cells begin to function as PS inducers only after their migration into the hypoblast, in preparation.

Hamburger, V., and Hamilton, H. L., 1951, A series of normal stages in the development of the chick embryo, J. Morph., 88:49-92.

Khaner, O., and Eyal-Giladi, H., 1986, The embryo-forming potency of the posterior marginal zone in stages X through XII of the chick, Develop. Biol., 115:275-281.

Khaner, O., and Eyal-Giladi, H., 1989, The chick's marginal zone and primitive streak formation. I. Coordinative effect of induction and inhibition, Develop. Biol., 134:206-214.

Kochav, S., Ginsburg, M., and Eyal-Giladi, H., 1980, From cleavage to primitive streak formation: A complementary normal table and a new look at the first stages of the development of the chick. II. Microscopic anatomy and cell population dynamics, Develop. Biol., 79:296-308.

Spratt, N. T., 1963, Role of substratum, supracellular continuity and differential growth in morphogenetic cell movements, Develop. Biol., 7:51-63.

Vakaet, L., 1962, Some new data concerning the formation of the definitive endoblast in the chick embryo, J.Embryol.Exp.Morph., 10:38-57.

Waddington, C. H., 1933, Induction by the endoderm in birds, Roux' Arch. Dev. Biol., 128:502-521.

# EVIDENCE FOR STEM CELLS IN THE MESODERM OF HENSEN'S NODE AND THEIR ROLE IN EMBRYONIC PATTERN FORMATION

Mark A.J. Selleck* and Claudio D. Stern

Dept. of Human Anatomy, South Parks Road, Oxford OX1 3QX, UK
*Present address: Developmental Biology Center, University of California Irvine, CA 92717, USA

In the chick embryo, both mesoderm and endoderm are derived from a single germ layer, the epiblast, through the primitive streak, which is the first axial structure to develop. The streak arises after about 8 hours' incubation, at stage 2 (Hamburger and Hamilton, 1951), as a thickening at the future caudal end of the blastoderm, just beneath the epiblast. It elongates cranially over the next 9 hours, while a groove appears along its length in the midline; cells migrate laterally from it to form the early mesoderm and the definitive (gut) endoderm. At the definitive streak stage (16-18 h of incubation; stage 4), the streak extends about three-quarters of the way across the area pellucida, with the primitive groove terminating rostrally in a pit. Anterior to this pit lies a mass of cells, termed Hensen's node, after Viktor Hensen who first described it in 1876.

Many lines of evidence suggested Hensen's node as the "organiser" of the amniote embryo, since it can induce the formation of a second embryonic axis if grafted into a host embryo, much like the dorsal lip of the blastopore in the amphibian (for reviews see Leikola, 1976; Hara, 1978; Nieuwkoop et al., 1985; Slack, 1991). Wetzel (1924) found that damage to Hensen's node prevented correct somitogenesis. His chorioallantoic grafts of portions of the early blastoderm showed that head or trunk structures developed only when the fragment of the blastoderm cultured contained Hensen's node. Furthermore, when a node is grown on the chorioallantoic membrane, its cells can autonomously differentiate into a number of cell types (Hunt, 1931; Willier and Rawles, 1931; Viswanath and Mulherkar, 1972; Leikola, 1975, 1978; Veini and Hara, 1975). Direct evidence that Hensen's node is the organiser of the avian embryo came from Waddington (1932; 1933) and Waddington and Schmidt (1933) who grafted Hensen's node from a donor embryo to the lateral margin of a host embryo and found that a second neural tube was induced from the host ectoderm. This result compares directly with the findings of Spemann and Mangold in their original organiser experiment, and has since been confirmed by many workers (Vakaet, 1965; Gallera, 1971; McCallion and Shinde, 1973; Dias and Schoenwolf, 1990). Furthermore, when the node is grafted into the anterior margin of the developing limb bud, it can induce supernumerary digits (Hornbruch and Wolpert, 1986; Stocker and Carlson, 1990).

The node has been studied by various fate mapping techniques such as marking with carbon particles, vital stains or tritiated thyimidine (Spratt, 1955; Rosenquist, 1983). However, these techniques are not always reliable because, for instance, one cannot be sure that the marker remains with those cells originally labelled. Furthermore, the resolution of these techniques is poor since only large groups of cells can be labelled. We have used new fate mapping techniques to generate more refined fate maps (Selleck and Stern, 1991). The carbocyanine dye, DiI, was used to label groups of cells in Hensen's node and the lineages of single cells within the node were also studied by intracellular iontophoretic injection of Lysine-Rhodamine-Dextran (LRD). The findings showed that at the definitive streak stage

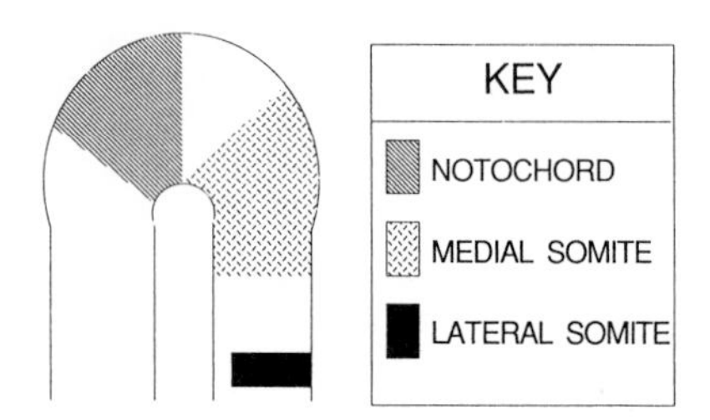


Fig. 1. Diagram illustrating the fate of mesoderm cells within different regions of Hensen's node and rostral primitive streak at the definitive primitive streak stage. In a ∇-shaped midline region of the node, mesoderm cells will contribute only to notochord, while in more lateral regions, cells will contribute to the medial halves of somites. The dorsal part of the lateral node sector also contains presumptive notochord cells (not shown). The lateral parts of somites are derived from more caudal regions of the primitive streak.

(stage 4), prospective notochord cells are contained within a ∇-shaped medial sector, and in the dorsal parts of lateral sectors (Fig. 1). The mesoderm of the lateral sectors contains prospective medial somite cells, while the lateral halves of somites are derived from a region of rostral primitive streak situated behind the node (Fig. 1). The results obtained from injection of LRD into single cells suggest that the regions intervening between the medial and lateral sectors contain 'intermediate' cells, whose progeny can contribute to both notochord and medial half somite. At stages 3-3+, the node also contains prospective endoderm cells. At later stages (5-9), the medial ∇-sector becomes narrower as presumptive notochord cells are lost from the node, and the proportion of prospective somite cells within it increases.

## Commitment of node mesoderm cells to their fates

To test the commitment of cells to their fates, it is necessary to challenge the fates of the cells in different environments. We have performed experiments to investigate whether the prospective notochord and somite cells within Hensen's node are committed by heterotopic grafting between various sectors of the node and rostral primitive streak (Selleck and Stern, 1992). The cells of the graft were labelled with DiI to follow their incorporation into different tissues; notochord was identified using a specific antibody (Not1; Yamada et al., 1991). At the definitive streak stage, prospective notochord cells appear to be committed to their fates because cells in the medial node sector always contribute to notochord, or form Not1-positive notochord-like structures when grafted into different positions in the embryo. In contrast, the presumptive somite cells are able to contribute to notochord and stain with Not1, showing that they are not committed to a somitic fate.

We were unable to determine the time at which notochord cells first become committed. It is difficult to perform grafting experiments to investigate this at stage 3 or earlier, since the notochord cells are spread throughtout the node, intermingled with other prospective cell types. However, three lines of evidence suggest that at least some prospective notochord cells become committed to their fate in Hensen's node. The first is that some individual cells in the epiblast of the node can populate both the notochord and the neural tube as late as stage 4, showing that they cannot yet be committed to either fate (Selleck and Stern, 1991). Second, the progeny of single cells lying between the medial and lateral node sectors contribute to both notochord and somite (Selleck and Stern, 1991). The third piece of evidence is that prospective somite cells in the lateral node sectors can become notochord cells when grafted into the medial sector, which suggests that mechanisms that specify cells as notochord are still operating at this stage (Selleck and Stern, 1992).

It seems likely that presumptive notochord cells become committed when they enter the medial sector of the node. What could be the mechanism? Amphibian embryos could provide part of the answer. In *Xenopus*, factors related to TGFβ, especially the activins, can induce ectoderm cells to become notochord (Green and Smith, 1990; Smith et al., 1990; Van den Eijnden-Van Raaij et al., 1990; Green and Smith, 1991). They operate in a concentration-dependent manner: the highest concentrations give rise to notochord (Green and Smith, 1990; Smith et al., 1991) and to cells that display 'organiser' activity when mixed with untreated cells (Green and Smith, 1990). It has been suggested (Smith et al., 1985; Stewart and Gerhart, 1991) that organiser cells are responsible for the induction of notochord. If the notochord is specified in a similar way in the chick, then one might expect that cells with organiser activity will be located in Hensen's node.

If 'organiser' cells are present in the node, and serve to commit notochord cells to their fate, where do they come from? Both fate maps and specification maps of the epiblast at pre-streak stages place the presumptive notochord cells at the centre of the blastoderm (Rudnick, 1948; Waddington, 1952; Balinsky, 1975), which may indicate that these cells are already committed to notochord. However, this finding is also consistent with the view that this region contains organiser cells. When isolated, some cells in this region would form notochord, under the influence of the organiser cells. This conclusion suggests that the tip of the elongating streak between stages 2 and 3 does not yet contain organiser cells, and therefore no prospective notochord cells. These cells can only be recruited when the tip of the streak reaches the centre of the blastoderm, at about stage 3. This agrees with the findings of Nicolet (1970), who found that the tip of the streak at stage 2 does not contribute to the notochord.

The organiser cells found at the centre of the blastoderm may originate from the posterior end of the embryo, at or near the posterior marginal zone, which also appears to have 'organiser' qualities such as controlling the orientation of the primitive streak (Eyal-Giladi and Spratt, 1965; Azar and Eyal-Giladi, 1979; 1981; Mitrani et al., 1983; Khaner et al., 1985; Khaner and Eyal-Giladi, 1986; 1989; Eyal-Giladi and Khaner, 1989; Stern, 1990). Carbon particle marking experiments (Spratt, 1946) show that before the primitive streak appears, cells at the posterior end of the blastoderm migrate rostrally. This finding is compatible with a movement of organiser cells in the epiblast towards the centre of the blastoderm prior to streak formation, and may indicate that organiser cells play different controlling roles at different stages of development.

## HENSEN'S NODE AND THE GENERATION OF PATTERN

### Periodic contribution of node-derived cells to notochord and somite

One striking finding obtained from our single cell lineage experiments (Selleck and Stern, 1991) is that in some cases, the progeny of single LRD-labelled cells were arranged in a periodic fashion in the notochord and somites (Fig. 2a). Several groups of labelled cells could be found along the length of the notochord, each separated by about 2 somite intervals. In specimens where single cells in Hensen's node populated the somites, about 2-3 consecutive somites were labelled (Fig. 2d), with a periodicity of about 5-6 somites. A similar finding was made when DiI labelled lateral node sectors were placed in the medial sector (Selleck and Stern, 1992; Fig. 2b,c): labelled cells populated the notochord in a periodic fashion. These findings are consistent with the idea that Hensen's node may contain presumptive notochord and somite founder cells, with stem-cell-like, self-renewing ability.

### Founder cells and stem cells in the node?

Further evidence that the node contains stem-cell-like cells that can give rise to notochord is the finding (Selleck and Stern, 1991) of a group of cells that can populate the entire length of the notochord, lying between the medial and lateral node sectors. These observations suggest that the node contains a population of 'notochord founder cells' ($FC_N$; Fig. 3), with stem cell properties: at each division of one of these cells, one daughter remains in the node and becomes the progenitor of subsequent groups of cells, whilst the other daughter gives rise to a single, more anterior group of cells. At each division of a founder cell, therefore, the ejected cell and its progeny occupy successively more posterior levels of the notochord, with clusters separated by about 2-3 somite lengths (400μm). The distance between successive clusters would be determined by the rate of division of the founder cell ($FC_N$), while the length of each cluster depends on the rate of division of its daughters.

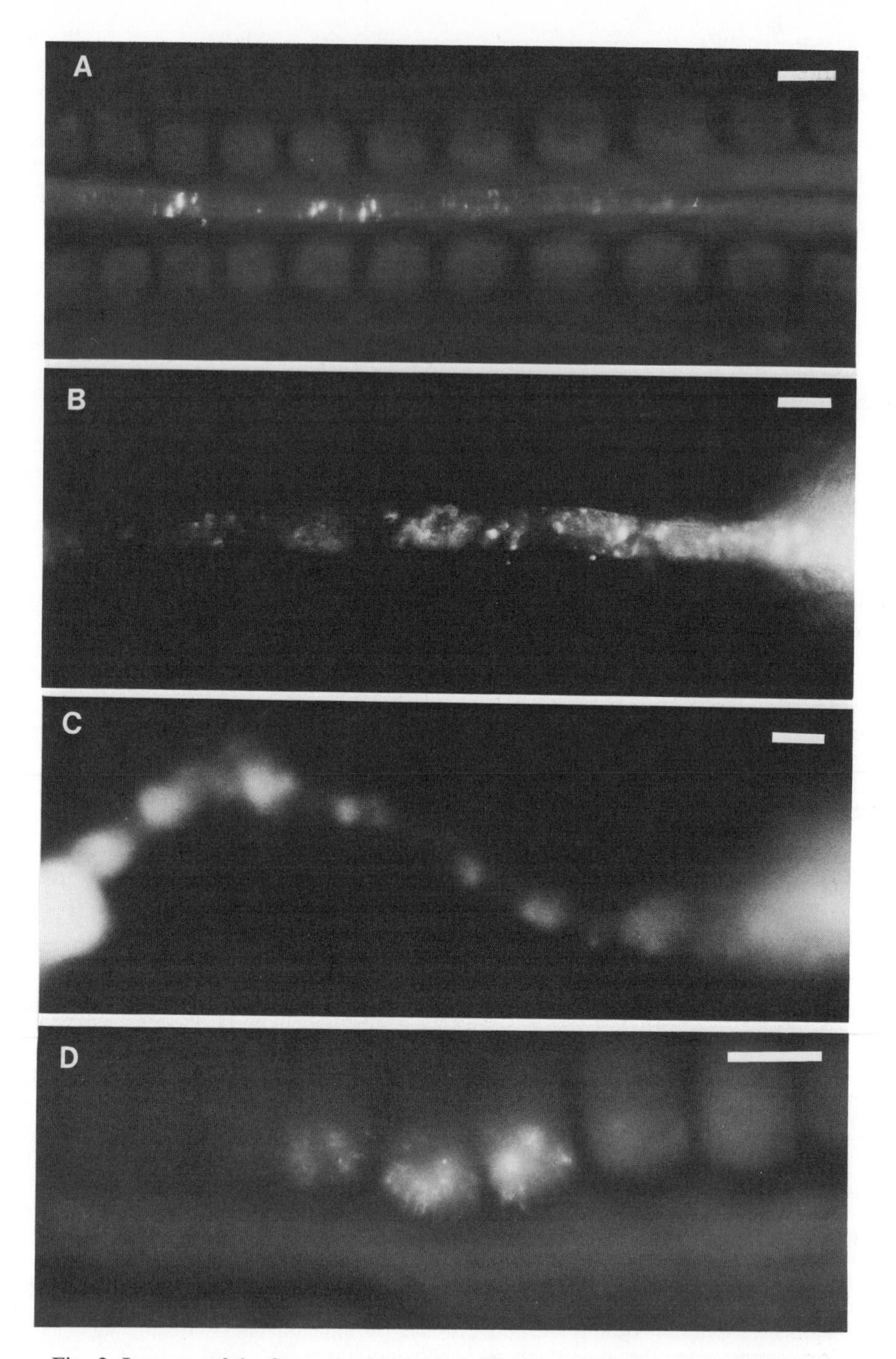


Fig. 2. In some of the fate mapping and grafting experiments, a periodicity can be seen in the allocation of cells to notochord and somite. A. Labelled progeny derived from a single node cell that has been injected with LRD can be found in several groups along the notochord. The groups are separated by about 2 somite-lengths. B, C. The same periodicity of fluorescent cells can be found when DiI-labelled medial sectors are placed into the lateral sector (B) or when lateral sectors are placed into the medial portion of Hensen's node (C). D. Single cells in the lateral node sectors typically populate 2-3 consecutive somites. Rostral to the right. Space bars= 100μm.

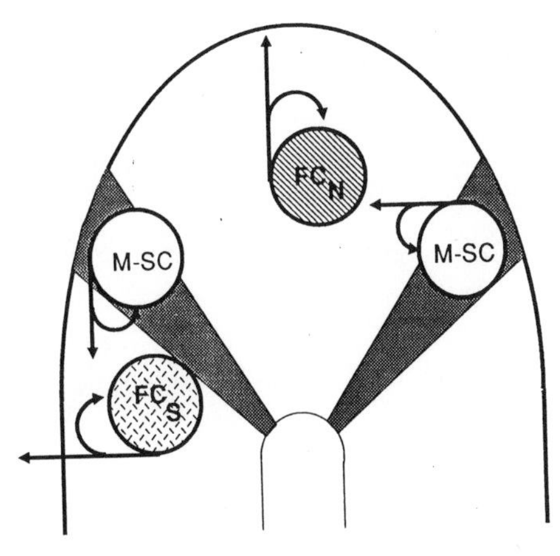


Fig. 3. Diagram illustrating a stem cell and founder cell mechanism that may control patterning of the notochord and somites. Stem cells (M-SC) lying between the medial and lateral node sectors give rise to daughter cells that populate both areas. These daughter cells are founders for prospective notochord or somite cells ($FC_N$ and $FC_S$ respectively). The cell cycle time of the notochord founder cells is different from that of somite founder cells. The cell cycle time of the founder cells will control the periodicty in allocation of cells to their mesoderm structures.

Medial somite cells also seem to be derived from founder cells in Hensen's node (*$FC_S$*; Fig. 3). Single labelled cells give rise to progeny that are similarly arranged into clusters in the medial halves of the somites, each cluster being some 2-3 somites long (about 300-500μm; Fig. 2d) and with a spacing of about 5-6 somites (800μm). The difference in spacing between clusters in notochord and somites indicates that each kind of founder cell has a characteristic mitotic rate.

Single cells lying between the medial and lateral node sectors are able to populate both notochord and somites. Therefore, these may represent a population of multipotent stem cells (*M-SC*; Fig. 3), which gives rise to the founder cells of both tissues ($FC_N$ and $FC_S$). The rate of division of the daughters of M-SC may be set according to whether they are specified as notochord ($FC_N$) or somite ($FC_S$) founder cells.

**Hensen's node and the control of segmentation**

Many workers have investigated the role of Hensen's node in somitogenesis, but have reached conflicting conclusions as to its importance (see Bellairs, 1963; Nicolet, 1971; Stern and Bellairs, 1984; Bellairs, 1986). This may be due to the absence of detailed information about the cellular contents of the node. Moreover, some workers have assumed that the node comprises only presumptive notochord cells.

The finding (Selleck and Stern, 1991; Ordahl and Le Douarin, 1992) that somites are subdivided into medial and lateral halves, derived from two separate sources (lateral node and anterior primitive streak, respectively) suggests that this dual origin may play an important role in some aspect of somite development. However, despite a difference in fate of medial and lateral somite cells in muscle formation, they appear to be equivalent in their commitment (Ordahl and Le Douarin, 1992). Perhaps the medial/lateral subdivision of somites reflects a property required for the generation of a metameric pattern rather than for the subsequent differentiation of different cell types. Bellairs and Veini (1984; see also Bellairs, 1986) have suggested that "somitogenic clusters" might be present in the embryo, which recruit cells from unsegmented paraxial mesoderm to form somites. They propose that the number of such clusters is correlated with the number of somites that will form, and that the recruited cells make up cell numbers so that sufficient cells gather to form individual somites.

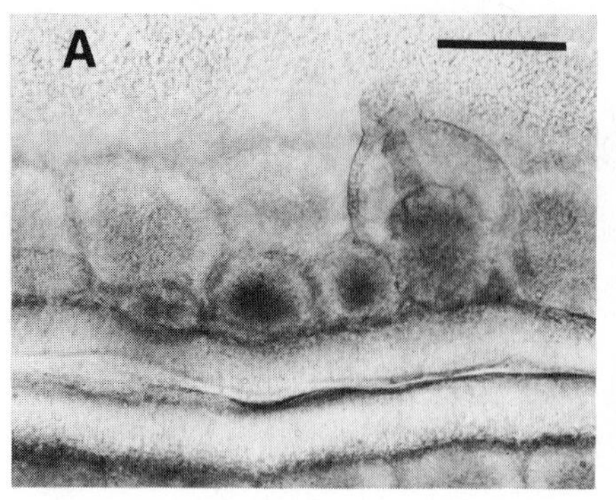

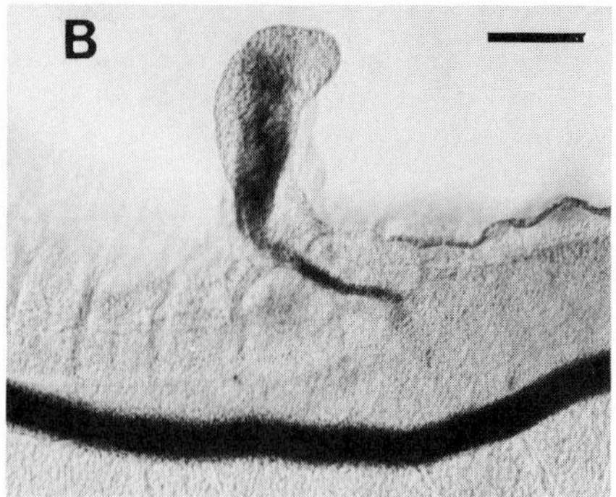


Fig. 4. When grafted into the segmental plate of a host embryo, medial and lateral node sectors behave very differently. A. Grafts of lateral node sector differentiate into small epithelial spheroids adjacent to, but not aligned with, the host somites. These spheres are like miniature epithelial somites. In contrast (B), medial node sectors autonomously differentiate into rod-like structures that stain with Not1, an antibody specific for notochord. Scale bars= 100μm.

Could the medial somite cells derived from the lateral sectors of the node contain these somitogenic clusters, and the presumptive lateral somite cells in the anterior primitive streak constitute the source of recruited cells? Transplantation experiments (Selleck and Stern, 1992) suggest that the cells of the lateral node sector are the ones that determine the spacing of the metameric pattern, because when transplanted into the segmental plate, they often form somite-like epithelial spheroids (Fig. 4a). In contrast, grafts of primitive streak or medial node sector (Fig. 4b) do not give rise to such spheres.

It therefore seems possible that the somitogenic cells in Hensen's node, which will contribute to the medial halves of the somites, are required for setting up a metameric pattern and the spacing of somites. What is the mechanism that sets up such a spacing pattern? If somite-stage embryos are subjected to brief heat-shock, periodic anomalies, separated by intervals of 5-8 somites, are seen in the somites that form after the shock (Primmett et al. 1988). Given that somites form at a rate of about one pair every 100 minutes (Menkes et al., 1961), this period corresponds to about 10 hours. Primmett et al. (1989) showed that this same period corresponds to the duration of the cell cycle of somite precursor cells. From these observations, they put forward a model to account for the periodicity of somitogenesis, based on a subdivision of the cell cycle into 7 periods, during each of which the cells destined for a somite are allocated. The last anomalies produced by a single heat shock in Primmett et al.'s (1988) experiments affect cells that segment 4-5 cell cycles after the time of the shock. Some of these cells may have been in Hensen's node (medial somite) at the time of the shock.

Thus, the medial somite precursor cells in Hensen's node may be responsible for setting up metameric pattern in the paraxial mesoderm, and the periodicity of this pattern may depend upon cell-autonomous timing mechanisms in these cells. This conclusion agrees with the above proposal that the node contains stem-cell-like founder cells ($FC_S$; Fig. 3). These arguments suggest that the somitic founder cells in the lateral node sectors ($FC_S$) are those whose cycle time is affected by heat shock, thereby affecting the periodicity of segmentation.

The conclusion that the lateral sectors of the node contain somitic founder cells equivalent to Bellairs and Veini's (1984) somitogenic clusters could also account for another result: Hornbruch et al. (1979) found that when quail Hensen's nodes were transplanted into the area pellucida of a chick host, some host cells formed somites. In contrast, grafts into the area opaca do not generate somites from the host, although small donor-derived somites develop. This finding can be explained by assuming that the area opaca, which does not contain mesoderm cells at this stage, cannot replace the cells normally derived from the anterior primitive streak (presumptive lateral half somite).

## REGULATION FOLLOWING NODE EXTIRPATION

The preceding discussion has suggested that Hensen's node contains organiser cells and stem cells retained within it as it regresses, rather than being merely a passageway for ingressing cells.

One problem with this view is that node extirpation or rostral primitive streak rotation at the definitive streak stage allows more or less normal development, provided that the wound heals (Waddington, 1932; Abercrombie, 1950; Spratt, 1955; Grabowski, 1956; Gallera, 1965, 1972, 1974a,b; Gallera and Nicolet, 1974; Hara, 1978). Under appropriate conditions, such embryos form a notochord and somites. In all of these experiments, the embryo shows a remarkable ability to compensate for the loss of Hensen's node or for a large disturbance in its position or orientation. If there are special cells with organiser and stem cell properties in the node, then these cells and properties must be maintained as a result of interactions with the neighbouring cells, and these interactions must continue at least until the definitive streak stage.

## CONCLUSIONS

Hensen's node appears to contain a population of organiser cells responsible for the commitment of notochord cells to their fates, and a population of stem/founder cells that contribute to somites and to the notochord. Somitic founder cells appear to be responsible for controlling the spacing of the segmental pattern of somites.


## ACKNOWLEDGEMENTS

The fate mapping and commitment experiments described in this paper were supported by a Wellcome Trust Prize Studentship.


### References


Abercrombie, M., 1950, The effects of antero-posterior reversals of lengths of the primitive streak in the chick, Phil. Trans. Roy. Soc. Lond. B., 234:317-338.

Azar, Y. and Eyal-Giladi, H., 1979, Marginal zone cells: the primitive streak-inducing component of the primary hypoblast in the chick, J. Embryol. exp. Morph., 52:79-88.

Azar, Y. and Eyal-Giladi, H., 1981, Interactions of epiblast and hypoblast in the formation of the primitive streak and the embryonic axis in the chick, as revealed by hypoblast-rotation experiments, J. Embryol. exp. Morph., 61:133-144.

Balinsky, B.I., 1975, "Introduction to embryology" 4th Edn., Saunders, Philadelphia.

Bellairs, R., 1963, The development of somites in the chick embryo, J. Embryol. exp. Morph., 11:697-714.

Bellairs, R., 1986, The tail bud and cessation of segmentation in the chick embryo, In: "Somites in developing embryos", R. Bellairs, D.A. Ede and J.W. Lash eds., Plenum Press, London.

Bellairs, R. and Veini, M ., 1984, Experimental analysis of control mechanisms in somite segmentation in avian embryos. II. Reduction of material in the gastrula stages of the chick, J. Embryol. exp. Morph., 79:183-200.

Dias, M.S. and Schoenwolf, G.C., 1990, Formation of ectopic neurepithelium in chick blastoderms: age-related capacities for induction and self-differentiation following transplantation of quail Hensen's node, Anat. Rec., 229:437-448.

Eyal-Giladi, H. and Khaner, O., 1989, The chick's marginal zone and primitive streak formation. II. Quantification of the marginal zone's potencies: temporal and spatial aspects, Dev. Biol., 134:215-221.

Eyal-Giladi, H. and Spratt, N.T. Jr., 1965, The embryo-forming potencies of the young chick blastoderm, J. Embryol. exp. Morph., 13:267-273.

Gallera, J., 1965, Excision et transplantation des différentes région de la ligne primitive chez le poulet, C.R. Ass. Anat. 49 Reunion, Madrid, 1964 In: Bull. Ass. Anat., 125:632-639.

Gallera, J., 1971, Primary induction in birds, Adv. Morphogen., 9:149-180.

Gallera, J., 1972, Alteration of the prospective fate and the inductive power of the definitive streak stage node in the chick, Experientia, 28:1217-1218.
Gallera, J., 1974a, Regulation des excédents nodaux dans les jeunes blastodermes de poulet. I. Implantation d'un deuxième nœud de Hensen dans la region directement postnodale de la ligne primitive, Arch. Biol. (Liège), 85:399-413.
Gallera, J., 1974b, Regulation des excédents nodaux dans les jeunes blastodermes de poulet.
Gallera, J., 1974b, Regulation des excédents nodaux dans les jeunes blastodermes de poulet. II. Implantation d'un deuxième nœud de Hensen a l'avant de la ligne primitive, Arch Biol. (Liège), 85:415.
Gallera, J. and Nicolet, G., 1974, Regulation in nodeless chick blastoderms. Experientia, 30:183.
Grabowski, C.T., 1956, The effects of the excision of Hensen's node on the early development of the chick embryo, J. exp. Zool., 133:301.
Green, J.B.A. and Smith, J.C., 1990, Graded changes in dose of a Xenopus activin A homologue elicit stepwise transitions in embryonic cell fate, Nature, 347:391.
Green, J.B.A. and Smith, J.C., 1991, Growth factors as morphogens: do gradients and thresholds establish body plan? Trends Genet, 7:245.
Hamburger, V. and Hamilton, H.L., 1951, A series of normal stages in the development of the chick, J. Morph., 88:49.
Hara, K., 1978, Spemann's organizer in birds, in: "Organizer: a milestone of a half-century from Spemann", O. Nakamura, and S. Toivonen eds., Elsevier, Amsterdam.
Hensen, V., 1876, Beobachtungen über die Befruchtung und Entwicklung des Kaninchens und Meerschweinchens, Z. Anat. EntwGesch., 1:353.
Hornbruch, A., Summerbell, D. and Wolpert, L., 1979, Somite formation in the early chick embryo following grafts of Hensen's node, J. Embryol. exp. Morph., 51:51.
Hornbruch, A. and Wolpert, L., 1986, Positional signalling by Hensen's node when grafted to the chick limb bud, J. Embryol. exp. Morph., 94:257.
Hunt, T.E., 1931, An experimental study of the independent differentiation of the isolated Hensen's node and its relation to the formation of axial and non-axial parts in the chick embryo, J. exp. Zool., 59:395.
Khaner, O. and Eyal-Giladi, H., 1986, The embryo-forming potency of the posterior marginal zone in stages X through XII of the chick, Dev. Biol., 115:275.
Khaner, O. and Eyal-Giladi, H., 1989, The chick's marginal zone and primitive streak formation. I. Coordinative effect of induction and inhibition, Dev. Biol., 134:206.
Khaner, O., Mitrani, E. and Eyal-Giladi, H., 1985, Developmental potencies of area opaca and marginal zone areas of early chick blastoderms, J. Embryol. exp. Morph., 89:235.
Leikola, A., 1975, Differentiation of quail Hensen's node in chick coelomic cavity, Experientia, 31:1087.
Leikola, A., 1976, Hensen's node- the 'organizer' of the amniote embryo, Experientia, 32:269.
Leikola, A., 1978, Differentiation of the epiblastic part of chick Hensen's node in coelomic cavity, Med. Biol., 56:339.
McCallion, D.J. and Shinde, V.A., 1973, Induction in the chick by quail Hensen's node, Experientia, 29:321.
Menkes, B., Midea, C., Elias, S. and Deleanu, M., 1961, Researches on the formation of axial organs. I. Studies on the differentiation of somites, Stud. Cercet. Stiint. Med., 8:7.
Mitrani, E., Shimoni, Y. and Eyal-Giladi, H., 1983, Nature of the hypoblastic influence on the chick embryo epiblast, J. Embryol. exp. Morph., 75:21.
Nicolet, G., 1970, Analyse autoradiographique de la localization des différentes ébauches présomptives dans la ligne primitive de l'embryon de poulet, J. Embryol. exp. Morph., 23:79.
Nicolet, G., 1971, The young notochord can induce somite genesis by means of diffusible substances in the chick, Experientia, 27:938.
Nieuwkoop, P.D., Johnen, A.G. and Albers, B., 1985, "The epigenetic nature of eaarly chordate development", Cambridge University Press, Cambridge.
Ordahl, C.P. and Le Douarin, N., 1992, Two myogenic lineages within the developing somite, Development, in press.
Primmett, D.R.N., Norris, W.E., Carlson, G.J., Keynes, R.J. and Stern, C.D., 1989, Periodic segmental anomalies induced by heat shock in the chick embryo are associated with the cell cycle, Development, 105:119.

Primmett, D.R.N., Stern, C.D. and Keynes, R.J., 1988, Heat shock causes repeated segmental anomalies in the chick embryo, Development, 104:331.
Rosenquist, G.C., 1983, The chorda center in Hensen's node of the chick embryo, Anat. Rec., 207:349.
Rudnick, D., 1948, Prospective areas and differentiation potencies in the chick blastoderm, Ann. N.Y. Acad. Sci., 49:761.
Selleck, M.A.J. and Stern, C.D., 1991, Fate mapping and cell lineage analysis of Hensen's node in the chick embryo, Development, 112:615.
Selleck, M.A.J. and Stern, C.D., 1992, Commitment of mesoderm cells in Hensen's node of the chick embryo to notochord and somite, Development, in press.
Slack, J.M.W., 1991, "From egg to embryo: determinative events in early development" 2nd edn., Cambridge University Press, Cambridge.
Smith, J.C., Dale, L. and Slack, J.M.W., 1985, Cell lineage labels and region-specific markers in the analysis of inductive interactions, J. Embryol. exp. Morph., 89(supplement):317.
Smith, J.C., Price, B.M.J., Van Nimmen, K. and Huylenbroeck, D., 1990, Identification of a potent Xenopus mesoderm-inducing factor as a homologue of activin A, Nature, 345:729.
Smith, J.C., Price, B.M.J., Green, J.B.A., Weigel, D. and Herrmann, B.G., 1991, Expression of a Xenopus homolog of Brachyury (T) is an immediate-early response to mesoderm induction, Cell, 67:79.
Spratt, N.T. Jr., 1946, Formation of the primitive streak in the explanted chick blastoderm marked with carbon particles, J. exp. Zool., 103:259.
Spratt, N.T. Jr., 1955, Analysis of the organizer center in the early chick embryo. I. Localization of prospective notochord and somite cells, J. exp. Zool., 128:121.
Stern, C.D., 1990, The marginal zone and its contribution to the hypoblast and primitive streak of the chick embryo, Development, 109:667.
Stern, C.D. and Bellairs, R., 1984, The roles of node regression and elongation of the area pellucida in the formation of somites in the avian embryo, J. Embryol. exp. Morph., 81:75.
Stewart, R.M. and Gerhart, J.C., 1991, Induction of notochord by the organizer in Xenopus, Roux's Arch. Dev. Biol., 199:341.
Stocker, K.M. and Carlson, B.M., 1990, Hensen's node, but not other biological signallers, can induce supernumerary digits in the developing chick limb bud, Roux's Arch. Dev. Biol., 198:371.
Vakaet, L., 1965, Résultats de la greffe de nœuds de Hensen d'âge différent sur le blastoderme de poulet, C.R. Soc. Biol., 159:232
Van den Eijnden-Van Raaij, A.J.M., Van Zoelen, E.J.J., Van Nimmen, K., Koster, C.H., Snoek, G.T., Durston, A.J. and Huylenbroeck, D., 1990, Activin-like factor from a Xenopus laevis cell line responsible for mesoderm-induction, Nature, 345:732.
Veini, M. and Hara, K., 1975, Changes in the differentiation tendencies of the hypoblast-free Hensen's node during "gastrulation" in the chick embryo, Wilhelm Roux's Arch., 177:89.
Viswanath, J.R. and Mulherkar, L., 1972, Studies on self-differentiating and induction capacities of Hensen's node using intracoelomic grafting technique, J. Embryol. exp. Morph., 28:547.
Waddington, C.H., 1932, Experiments on the development of chick and duck embryos, cultivated in vitro, Phil. Trans. Roy. Soc. Lond. B., 221:179.
Waddington, C.H., 1933, Induction by the primitive streak and its derivatives in the chick, J. exp. Biol., 10:38.
Waddington, C.H., 1952, "The Epigenetics of birds", Cambridge University Press, Cambridge.
Waddington, C.H. and Schmidt, G.A., 1933, Induction by heteroplastic grafts of the primitive streak in birds, Wilhelm Roux's Arch., 128:522.
Wetzel, R., 1924, Über den Primitivknoten des Hühnchens, Verh. Phys. Med. Ges. Würzburg, 49:227.
Willier, B.H. and Rawles, M.E., 1931, The relation of Hensen's node to the differentiating capacity of whole chick blastoderms as studied in chorioallantoic grafts, J. exp. Zool., 59:429.
Yamada, T., Placzek, M., Tanaka, H., Dodd, J. and Jessell, T.M., 1991, Control of cell pattern in the developing nervous system: polarizing activity of the floor plate and notochord, Cell, 64:635.

# EARLY MESODERM FORMATION IN THE MOUSE EMBRYO

Kirstie A. Lawson[1] and Roger A. Pedersen[2, 3]

[1]Hubrecht Laboratory, Netherlands Institute for Developmental Biology
Uppsalalaan 8, 3584 CT Utrecht, Netherlands
[2]Department of Radiobiology and Environmental Health and [3]Department of Anatomy, University of California, San Francisco, CA 94143, U.S.A.

## INTRODUCTION

The inaccessibility of the mammalian embryo to experimental manipulation during germ layer formation and early organogenesis, and the awkward geometry of the rodent embryo in particular, have meant that, compared with birds and amphibians, mammals have contributed little to our understanding of mesoderm formation in the vertebrates. Observations on morphology extend back more than 100 years (Snow, 1977) and have been supplemented with more refined investigation with transmission electron microscopy (Batten and Haar, 1979; Poelmann, 1981; Franke et al, 1983), and scanning electron microscopy (Tam & Meier, 1982; Tam et al., 1982; Hashimoto and Nakatsuji, 1989).

Blastocyst injection chimaeras established that all the foetal tissues are derived from the epiblast, which is set aside during the periimplantation period at about 4½ days gestation in the mouse (Gardner and Rossant, 1979). Primitive streak formation does not begin until two days later, but ectopic grafts have indicated that the epiblast retains its potency to form derivatives of all three germ layers until late in gastrulation (Svajger et al., 1986). Improved culture techniques for rodent embryos (New, 1990) were exploited to examine normal fate and potency of the epiblast during the second half of gastrulation (Beddington 1981, 1982; Tam and Beddington, 1987; Tam, 1989) and of the behaviour of isolated pieces and deficient embryos at mid- and late gastrulation (Snow, 1981). The results from these investigations led to somewhat conflicting interpretations, the results from heterotopic grafts demonstrating the ability of the epiblast to colonize a wider range of tissues than would have been expected from the behaviour of orthotopic grafts (Beddington, 1981, 1982), whereas the development of embryo pieces and deficient embryos showed marked regional autonomy. A compromise interpretation reconciling both sets of results is that gradual regionalization occurs in the epiblast during gastrulation; this is stabilized by cell interactions in the presence of mesoderm, which was present in all the partial embryos that developed recognisable structures.

Growth is very rapid during gastrulation in the mouse, but the pregastrulation epiblast has only about 600 cells (Snow, 1977) and so does not lend itself easily to the manipulations required for grafting. An additional disadvantage to grafting is that a clump of cells must be transplanted and any preexisting heterogeneity within the population may obscure results: it

*Formation and Differentiation of Early Embryonic Mesoderm*
Edited by R. Bellairs *et al.*, Plenum Press, New York, 1992

is not possible to distinguish between absence of commitment in all cells and selection within a mixed population of cells committed to different pathways. Clonal analysis of cells labelled in situ, while not necessarily providing information about potency and commitment, can give an accurate description of cell lineage, normal cell fate and morphogenetic movement. We have used such an approach to trace the origin and spread of mesoderm from the onset of gastrulation, immediately before primitive streak formation, up to the early somite stage in the mouse embryo.

## MATERIALS AND METHODS

Noninbred Swiss mice of the Dub: (ICR)strain were used. We have used a single-cell labelling technique (horseradish peroxidase (HRP) injected intracellularly by iontophoresis into the epiblast at 6.7 days gastrulation) (Beddington and Lawson, 1990) followed by embryo culture for 23 or 36 hr. In order to be able to localize accurately the position of the injected cell a fluorescent label (rhodamine dextran) was coinjected and detected using image intensification. An additional injection was made into a visceral extraembryonic endoderm in the same plane of longitude in the conceptus as the injected epiblast cell. Clones of these endoderm cells remain coherent and retain their position with respect to the embryonic axis and so can be used retrospectively to identify the angular distance from the embryonic axis of the injected epiblast cell (Lawson et al., 1991). HRP labelled descendants were identified by staining the intact embryo and their positions verified after histology and photographic reconstruction (Lawson et al., 1986, 1991). To aid classification of the results the initial epiblast was considered as three tiers (Fig. 1), the proximal and middle tier having an anterior axial, anterolateral, lateral, posterolateral and posterior axial zone (zones I-V and VI-X respectively) and the tier at the distal tip being a single, axial zone. Each zone consisted of approximately 35 epiblast cells in prestreak stage embryos and 75 cells at the early streak stage.

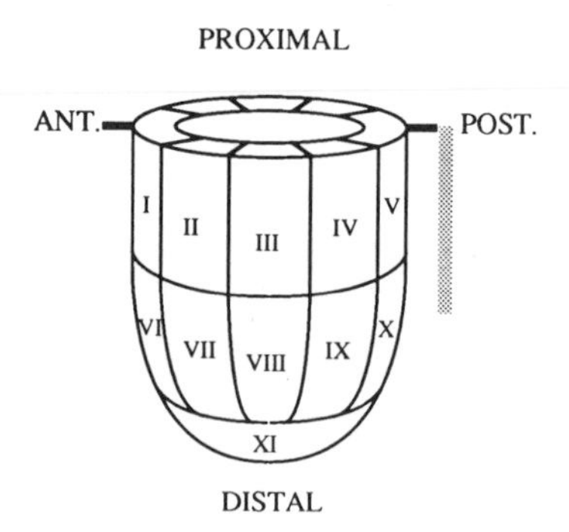


Fig. 1. Schematic representation of the epiblast of the early-streak stage mouse embryo. The anterior and posterior limits of the embryonic axis are marked by horizontal bars, the axis running from anterior (ANT) through the distal tip to posterior (POST). The extent of the primitive streak is indicated by the shaded bar. The injection zones are indicated (I-XI). Proximally the epiblast abuts the extraembryonic ectoderm and the whole is enveloped by visceral endoderm. From Lawson and Pedersen (1992).

## FATE MAPPING PRESUMPTIVE MESODERM

Comprehensive sampling of the epiblast of prestreak and early-streak stage embryos allowed the construction of a fate map valid for the first day of gastrulation, up to the neural plate stage (Fig. 2). Mesoderm (including extraembryonic mesoderm) is derived from most of the epiblast except for an axial strip anterior to the primitive streak which generates only ectoderm. The epiblast grows rapidly during gastrulation in the mouse (Snow, 1977) and rapid growth is maintained during the first day in culture: the population doubling time is 7.5 hr and more than 75% of labelled cells go through three to four cell cycles (Lawson et al., 1991). Expansion of the epiblast cup is achieved by succeeding generations of epithelial epiblast cells aligning towards the primitive streak (Lawson et al., 1991, Lawson and Pedersen, 1992), so that by the neural plate stage the ectoderm consists of descendants of epiblast originally in the axis anterior to the streak. This axial region appears to be relatively larger in the prestreak stage embryo (Fig. 2), but longer culture of such embryos to the neural plate stage showed that the anterior axial epiblast also has descendants in mesoderm (Lawson and Pedersen, 1992), indicating that transformation of epiblast to mesoderm is still continuing at midstreak and late-streak stages. The region mapped at the early-streak stage that exclusively produces ectoderm (Fig. 2) is not reduced by further culture to early somite stages (Lawson and Pedersen, 1992) and is therefore the only part of the epiblast that normally makes no contribution to mesoderm.

When different regions of presumptive mesoderm are traced through 22 hr culture a definite pattern emerges (Fig. 3). Epiblast cells nearest the origin of the primitive streak in

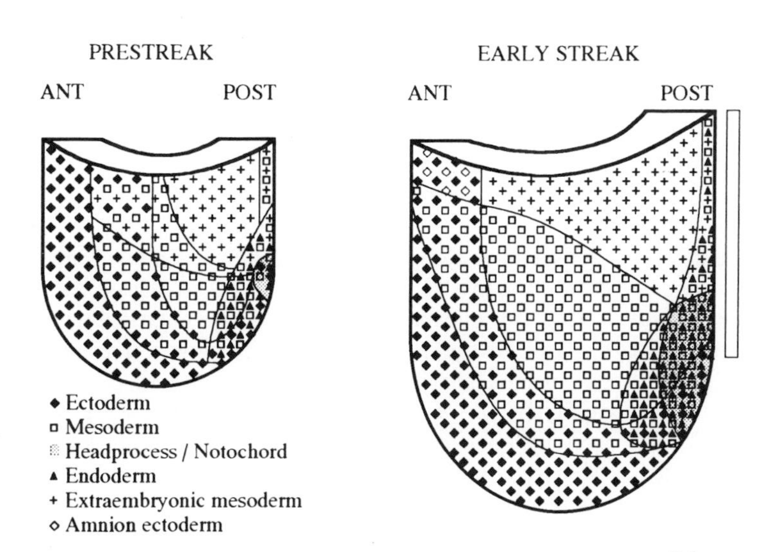


Fig. 2. Fate map of the prestreak and early-streak stage mouse epiblast to show the derivation of germ layers up to the mid/late streak and neural plate stages respectively. The left longitudinal half of the epiblast cup is viewed laterally. The primitive streak is represented by the vertical bar. ANT: anterior; POST: posterior. From Lawson et al. (1991).

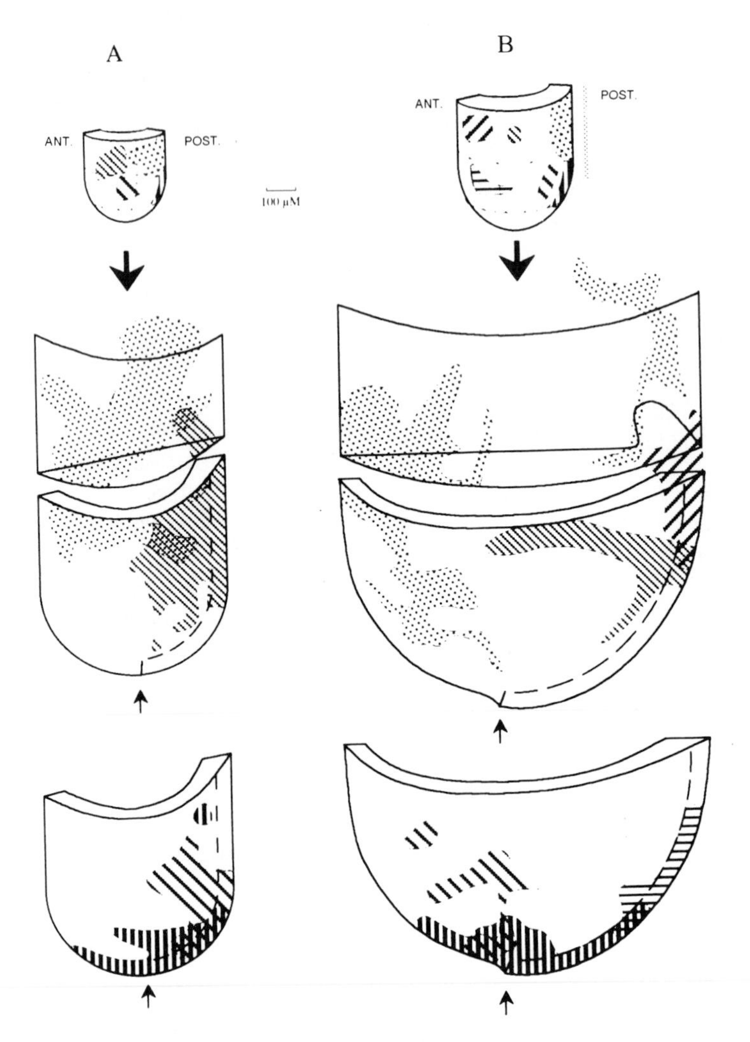


Fig. 3. Origin and localization of mesoderm during gastrulation. Lateral views of half embryos are represented. The two upper figures show the epiblast at the prestreak (A) and early-streak stage (B). Injection zones (Fig. 1) are indicated by thin lines; the primitive streak is indicated by a shaded bar. The middle and lower figures represent lateral views of the embryo 23 hr later at the midstreak/late streak (A) and neural plate stages (B). In the middle figures the amnion, yolk sac and allantoic bud are represented above the embryo. The primitive streak is indicated by a broken line, its anterior limit by a small arrow. Patterned areas cover the positions of mesoderm precursors in epiblast (upper figures) and their descendants in mesoderm and the primitive streak (middle and lower figures). (2-11 injected embryos/patterned area, m = 6, N = 60). The middle figures show the distribution of mesoderm descendants from the proximal half of the epiblast, the lower figures that of mesoderm descendants from the distal half. Data from Lawson et al., (1991).

the prestreak-stage embryo (i.e. zones IV and V in the proximal posterior part of the epiblast cup) contribute to yolk sac mesoderm and the most proximal and anterior embryonic mesoderm. Lateral and posterior mesoderm continuous with the primitive streak at the midstreak stage are derived from more anterior regions of epiblast, whereas axial mesoderm and mesoderm close to the anterior end of the streak, as well as the anterior part of the streak itself, are derived from the axial region (zone X) where the anterior end of the early streak would be expected (Fig. 3A). The distribution of mesoderm descendants from early-streak stage epiblast is similar to that from the prestreak stage (Fig 3B): initially proximal posterior cells pass through the streak early to form yolk sac and amnion mesoderm and the most anterior embryonic mesoderm, whereas proximal lateral and anterolateral epiblast contribute to lateral and posterior mesoderm, the posterior part of the streak and the base of the allantois. Cells from and near the anterior end of the streak (zone X) have descendants in the head process, neighbouring mesoderm and anterior portion of the primitive streak at the neural plate stage.

The regionalization of presumptive mesoderm at the early-streak stage is more clearly seen when embryos are cultured for 36 hr and have one to five somites and concomitant development of heart, foregut, notochordal plate, neural folds, allantois, and blood islands in the yolk sac. The proximal half of the epiblast is the source of the extraembryonic mesoderm i.e. blood islands, yolk sac mesoderm, amnion mesoderm and allantois and also contributes to the extreme posterior portion of the primitive streak (Fig 4A). The blood islands are derived from the most posterior part of this region (zones IV and V), presumably from some of the first cells to leave the epiblast. The basal part of the allantois and the posterior portion of the streak are derived from the more anterior, paraxial region of epiblast (mainly zone II). The absence of labelled cells in embryonic mesoderm from precursors in zone V suggests that the proximal and extremely anterior mesoderm of the late-streak stage embryo (Fig. 3) is later incorporated into mesoderm lining the extraembryonic coelom. Precursors from a slightly more caudal paraxial region in zones II and VII have mesoderm descendants in the posterior portion of the streak and in posterior mesoderm (Fig. 4B). The remaining epiblast that produces mesoderm (mainly zones VIII, IX and X) contributes to axial (head process/notochordal plate), paraxial and lateral mesoderm, heart, and the anterior portion of the primitive streak (Fig. 4B). Thus the bulk of the embryonic mesoderm at the early somite stage is derived from a relatively small region of the early-streak stage epiblast anterolateral to the anterior end of the streak.

More detailed analysis of the fate of epiblast cells in zones VIII, IX and X shows that there are regional differences in cell fate related to the distance from the anterior end of the early streak (Fig. 5). The region at or nearest the streak (Fig. 5C) makes the most extensive contribution to the axial mesoderm, the region furthest away (Fig. 5A) makes the smallest contribution; this is the continuation of a process already evident at the neural plate stage (Fig. 3). Members of clones within the axial mesoderm tend to be widely spread, anteriorly labelled cells having relatives scattered back to the node. There must therefore be extensive rearrangement of cells within the axial mesoderm, either during longitudinal expansion along the midline, or before insertion from the node, or both. The results are also consistent with the presence within the node of stem cells for axial mesoderm.

Paraxial mesoderm forming the cranial mesoderm and anterior somites is derived from all three injection zones, but there is no indication that the rostrocaudal position of this mesoderm is correlated with the position of epiblast precursors relative to the anterior end of the early streak (Fig. 5). Clonal descendants in paraxial mesoderm show less longitudinal spread than in axial mesoderm; descendants may be restricted to a single somitomere or somite, but are more normally spread through two or three, occasionally four, adjacent

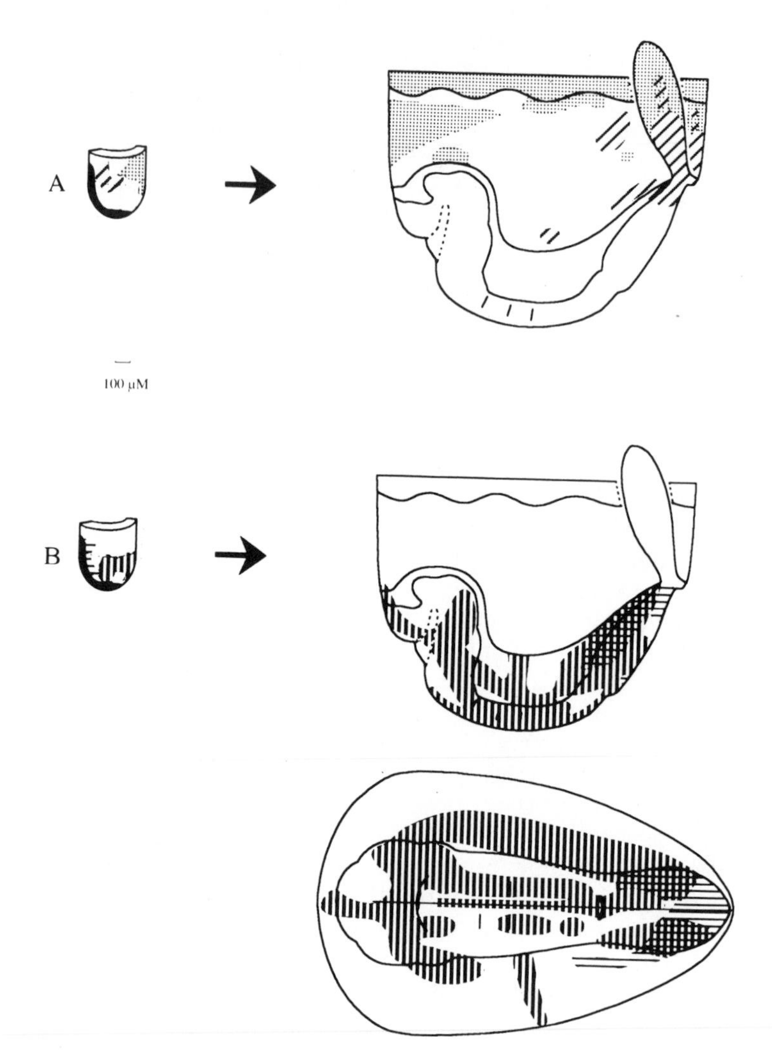


Fig. 4. Origin and localization of mesoderm at the early somite stage. The two figures on the left represent a lateral view of the early-streak stage epiblast. The injection zones are indicated by thin lines, the primitive streak by a vertical bar. The upper and middle figures on the right represent a lateral view of the embryo and yolk sac after 36 hr culture. The position of blood islands in the yolk sac is indicated by a wavy line; the allantois protrudes into the yolk sac from the posterior end of the embryo. The bottom figure represents the ventral view of an embryo flattened dorsoventrally and with its axis straightened. The oval outline indicates the junction between embryo and amnion. The bulges in the curved outline of the flattened neural folds represent the forebrain, midbrain and anterior hindbrain. The broad arc spanning the midline represents the anterior intestinal portal; the short arc represents the node. Patterned areas cover the positions of mesoderm precursors (figures on the left) and their descendants (figures on the right), (6-26 injected embryos/patterned area, m = 14, N = 60). The (solid-shaded axial region in the left hand figures produces only ectoderm. A: Clones contributing to extraembryonic mesoderm; B: Clones contributing to embryonic mesoderm.

metameres. Of 13 embryos with labelled paraxial mesoderm, 8 also had labelled cells in the node or anteriorly in the streak (eg Fig. 6). It is therefore possible that the supply of paraxial mesoderm is generated periodically from a stem cell population, possibly controlled by the cell cycle (Stern et al., 1988).

At the late streak stage the source of paraxial mesoderm for the following two days (i.e. for 18 somites and associated presomitic mesoderm) is the anterior end of the streak and the ectoderm immediately lateral to it (Beddington 1981, 1982; Tam and Beddington, 1987); this function of the streak is maintained at least until the early somite stage (3-7 somites) (Tam and Beddington, 1987). Zone X of the early streak stage epiblast colonizes the anterior part of the late streak (Fig. 3) and zones VIII, IX and X contribute to the adjacent ectoderm (Lawson et al., 1991). Therefore it is likely that this region of epiblast in the early-streak stage embryo is the source of the paraxial mesoderm of all later stages.

Lateral mesoderm at the early-somite stage is derived mainly from zone VIII (Fig. 5A). Since zone VIII colonizes the middle portion of the streak at the late-streak stage (Fig. 3) and orthotopic grafts in this region contribute mainly to lateral mesoderm (Tam and Beddington, 1987), there is good agreement between the results obtained using different labelling techniques. Descendants of cells labelled in zone VIII are also found in mesoderm posterior to the node (Fig. 5A), but the subsequent destination of these cells is unknown at present.

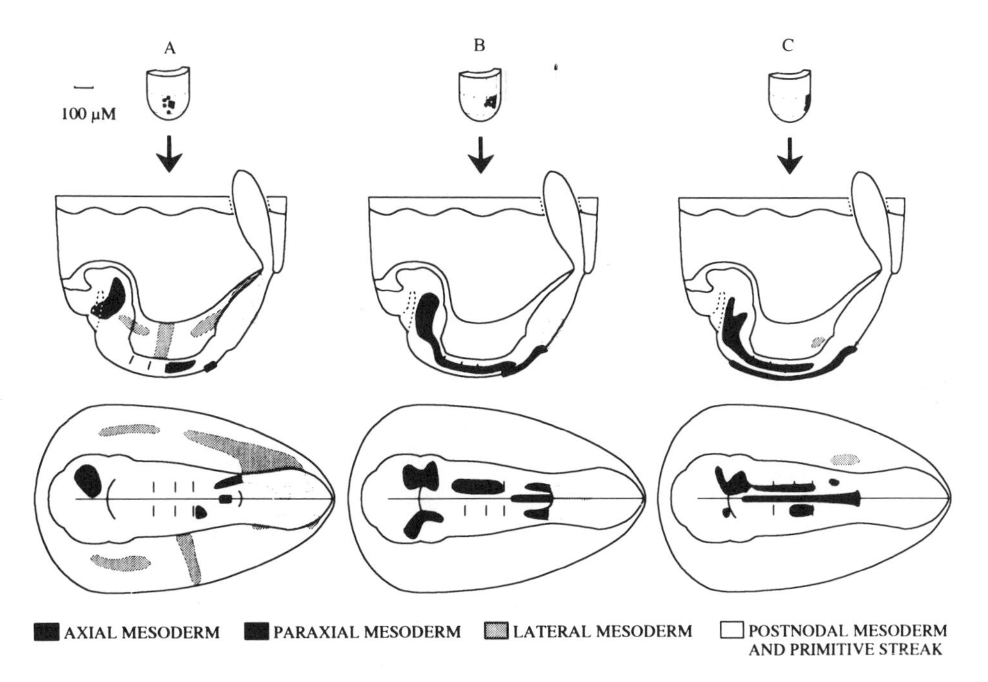


Fig. 5. Distribution of descendants in mesoderm of epiblast injected at and near the anterior end of the early streak. The positions of individual precursors (one injection/embryo)are shown by dots in the top row of figures. The positions occupied by descendants after 36 hr are enclosed by the shaded areas in lateral (middle row) and ventral (bottom row) views. A: injections in zone VIII; B: injections in zone IX; C: injections in zone X.

Clones that contribute to the heart originate in the boundary region overlapping presumptive extraembryonic and embryonic mesoderm (Fig. 7). Of the five labelled clones, four contributed only a few cells to myocardium or pericardium, with the major component in lateral mesoderm. The remaining clone was generated in a relatively young embryo which developed to the one somite stage; the large clone was fairly compact and restricted to the left half of the heart primordium, although initiated on the right side of the embryo. These results suggest that heart precursors may be allocated relatively early in gastrulation. Material in a similar region at the midstreak stage, and more distally at the late-streak stage, is essential for heart development (Snow, 1981). Although we made no injections into the presumptive heart epiblast of embryos that were cultured for one day (Fig. 3), comparison with neighbouring regions indicate that heart precursors would probably have gone through the streak by the midstreak stage and would certainly have done so by the late-streak stage. This suggests that the material essential for heart formation, identified by Snow in full-thickness explants, was already in the mesoderm layer at the late-streak stage and probably by the midstreak stage, and may indeed contain most, if not all, precursors of the future heart.

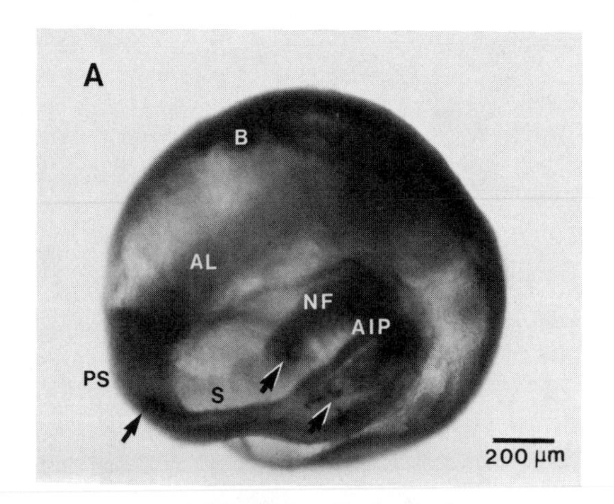


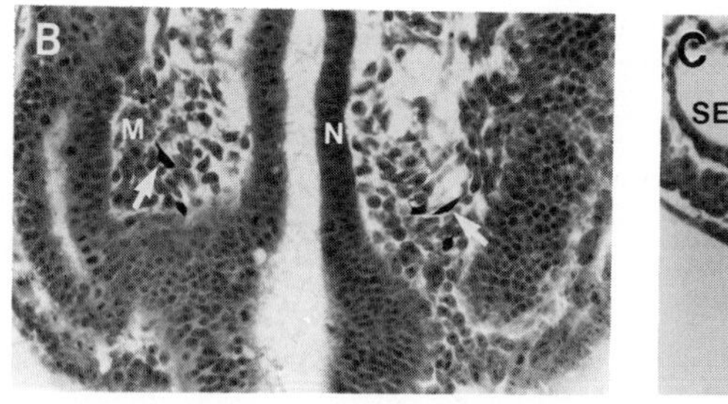


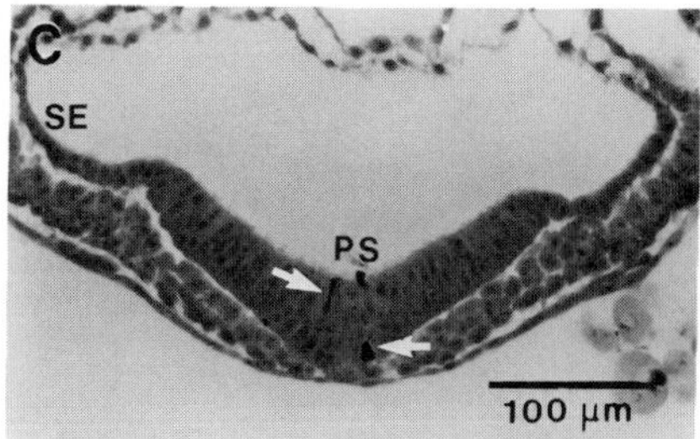


Fig. 6. Three somite stage embryo in which a single epiblast cell anterolateral to the anterior end of the early streak (zone IX) had been injected with HRP 36 hr earlier. A: Intact embryo showing labelled descendants (arrows) in cranial mesoderm and near the node. B and C: Frontal sections of the same embryo; B is a section through the neural folds of the presumptive hindbrain; C is a section immediately caudal to the node. Labelled cells indicated by arrows.
AL: allantois; AIP: anterior intestinal portal; B: blood islands; M: cranial mesoderm; N: lateral neurectoderm; NF: neural folds; PS: primitive streak; S: somite; SE: surface ectoderm.

Although presumptive mesoderm is sufficiently regionalized that a fate map of the epiblast can be drawn at the onset of gastrulation, this does not mean that lineage restriction has occurred. Indeed, the boundary region between presumptive ectoderm and the remainder of the epiblast consists mainly of cells that will have descendants in both ectoderm and mesoderm and is not a mixture of cells with descendants in only one or other of the germ layers (Lawson and Pedersen, 1992). Of the epiblast cells that traverse the primitive streak to form extraembryonic mesoderm, embryonic mesoderm and endoderm, less than 30% have descendants exclusively in one of the structures identifiable at the early somite stage; the majority will contribute to two or more structures (Lawson and Pedersen, 1992). Therefore the clonal analysis provides no evidence for cellular commitment or precursor allocation within the epiblast at the onset of gastrulation.

## THE CHANGING COMPOSITION OF THE PRIMITIVE STREAK

Primitive streak formation starts at about 6.5 d gestation in the mouse embryo, but may vary between 6.2 and 6.7 days, depending on the strain and variation within litters. For

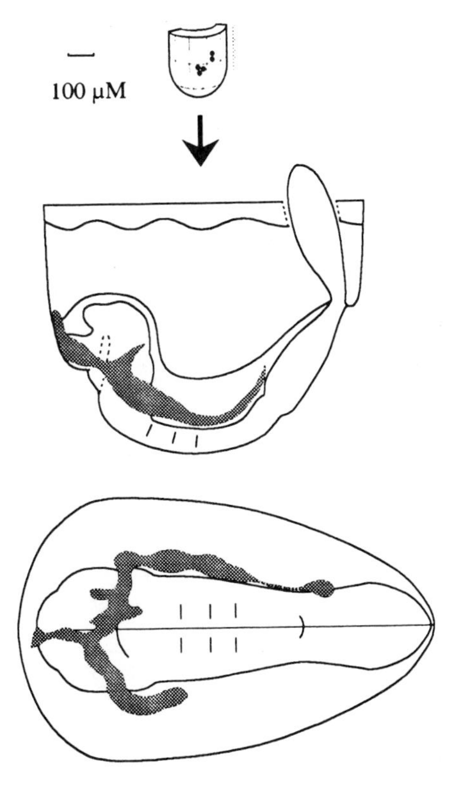


Fig. 7. Distribution of clones that contribute to the heart after labelling at the early-streak stage and culturing for 36 hr. The positions of precursors in epiblast are shown in the upper figure, the positions of their descendants in mesoderm are enclosed in the shaded areas in the two lower figures. The heart lies anterior to the foregut invagination which is indicated by a broken line in lateral view (middle figure) and by a broad arc spanning the midline in ventral view (bottom figure).

example, embryos of $(C_{57Bl} \times CBA)^2$ characteristically start primitive streak formation between 6.2 and 6.5 days and are smaller than embryos of noninbred Swiss mice which start between 6.5 and 6.7 days. Some vital statistics of embryo size and streak growth are given in Table 1.

The streak is initiated locally at the junction of the epiblast and extraembryonic ectoderm (Snell and Stevens, 1966; Tam and Meier, 1982; Hashimoto and Nakatsuji, 1989). It is presaged by slight embryo asymmetry and by thickening of the epiblast epithelium (Batten and Haar, 1979) (prestreak$^+$ in Table 1) which, according to Snow and Bennett (1978) is the result of an increase in frequency of mitotic spindles oriented perpendicular to the cell sheet in the streak region, thus producing several cell layers. During the early-streak stage the streak extends about 1/2 - 2/3 of the distance to the distal tip of the egg cylinder. It is not a well defined structure but is recognisable by the mesoderm which is produced from the time of its initiation, both proximally as extraembryonic mesoderm and laterally between the epiblast and endoderm. Not unexpectedly, the basal lamina between epiblast and endoderm of the streak is severely disrupted or absent (Batten and Haar, 1979; Poelmann, 1981; Franke et al., 1983). The anterior end of the streak is located at the distal tip of the embryo at the midstreak stage up to the neural plate stage and a recognisable node has formed by the late-streak stage. Laterally spreading mesoderm meets and overlaps axially spreading head process cells issuing from the anterior end of the streak, making the mesoderm layer continuous by the late-streak stage (Snell and Stevens, 1966).

Short term labelling is required to plot the cellular composition of and transitions in the streak, but in the absence of such data some conclusions can be drawn from the fate maps and clonal analysis already presented. Presumptive notochord is situated in the epiblast at the anterior end of the early streak, and is in the same relative position in the epiblast cup at the prestreak stage (Fig. 2). This suggests that cells may be being rapidly mobilized into the streak over approx. 100 $\mu$m during early-streak formation. Subsequent growth of the streak does not appear to be by anterior extension due to continued incorporation of axial epiblast: one day later the anterior portion of the streak still consists of cells descended from zone X (Figs 3, 8), while relatives and neighbours of these cells have passed into the mesoderm and endoderm. (Some descendants from 3/11 labelled precursors from zone X of the prestreak stage and 4/10 from the early-streak stage were found in the anterior portion of the streak after one day of culture). The composition of the rest of the late streak reflects its dynamic history: the main portion of the early streak (zone V) has been replaced; the middle portion is now occupied by descendants of cells originating in the lateral and anterolateral middle-tier injection zones; the posterior portion consists of cells derived from lateral, anterolateral and (in the neural plate stage) anterior proximal-tier injection zones (Figs 3, 8). Contributions from more anterior zones are located posterior in the streak relative to material derived from a more posterior zone in the same tier. These results are shown in more detail in Lawson et al. (1991).

Also taking into consideration the position of mesoderm descendants (Fig. 3) and the alignment of expanding epiblast clones towards the primitive streak (Lawson et al., 1991), the following statements can be made about the five-fold expansion of the streak (Table 1) between the early-streak and neural plate stages: 1) The anterior end of the streak is defined by a resident population of cells, descendants of which also contribute to axial and paraxial mesoderm. 2) The rest of the streak extends by the incorporation of descendants from progressively more anteriorly derived epiblast, later arrivals inserting more posterior to earlier ones. The incorporation of new material continues until at least the late-streak stage. 3) Cells continuously leave the growing streak to form mesoderm (and endoderm). All cells of the

Table 1. Linear dimensions of the embryonic portion of the mouse conceptus during gastrulation.

| Age | n | Stage[1)] | h[2)] (μm) m ± sd | d[3)] (μm) m ± sd | primitive streak[4)] (μm) |
|---|---|---|---|---|---|
| *C57BL x CBA, F2 embryos* | | | | | |
| 6 d | 93 | prestreak | 133 ± 15 | 137 ± 16 | - |
| 6½ d | 81 | prestreak | 164 ± 23 | 162 ± 18 | - |
| | 62 | prestreak$^+$ | 188 ± 17 | 185 ± 19 | - |
| | 103 | early streak | 212 ± 26 | 211 ± 20 | 110 (50 - 140) |
| | 15 | early streak$^+$ | 240 ± 28 | 235 ± 16 | 216 |
| 7 d | 35 | mid streak | 281 ± 46 | 266 ± 34 | 357 |
| | 7 | mid streak$^+$ | 350 ± 52 | 333 ± 20 | 445 |
| 7½ d | 84 | late streak | 383 ± 47 | 417 ± 89 | 502 |
| | 40 | late streak$^+$ | 404 ± 64 | 536 ± 73 | 557 |
| | 38 | neural plate | 423 ± 57 | 552 ± 112 | 581 |
| *Noninbred Swiss* | | | | | |
| 6¾ d | 71 | prestreak | 168 ± 31 | 167 ± 22 | - |
| | 18 | early streak | 255 ±30 | 227 ± 26 | 130 |
| 7½ d | 7 | mid streak | 337 ± 37 | 259 ± 34 | 411 |
| | 9 | late streak | 443 ± 63 | 410 ± 78 | 560 |
| 7¾ d | 4 | neural plate | 456 ± 78 | 557 ± 142 | 615 |

1. For criteria see Lawson et al. (1987, 1991). Prestreak$^+$ embryos had slightly thickened posterior epiblast. Early streak$^+$ embryos had the first indication of exocoelom formation; the primitive streak had not reached the distal tip of the egg cylinder.
2. The distance from the junction of epiblast and extraembryonic ectoderm to the distal tip of the egg cylinder.
3. Diameter of the egg cylinder at the junction of epiblast and extraembryonic ectoderm.
4. Calculated from mean h and d values considering the embryonic portion as a cylinder closed distally with a semicircular plug.

initially posterior part, derived from posterolateral and posterior epiblast, have left the streak by the midstreak stage and populate mainly extraembryonic mesoderm.

The changes between the neural plate and early-somite stages seem dynamically more simple (Figs 4, 8). The anterior end of the streak continues to generate axial and paraxial mesoderm, but descendants of the original resident population are rarely found there; their place has been taken by descendants from the neighbouring zone (IX) of the early-streak stage embryo. Although zone IX was not represented in the late streak, it had contributed to axial mesoderm and so some cells must have passed through the streak. Relatively few data were obtained for the behaviour of this zone in embryos cultured for one day, and representation in the streak could have been missed. Comparison of the composition of the rest of the streak of neural plate and early-somite stage embryos shows that a relatively posterior shift of streak material takes place (Fig. 8). This is presumably associated with the continued incorporation of posterior streak into allantois and other extraembryonic structures (Fig. 4), a process which is complete by the early somite stage (Tam and Beddington, 1987), and is evidence for the conclusion that shortening of the streak between 7.5 and 8.5 d is entirely due to loss from the posterior half of the streak of allantois and PGC precursors (Snow, 1981).

## SUMMARY AND CONCLUSIONS

The presumptive embryonic and extraembryonic mesoderm can be mapped onto the mouse epiblast at the onset of gastrulation, in spite of fast and noncoherent growth of the epiblast. There is no lineage restriction of the germ layers or of any of the structures evident at the early somite stage; therefore the predictability of regional fate is presumably due to the emphatic alignment of clonal descendants within the epiblast towards the developing primitive streak.

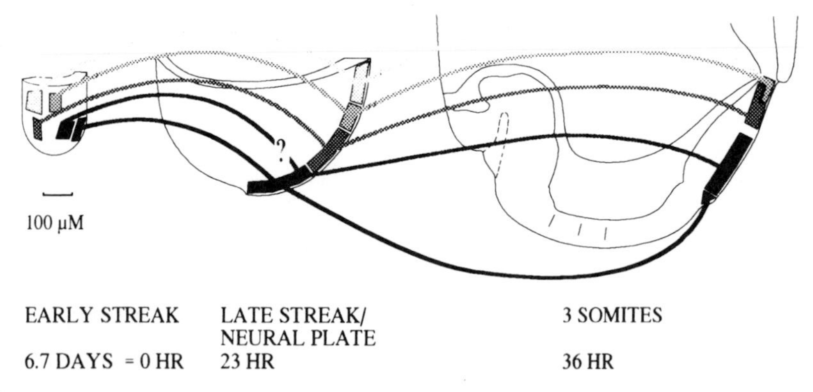


Fig. 8. Schematic summary of the composition of the primitive streak. The three figures represent lateral views of the embryonic portion of the conceptus at different stages of development. The extent of the primitive streak is indicated by a bar. Regions of the epiblast that have descendants in the streak at later stages are indicated by different intensities of shading and are joined by shaded lines to the appropriate region of the streak at later stages.

After the early streak stage the streak extends, not by mobilization of more anterior axial epiblast, but partly by growth of a resident population already in the anterior portion of the streak and mainly through expansion of its posterior half by incorporation of descendants from lateral and anterior epiblast. Mesoderm formation and primitive streak extension are concurrent processes, the posterior half of the streak being formed by the cells travelling through it into extraembryonic and lateral mesoderm. This activity may have slowed down by the late streak stage, after which much of the streak is incorporated directly into posterior structures. The anterior portion remains responsible for generating axial and paraxial mesoderm.

The mouse embryo has a fate map of topological identity with those of avian and amphibian embryos (Lawson et al., 1991). The process of gastrulation converts an epithelium, whether it be in the form of a flat sheet (chick and most mammals), a cup (rodents) or a hollow ball (amphibians) into a three layered structure by invagination through a primitive streak or a blastopore. It remains to be seen whether the different vertebrate groups employ similar cellular and molecular mechanisms to attain this end, and whether the rapid growth of the epiblast of mammals is a modifying or an additional feature.

## ACKNOWLEDGEMENTS

The work described here was supported by a NATO grant for collaborative research and by US DOE/OHER no. DEAC 03-76-SFO 1012. We are grateful to S. Kerkvliet and M. Flannery for histology, to W. Hage for computer graphics and to E. Cohen and M. Nortier for help in preparing the manuscript.

## REFERENCES

Batten, B.E. & Haar, J.L. (1979). Fine structural differentiation of germ layers in the mouse at the time of mesoderm formation. *Anat.Rec.* **194**, 125-142.

Beddington, R.S.P. (1981). An autoradiographic analysis of the potency of embryonic ectoderm in the 8$^{th}$ day postimplantation mouse embryo. *J.Embryol.exp.Morph.* **64**, 87-104.

Beddington, R.S.P. (1982). An autoradiographic analysis of tissue potency in different regions of the embryonic ectoderm during gastrulation in the mouse. *J.Embryol.exp.Morph.* **69**, 265-285.

Beddington, R.S.P., & Lawson, K.A. (1990). Clonal analysis of cell lineage. In *Postimplantation Mammalian Embryos: A Practical Approach.* (ed. A.J. Copp & D.L. Cockcroft) pp. 267-316. Oxford: IRL Press.

Franke, W.W., Grund, C., Jackson, B.W. & Illmansee, K. (1983). Formation of cytoskeletal elements during mouse embryogenesis. IV. Ultrastructure of primary mesenchymal cells and their cell-cell interactions. *Differentiation* **25**, 121-141.

Gardner, R.L. & Rossant, J. (1979). Investigation of the fate of 4.5 day *post coitum* mouse inner cell mass cells by blastocyst injection. *J.Embryol.exp.Morph.* **52**, 141-152.

Hashimoto, K. & Nakatsuji,H. (1989). Formation of the primitive streak and mesoderm cells in mouse embryos - Detailed scanning electron microscopical study. *Dev.Growth Differ.* **31**. 209-218.

Lawson, K.A., Meneses, J.J. & Pedersen, R.A. (1986). Cell fate and cell lineage in the endoderm of the presomite mouse embryo, studied with an intracellular tracer. *Devl Biol.* **115**, 325-339.

Lawson, K.A., Meneses, J.J. & Pedersen, R.A. (1991). Clonal analysis of epiblast fate during germ layer formation in the mouse embryo. *Development* **113**, 891-911.

Lawson, K.A., & Pedersen, R.A. (1992). Clonal analysis of cell fate during gastrulation and early neurulation in the mouse. In Postimplantation in the Mouse *Ciba Foundation Symposium* **165**, pp. 3-26.

New, D.A.T. (1990). Introduction. In *Postimplantation Mammalian Embryos: A Practical Approach* (eds A.J. Copp and D.L. Cockcroft). pp. 1-14. Cambridge: Cambridge University Press.

Poelmann, R.E. (1981). The formation of the embryonic mesoderm in the early post-implantation mouse embryo. *Anat.Embryol.* **162**, 29-40.

Snell, G.D. & Stevens, L.C. (1966). Early embryology. In *Biology of the Laboratory Mouse*, 2nd edn. (ed. E.L. Green) pp 205-245. New York: McGraw-Hill.

Snow, M.H.L. (1977). Gastrulation in the mouse: growth and regionalization of the epiblast. *J.Embryol.exp.Morph.* **42**, 293-303.

Snow, M.H.L. (1981). Autonomous development of parts isolated from primitive-streak-stage mouse embryos. Is development clonal? *J.Embryol. exp.Morph.* **65** (Suppl.), 269-287.

Snow, M.H.L. & Bennett, D. (1978). Gastrulation in the mouse: assessment of cell populations in the epiblast of $t^{w18}/t^{w18}$ embryos. *J.Embryol.exp.Morph.* **47**, 39-52.

Švajger, A., Levak-Švajger, B. & Škreb, N. (1986). Rat embryonic ectoderm as renal isograft. *J. Embryol. exp. Morph.* **94**, 1-27.

Tam, P.P.L. (1989). Regionalisation of the mouse embryonic ectoderm: allocation of prospective ectodermal tissues during gastrulation. *Development* **107**, 55-67.

Tam, P.P.L. & Beddington, R.S.P. (1987). The formation of mesodermal tissues in the mouse embryo during gastrulation and early organogenesis. *Development* **99**, 109-126.

Tam, P.P.L. & Meier, S. (1982). The establishment of a somitomeric pattern in the mesoderm of the gastrulating mouse embryo. *Amer. J. Anat.* **164**, 209-225.

Tam, P.P.L. & Meier, S., & Jacobson, A.G. (1982). Differentiation of the metameric pattern in the embryonic axis of the mouse, II. Somitomeric organization of the presomitic mesoderm. *Differentiation* **21**, 109-122.

# MEDIOLATERAL INTERCALATION OF MESODERMAL CELLS IN THE *XENOPUS LAEVIS* GASTRULA

Ray Keller and John Shih

University of California, Berkeley
Berkeley, CA 94720

## Introduction

The prospective mesoderm is important in nearly every aspect of early vertebrate embryogenesis because of its morphogenetic functions and because of the pattern-forming, inductive functions of the Spemann "organizer" contained in its dorsal sector (Spemann, 1938). The objective of this paper is to focus on the morphogenesis of the mesoderm in early embryogenesis, specifically on the powerful convergence and extension movements of the prospective dorsal mesoderm. The two major morphogenetic processes in the gastrula mesoderm are the migration of the leading edge mesoderm across the roof of the blastocoel (see Nakatsuji, 1984; Keller and Winklbauer, 1991) and the convergence (narrowing) and extension (lengthening) movements of the prospective dorsal, axial and paraxial mesoderm (see Keller, 1986). The function, the cellular basis, and the tissue interactions involved in convergence and extension of the dorsal mesoderm of *Xenopus* have been investigated extensively in this laboratory and reviewed with emphasis on history and function (Keller, 1986), on the underlying motility (Keller et al.,1991a), and on cell interactions regulating cell motility and behavior (Keller et al.,1991b). Our objective here will be to summarize and simplify the fundamental features of the convergence and extension movements, including their cellular basis, their function, and their regulation by tissue interactions in *Xenopus laevis,* the African clawed frog.

## What are convergence and extension?

*Convergence* is narrowing of tissue towards the dorsal side of the embryo and *extension* is elongation of the tissue in the anterior-posterior direction as it converges toward the dorsal midline (see Keller et al., 1991b). The two processes often occur together. In these cases they are probably the result of a single underlying cellular behavior and thus are referred to as *convergent extension* (Keller et al., 1991b). However, convergence of tissue may also result

*Formation and Differentia ion of Early Embryonic Mesoderm*
Edited by R. Bellairs *et al.*, Plenum Press, New York, 1992

in *thickening* rather than extension (Wilson et al., 1989), and extension may be produced by *thinning* rather than convergence (Wilson and Keller, 1991;Keller et al. 1991b). In these mass movements of tissue, change in one dimension of the tissue results in compensating changes in one or both of the other dimensions with conservation of tissue volume.

**How were convergence and extension discovered?**

Vogt (1929) and others before him (see Keller et al., 1991b, for a history) found that vital dye marks placed in the prospective dorsal mesodermal and neural tissues of amphibian early gastrula elongate in the anterior-posterior direction and narrow toward the dorsal midline during gastrulation and neurulation. Recent work using vital dyes (Keller, 1975, 1976), cell lineage tracers (Keller and Tibbetts, 1989; Keller et al., 1991a), and time-lapse videomicroscopy (Jacobson and Gordon, 1976; Keller, 1978; Wilson and Keller, 1991; Wilson et al., 1989; Keller et al., 1989a) have characterized these movements in detail.

**What tissues show autonomous convergence and extension?**

A major problem in analysis of any local distortion of tissue is to determine whether it is a passive response to forces generated elsewhere, or whether it is an active process, produced by forces generated locally. A second major problem is to learn whether these behaviors are specific to a particular region or tissue. Explants of the early gastrula show that convergence and extension are locally autonomous and not dependent on forces generated elsewhere (Schechtman, 1942; Keller and Danilchik, 1988). Sandwich explants made by sandwiching two dorsal marginal zones of the early *Xenopus* gastrula together [Fig. 1], converge and extend in culture, using forces developed internally and without exerting traction on the external substratum. These explants of the dorsal sector of the gastrula show two regions of convergence and extension; one occurs in the dorsal involuting marginal zone (DIMZ) and the other occurs in the dorsal noninvoluting marginal zone (DNIMZ) [Fig. 1]. The DIMZ involutes during gastrulation to form the notochordal and somitic mesoderm and the endodermal roof of the archenteron (Keller, 1975, 1976]. In cultured explants, the DIMZ converges, extends and forms somites and notochord in their normal relationship [Fig. 1]. In the embryo, the DNIMZ converges and extends from the midgastrula stage through neurulation and forms the posterior part of the nervous system, specifically the spinal cord and hindbrain (Keller et al.,1991c). The prospective spinal cord and hindbrain also converge and extend in sandwich explants (Keller and Danilchik, 1988) [Fig. 1]. Neural convergence and extension are induced by planar signals emanating from the DIMZ (Keller et al., 1991d). The DNIMZ differentiates to form a tangle of neurons but does not form a normal neural tube (Keller and Danilchik, 1988). In normal embryogenesis, vertical signals from the underlying mesoderm modify the development of the converging and extending prospective spinal cord and hindbrain, resulting in floorplate formation (see Holtfreter and Hamburger, 1955) and other aspects of

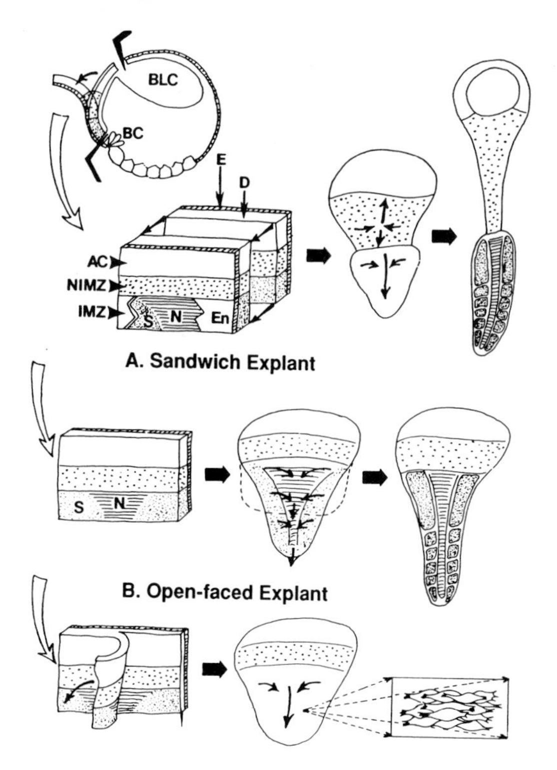


Figure 1. A diagram shows the development of sandwich explants (A), open-faced explants (B), and shaved open-faced explants (C). The dorsal sector of the early gastrula, seen in sagittal section at the upper left, is excised on both sides about 60 degrees from the midline and extending from the bottle cells (BC) to near the animal pole. The explant parts are: a superficial epithelium (E), a deep, nonepithelial (mesenchymal) region (D), an animal cap region (AC), a noninvoluting marginal zone (NIMZ), and an involuting marginal zone (IMZ). The AC and NIMZ regions are prospective neural ectoderm, the NIMZ making the ventral and ventrolateral parts of the spinal cord and hindbrain. The epithelial IMZ consists of prospective endoderm (En) and the deep region forms notochordal (N) and somitic (S) mesoderm. Sandwich explants show two regions of extension, one in the NIMZ and one in the IMZ (arrows in A). The open-faced explants are cultured in Danilchik's solution under a coverslip. Under these conditions, the IMZ converges, extends, and differentiates into somites and notochord (B). In open-faced explants, the NIMZ does not converge or extend (Wilson and Keller, 1991). Shaved explants (C) are made and cultured like open-faced explants but the inner (deepest) layers of cells are shaved off with an eyebrow knife, exposing the deep cells next to the epithelium to observation, videorecording, and analysis of cell behavior at high resolution (box, lower right). Figure and legend modified from Keller and others, 1991a.

neural patterning (see Discussion in Keller et al.,1991d). Since these vertical interactions do not occur in sandwich explants, many features of normal spinal cord and hindbrain development do not occur in sandwich explants (see Keller and Danilchik, 1988; Keller et al., 1991d). After the planar interactions have occurred, the neural region undergoes locally autonomous convergence and extension, beginning in the midgastrula stage (Keller and Danilchik, 1988; Keller et al., 1991d)[Fig. 1]. Convergence and extension of these explants are not dependent on attachment to or traction on an external substratum. In fact, adhesion to a substratum retards convergence and extension (see Keller and Danilchik, 1988).

**Which cells produce the forces driving convergence and extension?**

Does the epithelial (endodermal) monolayer or the deep mesenchymal (mesodermal) region of the marginal zone produce the forces driving convergence and extension? The superficial layer of the marginal zone consists of an epithelial monolayer of cells connected circumapically by a junctional complex, whereas the deep region consists of several layers of mesenchymal cells, locally attached to one another by focal adhesions(see Keller and Schoenwolf, 1977). The evidence shows that the deep mesodermal cells produce most of the forces driving convergence and extension. Manipulation of the deep mesodermal cells has major effects on marginal zone movements whereas manipulation of superficial epithelial cells has much less effect (Keller, 1981; Keller and Danilchik, 1988). More convincing is the fact that explants of deep DIMZ cells of the midgastrula, without epithelium, will converge and extend (Shih and Keller, 1991a) whereas the epithelium alone does not make a stable explant and does not converge and extend. We do not know whether the deep or superficial layer of the DNIMZ is mechanically active.

**How do the epithelial cells accomodate convergence and extension?**

The endodermal epithelial cells do not actively converge and extend and thus are mechanically passive. They appear to be passively distorted by the activity of the underlying mesodermal cells. Time-lapse recordings of the epithelial cells show that they accomodate convergence and extension by intercalating between one another along the mediolateral axis to form a longer but narrower array (Keller, 1978). This process is called *mediolateral intercalation*. Despite being connected circumapically by junctional complexes, epithelial cells can exchange neighbors while maintaining a physiological barrier (see Keller and Trinkaus, 1987). Although junctional complexes, including tight junctions and desmosomes, appear immutable in electron micrographs, they must be very dynamic to allow rearrangement. Cell rearrangement, including rearrangement of epithelial cells, occurs in morphogenesis throughout the Metazoa (Keller, 1987).

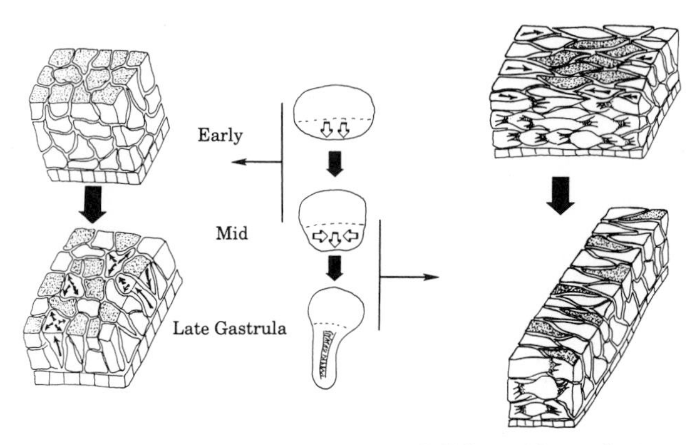


Figure 2. The behaviors of deep mesodermal cells during convergence and extension are illustrated as viewed from the inner (deep) surface of open-faced explants of the involuting marginal zone (IMZ). Convergence and extension of open-faced explants at early, middle and late gastrula stages are shown in the center. Deep cell behaviors are shown at the sides. Between the early and middle gastrula stages, the IMZ (area below the dashed lines) extends considerably but converges very little (arrows, center). High-resolution videorecordings show that this extension is produced by radial intercalation (left) in which several layers of deep cells intercalate along the radius of the embryo to occupy a greater area. From the middle gastrula stage onward, the IMZ continues to extend but also converges. Recordings of cell behavior show that this occurs by mediolateral intercalation in which individual cells move between one another along the mediolateral axis to form a longer, narrower array (right).Modified from Wilson and Keller, 1991.

The epithelial cells of the dorsal marginal zone may be specialized to allow rearrangement. As the DIMZ and DNIMZ elongate, the epithelial cells first lengthen to accommodate this distortion, and then they rearrange, returning to their original shape as they do so (see Keller, 1978). This suggests that they are stretched by movements of the underlying mesoderm, and then they return to their original shape as they intercalate between one another to form a longer array. In contrast, animal cap cells grafted into the marginal zone elongate much more than the native cells and do not rearrange (see Keller, 1981). The junctions of marginal zone cells may be specialized in a way that allows them to slide past one another.

**What do the deep mesodermal cells do to produce thinning/extension and convergence/extension?**

Time-lapse video recordings of cells in explants (Wilson et al., 1989; Wilson and Keller, 1991), tracing of cells with lineage tracers in embryos (Keller and Tibbetts, 1989) and explants (Keller et al., 1991c), and morphometric analysis of cells in embryos prepared for SEM (Keller, 1981) show various aspects of the cell behavior underlying convergence and extension.

However, the most direct evidence on deep cell behavior comes from high resolution video recordings of "open-faced" explants. Since the amphibian embryo is opaque, we devised a method of culturing the DIMZ and DNIMZ with the deep mesodermal cells exposed to observation, using "open faced" explants (Keller et al., 1989a; Wilson et al., 1989; Wilson and Keller, 1991) [Fig. 1B]. Recordings of these explants show that covergence and extension actually occurs in two phases. In the first half of gastrulation, the DIMZ thins and extends, but does not converge greatly [Fig. 2]. These *thinning* and *extension* movements are driven by *radial intercalation* of multiple layers of cells to form fewer layers of greater area [Fig. 2] (Keller, 1981; Wilson et al.,1989; Wilson and Keller, 1991). For reasons not yet understood, all the area produced by radial intercalation is expressed as increased length (extension) rather than as length and width (Wilson and Keller, 1991). Radial intercalation is so named because it involves intercalation of cells along radii of the spherical embryo (perpendicular to the surface) (Keller et al., 1991b).

In the second half of gastrulation, convergence and extension occur together, with some of the convergence being expressed as thickening [Fig. 2]. The convergence and extension occurring in the second half of gastrulation and through neurulation occurs by *mediolateral intercalation* of several rows of cells to form a longer, narrower array [Fig. 2] (Wilson et al., 1989; Keller et al., 1989; Wilson and Keller, 1991). Tracings of cells with fluorescent markers shows that the same sequence of radial and mediolateral cell intercalations occur in the embryo (Keller and Tibbetts, 1989; Keller et al., 1991c), and thus these behaviors are not a artifact of explantation and culture. Mediolateral intercalation is so named because the cells move between one another parallel to the mediolateral axis of the embryo.

A similar pattern of thinning and extension and convergence and extension, also driven by radial and mediolateral intercalation of deep cells, occurs in the neural region (Keller et al., 1991c).

**What kind of protrusive activity drives the mediolateral intercalation of mesodermal cells during convergence and extension?**

The terms radial and mediolateral intercalation describe what happens to cells as the tissue gets thinner and longer or narrower and longer, but they do not explain what moves the cells between one another. To move, cells must make protrusions, attach to a substratum, and either pull themselves forward or flow the cell body into the protrusion. High resolution, time-lapse videomicroscopy of epi-illuminated or fluorescently labeled deep mesodermal cells in cultured explants reveals the cell motility and protrusive activity underlying mediolateral intercalation in the gastrula(Shih and Keller, 1991a) and neurula stages (Keller et al., 1989a).

The motility driving mediolateral intercalation is best visualized in two types of explants that leave only one layer of deep cells for observation. With multiple layers of cells in the open-faced explant, the behavior observed in the inner, deepest layer may not be what is driving convergence and extension. The important behavior could be occurring in a layer closer to the epithelium. To avoid this difficulty, we made a "shaved" explant by removing all but the layer of deep cells immediately beneath the epithelium [Fig. 1C]. Thus any cell behavior observed in this layer is likely to be the "motor" of mediolateral intercalation (Shih and Keller, 1991a) [Fig. 2C]. In a second type of explant,the dorsal mesoderm is excised and cultured at the midgastrula stage, after it has thinned by radial intercalation; under these conditions, the explant consists of about one layer of deep cells and shows all the behavior associated with convergence and extension.

Based on extensive video recordings of cell behavior in the shaved explant and the deep cell explant, we have learned that mediolateral intercalation occurs as follows [Fig. 3]:

(1) The deep mesodermal cells divide once in the early part of gastrulation. Division occurs during radial intercalation in the embryo or in a thick explant, but little radial intercalation is seen in shaved explants.

(2) In the above period, protrusive activity, in the form of filiform and lamelliform protrusions, is rapid and directed randomly, so that the cells are multipolar.

(3) At the midgastrula stage, protrusive activity slows as stable protrusions are formed. These protrusions are filiform and lamelliform in shape. They are highly polarized, being directed medially and laterally, and they are applied directly to the surfaces of adjacent cells.

(4) The medially and laterally directed protrusive activity elongates the cells mediolaterally and at the same time brings all cells into parallel alignment with one another and the mediolateral axis.

(5) The continued protrusive activity at their medial and lateral ends results in intercalation of the elongated cells between one another, along the mediolateral axis (mediolateral intercalation).

**What is the evidence that the bipolar protrusive activity actually produces mediolateral intercalation?**

Because there is usually only one layer of deep cells in these explants, the protrusive activity observed among them is what probably moves them between one another. Most cells do not move without protrusive activity and the bipolar,

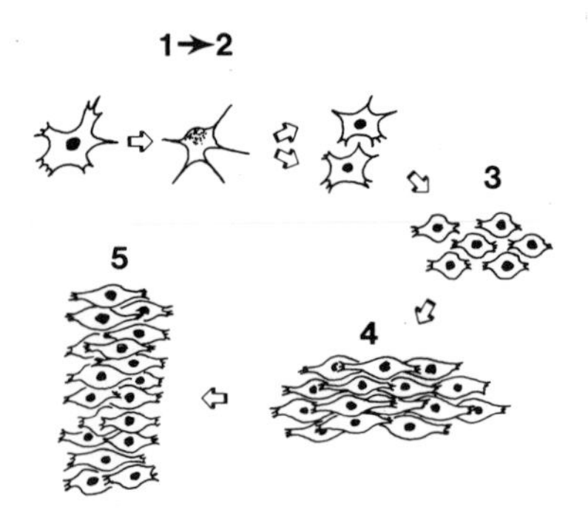


Figure 3. Diagrams illustrate the sequence of motile behavior shown by deep mesodermal cells before and during mediolateral cell intercalation.

mediolaterally directed protrusive activity is the only motile activity observed. Moreover, the relationship between protrusive activity and cell intercalation suggests that the two are causally related. The cells do not elongate and intercalate without bipolar protrusive activity. When the bipolar protrusive activity is unbalanced, the cells move as one would expected if these activities were controlling cell movement. For example, when one end of an elongated cell detaches, the cell quickly retracts and shortens. In poorly-extending explants, the cells show less intercalation and much greater elongation in the mediolateral direction. This suggests that medially and laterally directed protrusive activity normally pulls the cells between one another along the mediolateral axis. But if intercalation is resisted, the cells become more elongated in the mediolateral direction as they walk along the long sides of their neighbors.

**What are the deep cells using as a substratum during intercalation?**

The deep mesodermal cells, like the explant as a whole, do not require an external substrate to support their migration between one another. Instead, they appear to use one another's surfaces. The bipolar protrusions are lamelliform or filiform and they are closely applied to the surfaces of adjacent cells in video recordings (Shih and Keller,1991a) and in scanning electron micrographs (Keller et al., 1989b). Thus we believe that they use the surfaces of other cells as a substrate. Moreover, their protrusive activity does not appear to be contact inhibited. Thus they appear to be able to crawl between one another, wedging themselves into small spaces and expanding them, regardless of how tightly the cells are packed.

**What are the spatial and temporal patterns of expression of mesodermal cell behavior in the involuting marginal zone?**

High resolution videorecordings of thin, shaved explants show that the behavior of the mesodermal cells during mediolateral intercalation is highly organized with reference to the anterior-posterior direction, with reference to the lateral-medial direction, and with reference to the notochord-somite boundary (Shih and Keller, 1991b) [Fig. 4]:

(1) The bipolar protrusive activity and resulting cell elongation, alignment, and intercalation begin near the vegetal end of the DIMZ, forming an arc initially about 4 cells inward from the anterior end of the IMZ [#1 in Fig. 4]. This is called the *vegetal alignment zone* (VAZ).

(2) Then the *notochord-somite boundary* forms on either side, within the arc of elongated, aligned cells [asterisks, Fig. 4]. The boundary always begins as a shadowing between several cells, and thus is very short.

(3) Formation of the notochord-somite boundary then progresses vegetally (prospective anterior) and animally (prospective posterior) at the same time [#2, Fig. 4]. The posterior progression arcs laterally and animally, such that the posterior part of the notochord is much wider than the anterior, matching the outline of the prospective notochord on the fate map (Keller, 1976) [Fig.4].

(4) Cell elongation, alignment, and intercalation then progress inward from the notochord-somite boundary and posteriorly from the original VAZ [#3, Fig. 4]. Thus the last cells to undergo mediolateral intercalation are the ones in the posterior, medial notochord.

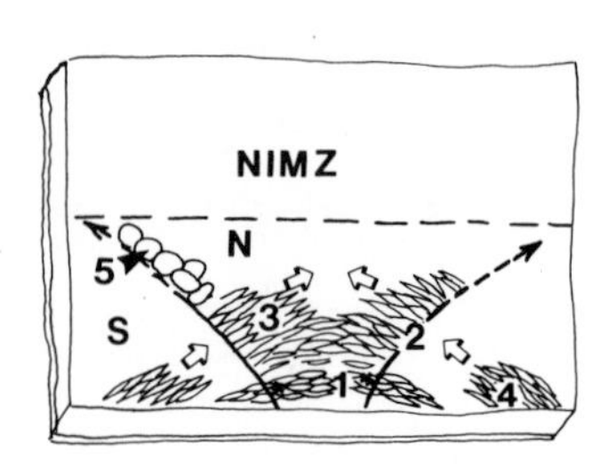


Figure 4. The diagram shows the spatial and temporal patterning of the cell behavior involved in mediolateral intercalation as it is displayed on the inner surface of a shaved explant of the dorsal marginal zone. N, prospective notochord; S, prospective somitic mesoderm.

(5) Cell elongation, alignment, and intercalation then progress in an arc from the prospective lateral somite boundary, dorsally and medially toward the notochord, and also from the prospective anterior somitic mesoderm toward the posterior somitic mesoderm [#4, Fig. 4]. Thus there is a progression of cell behavior from lateral to medial and from anterior to posterior in both notochordal and somitic tissues. This progression begins in the midgastrula and continues through neurulation when the posterior somitic mesoderm is still undergoing the early steps in the sequence (see Wilson et al., 1989; Keller et al., 1989a).

(6) Vacuolation of the notochord cells procedes from anterior to posterior along the boundary and inward from the boundary [#5, Fig. 4].

The progression of these behaviors is best visualized in shaved explants that do not converge and extend (Shih and Keller, 1991b). Normally, the same progression occurs in converging and extending explants and whole embryos but is altered by the movements themselves (Shih and Keller, 1991b).

## How does the posterior progression of mediolateral cell intercalation function in gastrulation?

The anterior-posterior progression of the arc of cell behaviors described above may have a function in involuting the marginal zone during gastrulation. The convergence and extension movements of the marginal zone can produce the involution of this region without the aid of pulling forces from the migration of the mesodermal cells on the inner surface of the animal cap and without the benefit of pushing forces from epiboly of the animal cap. The animal cap can be removed, depriving the mesoderm of a substrate for migration and also removing epibolic forces, but involution still

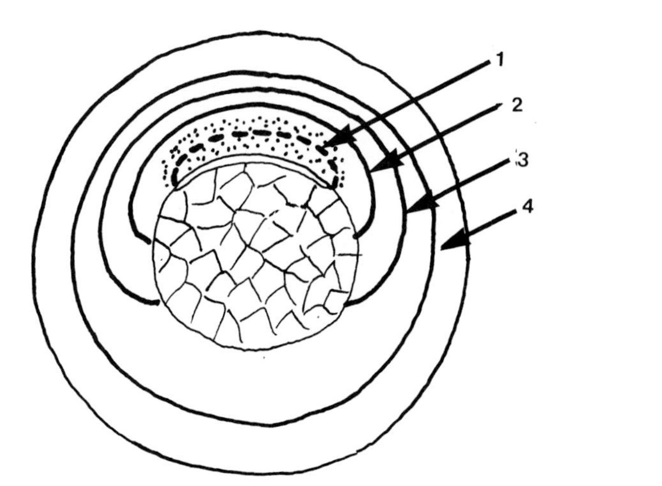


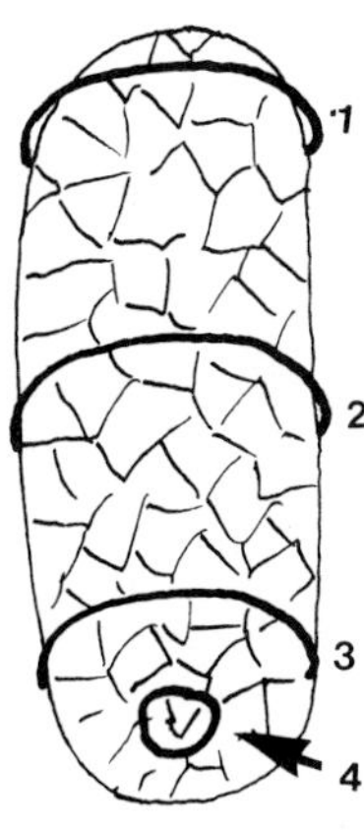


Figure 5. The role of the posteriorly progressing arc of mediolateral cell intercalation in involuting the marginal zone is illustrated. The involuting marginal zone of the early gastrula can be visualized as a series of hoops that will be progressively shortened (#1-4, right). As mediolateral intercalation progresses posteriorly, these hoops are shortened, one after the other, and the involuting marginal zone is pulled over the yolk plug to form the archenteron roof.

occurs (Holtfreter, 1933; Keller et al., 1985a, b). We believe that the mechanism by which convergence and extension can produce involution is related to the anterior-posterior progression of the cell behaviors driving mediolateral cell intercalation.

The way convergence and extension drive involution can be visualized as follows. Early in gastrulation, two processes conspire to produce a short, blind pocket, the incipient archenteron [shaded area, Fig.5, left]. The bottle cells bend the involuting marginal zone outward at the onset of gastrulation (Hardin and Keller, 1988), and in the early part of gastrulation the involuting marginal zone undergoes radial intercalation, and spreads vegetally, as described above. This tends to roll the IMZ over the bottle cells, forming a substantial groove or pocket mid-dorsally [Fig. 5].

Then shortly before stage 10.5 (early midgastrula), the VAZ forms in an arc extending from near the vegetal endoderm on one side, across the blastoporal lip to the other side [dashed line, Fig. 5, left].As the cells of the VAZ participate in convergence and extension by mediolateral intercalation, the arc shortens, forming a constricting hoop that tends to involute the IMZ. Progression of this arc of mediolateral cell intercalation animally (toward the prospective posterior) can be visualized as the progressive shortening of more hoops drawn at progressively more posterior levels in the IMZ, the last one being the limit of involution, which constricts around the blastopore [#4, Fig. 5, right] (see Wilson and Keller, 1991). Note that the VAZ forms in a postinvolution position, across the roof of the short archenteron of the early midgastrula. As the cell behavior initiated in the VAZ progresses posteriorly, the shortening arcs it produces would have a tendency to pass through the zone of involution, and into the preinvolution mesoderm. But the shortening arcs pull the IMZ over the lip, producing involution instead. The coordinate extension accounts for the increased anterior-posterior difference between the hoops as development proceeds [Fig. 5, right].

## Does the epithelial endoderm play a role in mediolateral intercalation?

The epithelial endodermal layer does not actively produce the forces causing the convergence and extension movements to occur, but there is evidence that the endoderm organizes some aspects of the mediolateral cell intercalation behavior among mesodermal cells. The deep mesodermal cells alone do not converge and extend when isolated at the early gastrula stage. However, by the midgastrula stage, they will converge and extend by mediolateral intercalation, using the protrusive activity described above (Shih and Keller, 1991a). Grafting experiments show that the epithelium of the dorsal marginal zone has the capacity to organize dorsal morphogenetic movements and tissue differentiation among deep mesodermal cells from the ventral side of a normal animal or from UV-ventralized embryos (Shih and Keller, 1991c). Thus in the first part of gastrulation, the epithelium appears to organize the deep mesodermal cells to do the mediolateral intercalation behaviors described above.

## Are radial and mediolateral intercalation common features of early morphogenesis ?

The teleost fish embryo shows a similar sequence of radial and mediolateral cell intercalation behaviors during early embryogenesis (Warga and Kimmel, 1990). Analysis of deep cell convergence on the embryonic shield in *Fundulus* shows that these cells intercalate but the protrusive activity and motility involved is different from that seen in *Xenopus*, in two major ways: the cells move into spaces rather than wedge themselves between closely packed cells, and they show contact inhibition of movement (Trinkaus et al., 1991). Mediolateral intercalation occurs in avian neurulation but oriented cell division and change in cell shape also play a role in this system (see Schoenwolf and Alvarez, 1989). Although cell intercalation is involved in many types of convergence and extension movements (Keller, 1987) other elongating and narrowing tissues may utilize other processes, such as the cell division seen in the avian embryo and change in cell shape (Condic et al., 1991).

## References


Condic, M., Fristrom, D., and Fristrom, J.,1991, Apical cell shape changes during Drosophila imaginal leg disc elongation: a novel morphogenetic mechanism. *Development* 111:23-33.

Hardin, J.and Keller, R., 1988, The behavior and function of bottle cells during gastrulation of *Xenopus laevis*. *Development* 103:211-230.

Holtfreter, J., 1933, Die total Exogastrulation, eine Selbstablosung des Ektoderms von Entomesoderm. *Arch. Entwicklungsmech. Org.* 129: 669-793.

Jacobson, A, Gordon, R., 1976, Changes in the shape of the developing vertebrate nervous system analyzed experimentally, mathematically, and by computer simulation. *J. exp Zool.* 197:191-246.

Keller, R.E.,1975, Vital dye mapping of the gastrula and neurula of *Xenopus laevis* . I. Prospective areas and morphogenetic movements of the superficial layer. *Develop. Biol.* 42:222-241.

Keller, R.E.,1976, Vital dye mapping of the gastrula and neurula of *Xenopus laevis*. II. Prospective areas and morphogenetic movements of the deep layer. *Develop. Biol.* 51:118-137.

Keller, R. E.,1978, Time-lapse cinemicrographic analysis of superficial cell behavior during and prior to gastrulation in *Xenopus laevis* . *J. Morph.* 157:223-248.

Keller, R. E.,1980, The cellular basis of epiboly: An SEM study of deep cell rearrangement during gastrulation in *Xenopus laevis. J. Embryol. exp. Morph.* 60: 201-234.

Keller, R. E.,1981, An experimental analysis of the role of bottle cells and the deep marginal zone in gastrulation of *Xenopus laevis. J. Exp. Zool.* 216:81-101.

Keller, R. E.,1984, The cellular basis of gastrulation in *Xenopus laevis*: active post-involution convergence and extension by medio-lateral interdigitation. *Am. Zool.* 24:589-603.

Keller, R. E.,1986, The cellular basis of amphibian gastrulation. In "Developmental Biology: A Comprehensive Synthesis" Vol. 2. Browder, L.(ed.) The Cellular Basis of Morphogenesis. Plenum Press, New York.

Keller, R. E.,1987, Cell rearrangement in morphogenesis. *Zool. Sci.* 4:763-779.

Keller, R. E. and Schoenwolf, G. C.,1977, An SEM study of cellular morphology, contact, and arrangement, as related to gastrulation in *Xenopus laevis. Wilhelm Roux's Arch.* 182:165-186.

Keller, R.E.,Danilchik, M., Gimlich, R., and Shih, J.,1985a, Convergent extension by cell intercalation during gastrulation of *Xenopus laevis*. In "Molecular Determinants of Animal Form" Edelman, G,M. (ed,) Alan R. Liss, Inc., New York. pp. 111-141.

Keller, R.E.,Danilchik, M., Gimlich, R., and Shih, J.,1985b, The function of convergent extension during gastrulation of *Xenopus laevis. J. Embryol. exp. Morph.* 89 Suppl: 185-209.

Keller, R.E., and Trinkaus, J.P.,1987, Rearrangement of enveloping layer cells without disruption of the epithelial permeability barrier as a factor in *Fundulus* epiboly. *Develop. Biol.* 120:12-24.

Keller, R. E.and Danilchik, M.,1988, Regional expression, pattern and timing of convergence and extension during gastrulation of *Xenopus laevis. Development*. 103: 193-210.
Keller, R. E. and Tibbetts, P. 1989. Mediolateral cell intercalation in the dorsal axial mesoderm of *Xenopus. laevis. Develop. Biol.*,131: 539-549
Keller, R. E., Cooper, M.S., Danilchik, M., Tibbetts,P. and Wilson, P.A.,1989a, Cell intercalation during notochord development in *Xenopus laevis. J. exp. Zool.* 251:134-154.
Keller, R. E., Shih, J. and Wilson, P.A.,1989b, Morphological polarity of intercalating deep mesodermal cells in the organizer of *Xenopus laevis* gastrulae.*Proceedings of the 47th annual meeting of the Electron Microscopy Society of America*. San Francisco Press. p. 840,
Keller, R.E., Shih, J.and Wilson, P.,1991a, Cell Motility, Control and Function of Convergence and Extension During Gastrulation in *Xenopus*. In "Gastrulation: Movements, Patterns, and Molecules". Keller, R., Clark, W. and Griffen. F, (eds.) Plenum Press. In press.
Keller, R.E., Shih, J., Wilson, P., and Sater, A.,1991b, Pattern and function of cell motility and cell interactions during convergence and extension in Xenopus. In "Cell-cell interactions in early development" *49th Symp. Soc. Develop. Biol.* Gerhart, J.C.(ed.) In press.
Keller, R.E., Shih, J. and Sater, A.,1991c, The cellular basis of convergence and extension of the *Xenopus* neural plate. Submitted.
Keller, R.E., Shih, J., Sater, A. and Moreno, C.,1991d, Planar induction of convergence and extension of the neural plate by the organizer of *Xenopus*. Submitted.
Keller, R. and Winklbauer, R.,1991, The cellular basis of amphibian gastrulation. In "Current Topics in Developmental Biology" Pedersen, R.(ed). In press.
Nakatsuji, N.,1984, Cell locomotion and contact guidance in amphibian gastrulation. *Am. Zool.* 24:615-627.
Schechtman, A.M.,1942, The mechanics of amphibian gastrulation. I. Gastrulation-producing interactions between various regions of an anuran egg (*Hyla regilia*). *Univ. Calif. Publ. Zool.* 51:1-39.
Schoenwolf G.C., Alvarez, I., 1989, Roles of neuroepithelial cell rearrangement and division in shaping the avian neural plate. *Development* 106: 427-439.
Shih, J. and Keller, R.E.,1991a, The mechanism of mediolateral intercalation during *Xenopus* gastrulation: directed protrusive activity and cell alignment. Submitted.
Shih, J. and Keller, R.E.,1991b, Patterns of cell motility in the organiser of *Xenopus*. Submitted.
Shih, J. and Keller, R.E.,1991c, The Epithelium of the dorsal marginal zone of *Xenopus* has organiser activity. Submitted.
Spemann, H.,1938, "Embryonic Development and Induction" Yale University Press, New Haven.
Trinkaus, J.P., Trinkaus, M. and Fink, R.,1991, On the convergent cell movements of gastrulation in *Fundulus*. J. Exp. Zool., in press.

Vogt, W.,1929, Gestaltanalyse am Amphibienkein mit ortlicher Vitalfarbung. II. Teil. Gastrulation und Mesodermbildung bei Urodelen und Anuren. *Wilhelm Roux Arch. EntwMech. Org.* 120:384-706.
Warga, R, Kimmel, C.,1990, Cell movements during epiboly and gastrulation in zebrafish. *Development* 108:569-580.
Wilson, P. A., Oster, G.and Keller, R., 1989, Cell rearrangement and segmentation in *Xenopus*: direct observation of cultured explants. *Development* 105:155-166.
Wilson, P.and Keller, R.,1991, Cell Rearrangement During Gastrulation of *Xenopus*: Direct Observation of Cultured Explants. *Development* 112:289-300.

# DIFFERENTIATION CAPABILITIES OF THE AVIAN PRECHORDAL HEAD MESODERM

Roswitha Seifert, Heinz Jürgen Jacob, and Monika Jacob

Ruhr-Universität Bochum
Abteilung für Anatomie und Embryologie
Universitätsstr. 150
D-4630 Bochum 1, Germany (FRG)

## INTRODUCTION

The early development of vertebrate embryos is characterized by a bilateral symmetry and a metameric segmentation of the paraxial trunk mesoderm in the form of the somites, which are considered to be the only sources of skeletal trunk musculature (Christ et al., 1986, 1990).

Apart from the occipital somites at its most posterior end the head region lacks a serial repetition of morphological units along the anteroposterior axis. Some swellings within the hindbrain called rhombomeres are sometimes considered to be segmental units (Guthrie and Lumsden, 1991). Holland (1988) and Fraser et al.(1990) described the appearance and distribution of homeobox expression which indicates a regional specification within the hindbrain corresponding to the rhombomeres. Couly and Le Douarin (1990) showed experimentally that the migration and spreading of the cephalic ectoderm corresponds with the expansion of the neuromeres.

Some structures in the paraxial mesoderm are still discussed as somite-homologues. These are the head cavities, the somitomeres and the visceral arches. Adelmann (1926) showed that one pair of premandibular head cavities exists in the chick embryo within the lateral parts of the prechordal head mesenchyme at the end of the second day of incubation. Jacob et al. (1984, 1986), who investigated these head cavities using structural and ultrastructural methods, concluded that they should not be described as somite-homologues.

Somitomeres were first described by Meier (1979) within the paraxial trunk mesoderm, as structures which precede the somites, and also within the paraxial head mesoderm (Meier, 1981), where no further segmentation follows. But their existence is not generally accepted.

The segmental nature of the visceral arches has often been discussed, but their development shows a great variability, so that they probably evolved secondarily to mesodermal segmentation.

Like the question of metamerism of the head mesenchyme, the origin of myogenic cells of head and visceral musculature is not yet settled, although the sources and distribution of mesenchymal and chondrogenic tissues are more firmly established (Le Lievre and Le Douarin, 1975, Le Douarin, 1982, Noden,

1978a+b, 1984). The muscles of the head mesoderm have been attributed to two sources, the axially located prechordal mesoderm and the paraxial mesoderm.

According to Adelmann (1922, 1926) there is a mesodermal proliferation immediately anterior to the notochord which he called the prechordal plate. With the formation of the foregut, this comes to lie on the dorsal wall of the gut between the anterior end of the notochord and the oral membrane. Since the term prechordal plate is also used for the site of the prospective oral membrane, to avoid confusion we prefer the term prechordal mesoderm which was introduced by Jacob et al.(1984). In line with Nicolet (1970) they took the view that in the chick embryo the prechordal mesoderm arises from Hensen's node, and they regarded it as the rostral part of the head process. According to Hinrichsen (1990) the rostral part of the chorda process in human embryos appears transiently in continuity with the endoderm forming the prechordal plate. By contrast, however, Jacob et al. (1984, 1986) considered that the prechordal mesoderm of the chick always exists as an independent mesodermal condensation anterior to the head process. From stage 8 HH onwards mediolaterally orientated cells can be observed detaching from the lateral margins of the prechordal mesoderm and perhaps migrating into the paraxial head mesoderm. According to the experimental results of Wachtler et al.(1984), Wachtler and Jacob (1986) and the observations of Jacob et al.(1984) these cells probably take part in the establishment of the extrinsic eye muscles. Moreover the experimental studies of Seifert and Christ (1990) showed that beside this the prechordal mesodermal cells of a quail, orthotopically grafted into a chick embryo, are able to migrate into the visceral arches. There they take part in the central core from which the branchiomeric musculature arises as McClearn and Noden (1988) described.

Johnston et al.(1979) and Noden (1983, 1984, 1986) who investigated the origins of avian cephalic muscles using the quail-chick marker technique concluded that the primordia of the extrinsic eye muscles derive from the paraxial head mesoderm, but their investigations were carried out on relatively late embryonic stages (9-11 HH). Although myogenic potency of prechordal mesoderm is found from stage 5 HH onwards (Wachtler et al. 1984, Krenn et al. 1988), this is true for the paraxial head mesoderm only from stage 8 HH onwards. Taking into account that in stages 8-10 HH Jacob et al.(1986) described a colonization of the paraxial head mesoderm by prechordal mesodermal cells it can be assumed that the myogenic potency of the paraxial head mesoderm is a secondary one.

The aim of this study is the further investigation of the prechordal head mesoderm:
- The development of the prechordal mesoderm in quail embryos is described and compared with that of the chick.
- Using the quail-chick marker technique the behaviour of prechordal mesoderm in heterotopic grafting sites is studied.
- The prechordal head mesoderm is compared with the paraxial trunk mesoderm with regard to myogenic potency, migratory ability, and pattern formation.

## LOCATION AND MORPHOLOGY OF THE PRECHORDAL MESODERM

Although the development and structure of chick prechordal mesoderm is well studied (Adelmann, 1922, 1926, Nicolet, 1970, Jacob et al., 1984, 1986) that of the quail is not,even though it is used in many experimental studies (Wachtler et al., 1984, Krenn et al., 1988).

According to Hamburger and Hamilton HH (1951) the early chick embryos are staged by their number of somites:

3 somites mean stage 8-HH, 4 somites mean stage 8 HH
5 somites mean stage 8+HH, 7 somites mean stage 9 HH.

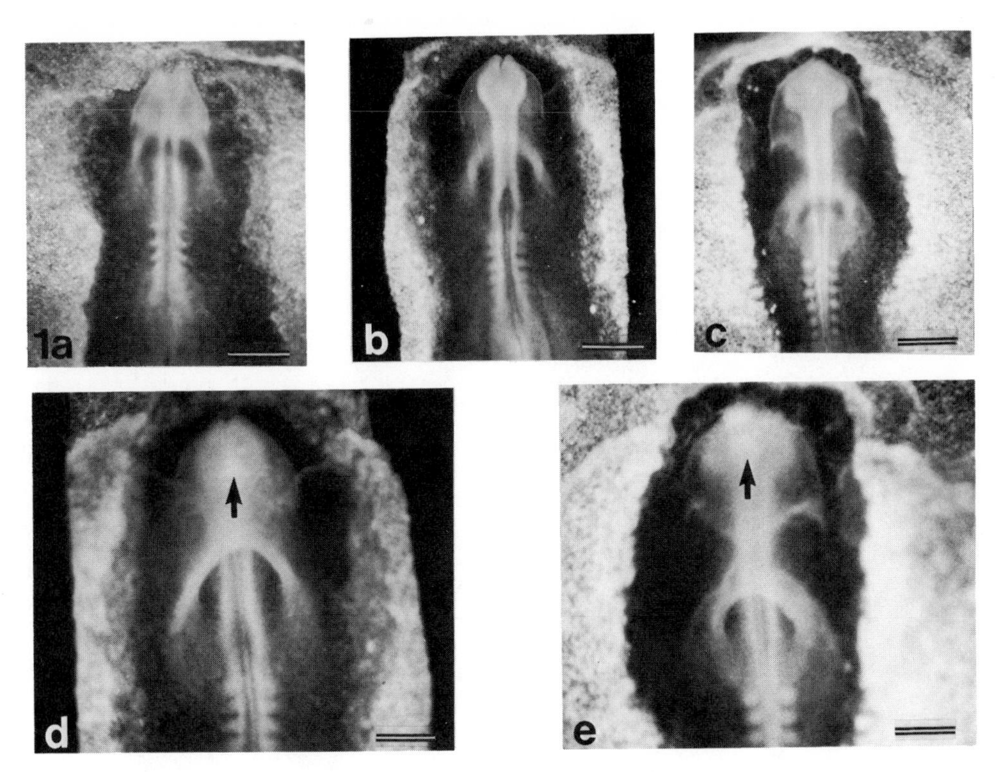


Fig.1. Dorsal view of a) a 4-somite chick embryo, b) a 4-somite quail embryo, c) a 7-somite chick embryo (bars = 500μm). Note the advanced head development of the quail compared with that of the chick. From a ventral view a round-shaped condensation can be recognized rostral to the notochord d) in the 4-somite quail embryo (arrow) and e) in the 7-somite chick embryo (arrow) (bars = 300μm).

Using these criteria also for staging quail embryos it has to be considered that the corresponding head development precedes that of the chick. As seen in Fig.1a-c the head of a 4-somite quail resembles that of a 7-somite chick more than that of a 4-somite chick. This means that the head development of the quail is about four hours in advance. Special attention has to be paid to a mesodermal condensation rostral to the head process (Fig.1d-e), which is distinctive in a 4-somite quail and a 7-somite chick embryo. This condensation appears to represent the prechordal head mesoderm; this has been studied in semithin sections.

<u>3-somite quail embryo</u>

In Fig.2a the prechordal mesoderm of a 3-somite quail embryo can be seen to be clearly delimited from the head process, which already shows notochordal features. In the quail embryo the prospective brain juts beyond the underlying prechordal mesoderm and the foregut, so that a space appears between the foregut and the medullary plate, especially in paramedian sections (Fig.2b). In line with Hamilton (1965) the term medullary plate is kept for the floor of the anterior neuropore until the further differentiation of the prosencephalon. Into this space cells which detach from the tip of the prechordal mesoderm seem to migrate into a caudo-rostrally direction (Fig.2c). Although the demarcation of the prechordal mesoderm to the adjacent dorsal neural floor is obvious, the one to the underlying endoderm is not clear, even though Jacob et al.(1984) observed the formation of an endodermal basal lamina from stage 7 HH onwards in the chick embryo. At its anterior end the endoderm of the foregut is thickened and seems to embrace the rostral part of the prechordal mesoderm from the rostro-ventral side, while the anterior mesodermal cells seem to be condensed at this point. Where the prechordal mesoderm and the columnar endodermal epithelium join each other an accumulation of cells can be observed. Whether these cells are of mesodermal or endodermal origin cannot be distinguished.

A remarkable feature is a pit, probably formed by the dorsal wall of the foregut bulging into the prechordal mesoderm (Fig.2d). This pit seems to separate the prechordal cell mass extending about 200 μm into a posterior and an anterior part. The posterior part resembles features of the presumptive notochord, whereas the anterior one takes part in the cell accumulation within the anterior part of the foregut. From the anterior part cells can be observed detaching and probably colonizing the space between foregut and neuroectoderm.

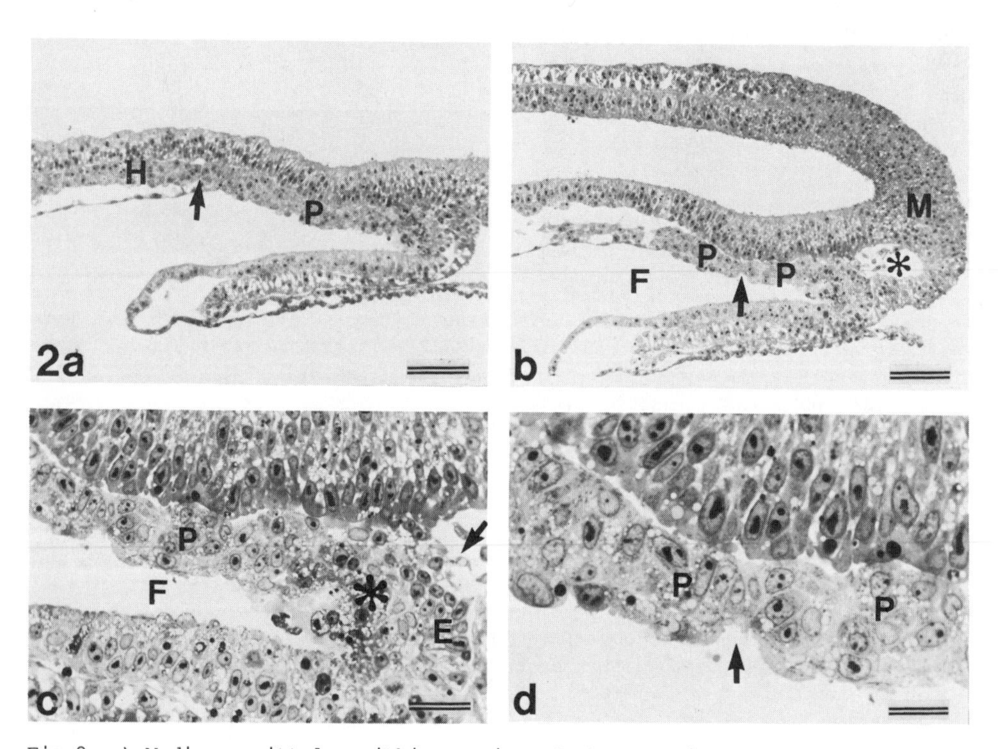


Fig.2. a) Median sagittal semithin section of the anterior head region of a 3-somite quail embryo. Note the demarcation between prechordal mesoderm P and head process H (arrow), bar = 70μm. b) Slightly paramedian section of the same embryo showing a clear space (asterisk) between the medullary plate M and the foregut F. Note the pit (arrow) that seems to divide the prechordal mesoderm P into two parts, bar = 70μm. c) At the anterior end of the foregut F the endodermal layer E becomes high columnar with distinctive apical cell contacts. Where prechordal mesoderm P and columnar endoderm join each other there arises a cell accumulation which is not clearly demarcated (asterisk). Dorso-rostrally there are cells visible which seem to detach into a rostral direction (arrow), bar = 20μm. d) The pit (arrow) is bulging into the prechordal mesoderm P, which appears not to be clearly separated from the underlying endoderm of the foregut, bar = 15μm.

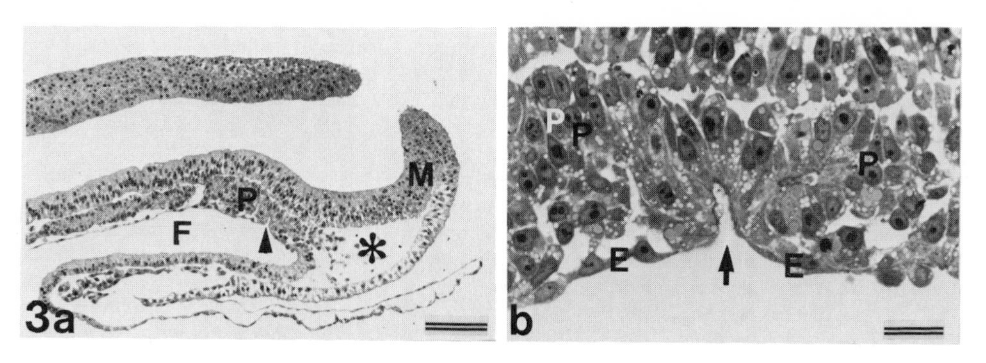


Fig.3. a) Median sagittal semithin section of the rostral head region of a 5-somite quail embryo. Compared with fig.2 the space between foregut F and medullary plate M has widened and the pit has deepened (arrowhead). Cells seem to detach from the prechordal mesoderm P in a rostral direction (asterisk), bar = 95µm. b) An enlarged detail from fig. a) shows the pit (arrow) that appears to be formed of high columnar endodermal cells. The borders of the prechordal mesoderm P to the underlying flat endoderm E of the foregut have become more obvious, bar = 18µm.

5-somite quail embryo

Sagittal sections give results similar to the 3-somite embryo (Fig.3), but the endodermal pit has more deepened and the space between the foregut and the medullary plate has widened. The prechordal mesoderm appears to be more separated from the underlying endoderm, and the amassing of cells within the anterior part of the foregut has been reduced to some single cells.

3-somite chick embryo

Since the endodermal pit, that seems to separate the prechordal mesoderm in the quail embryo, has never been described before, we also investigated semithin sections of chick embryos (Fig.4).

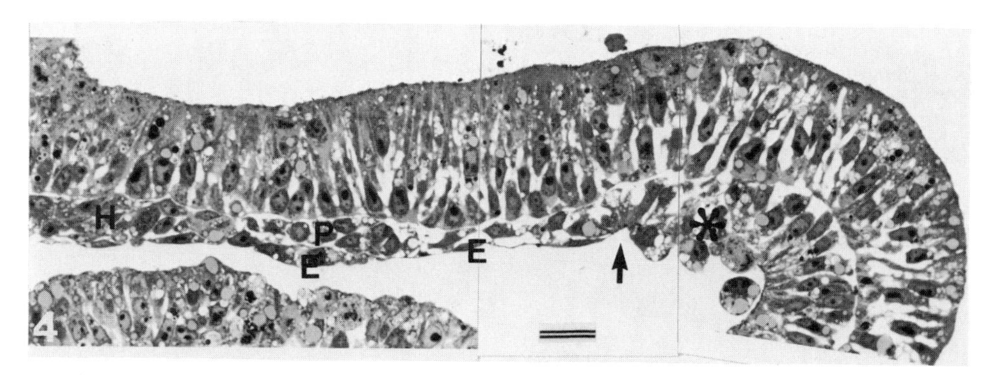


Fig.4. Median sagittal semithin section of the head region of a 3-somite chick embryo. The head process H continues into the prechordal mesoderm P which is obviously distinguishable from the underlying endoderm E. Note the pit (arrow) at the tip of the well recognizable prechordal mesoderm, bordered by high columnar endodermal cells, and a cell mass (asterisk) which cannot be assigned to the mesodermal or endodermal germ layer, bar = 20µm.

Surprisingly a comparable structure can be found, which seems to be bordered by endodermal cells. The endodermal cell layer at this point changes from a flat epithelium into a high columnar one. The posterior part of the prechordal mesoderm is continued with the following head process and has previously been described by Jacob et al. (1984, 1986). Rostral to the pit the continuation of the prechordal mesoderm and its border with the underlying endoderm is obscure, although the demarcation of the high columnar endoderm of the anterior part of the foregut to the prechordal mesoderm is obvious again. Some single cells can be found within the anterior end of the foregut. Cells which seem to detach in a rostral direction as visible in the quail embryo, have not been observed in the 3-somite chick embryo, neither in median nor in paramedian sections. Perhaps this is because there is no space between foregut and the neuroectoderm of the medullary plate in the chick. Further investigation is needed to determine whether the absence of this space is due to the fact that the chick head is less advanced than the quail head at the stages studied, or due to intrinsic differences in head development in the two species.

Summarizing, it can be stated that the macroscopically discernible condensation at the anterior end of the head process/notochord seems to coincide with the part of the prechordal mesoderm anterior to the endodermal pit.

For experimental procedures the prechordal mesoderm in its whole length of about 200 µm including the adjacent dorsal neural floor was taken as graft. In line with Couly and Le Douarin (1985, 1988) the rostral part of the neural primordia about a length of 450 µm has to be regarded as prosencephalon, the neural floor especially as prospective diencephalon.

## EXPERIMENTAL RESULTS

Using the quail-chick marker technique three series of experiments were performed (Fig.5).

### Grafting of prechordal mesoderm into a wing bud

In the first experimental series the prechordal mesoderm of quail embryos with 0 to 7 somites was heterotopically grafted into the wing bud of a four-day-

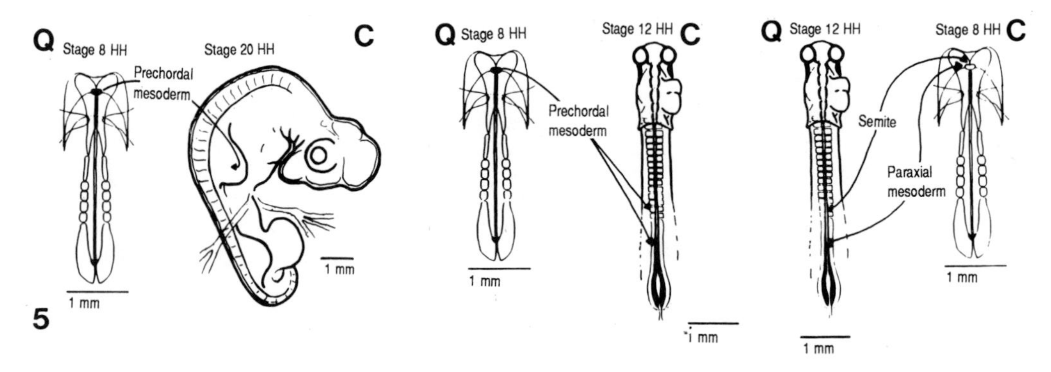


Fig.5. Experimental procedures: a) In the first series the prechordal head mesoderm of a quail (0-7 somites) was grafted into the wing bud of a 4 days incubated chick embryo. b) In the second series the prechordal head mesoderm of the quail was transplanted into the thoracic paraxial mesoderm of a 2-day-old chick embryo. c) In the third series paraxial trunk mesoderm of 2-day-old quail embryos was implanted to the prechordal mesoderm of chicks stages 6-9 HH. C: chick; Q: quail

old chick embryo. The graft included the adjacent neural ectoderm to stabilize the mesodermal tissue and to mark the site of implantation. Most of the operated wings developed normally, and the graft is located within the stylopodium, recognizable because of the existence of neural tissue with the typical quail nuclear marker at this site (Fig.6a). Quail cells can be found detached from the graft and distributed within the surrounding host tissue apparently at random (Fig.6b). They seem to differentiate into myotubes, if they contact muscle blastema or a suitable mesenchymal area which is of somatopleural origin (Fig.6c). In some cases graft-derived muscle cells have developed which participate in the normally patterned limb musculature (Fig.6d-e). Indications of a further distribution/migration have not been observed, perhaps because it does not take place or because the low number of grafted myogenic cells does not allow it. Other kinds of tissues made up of grafted cells cannot be detected.

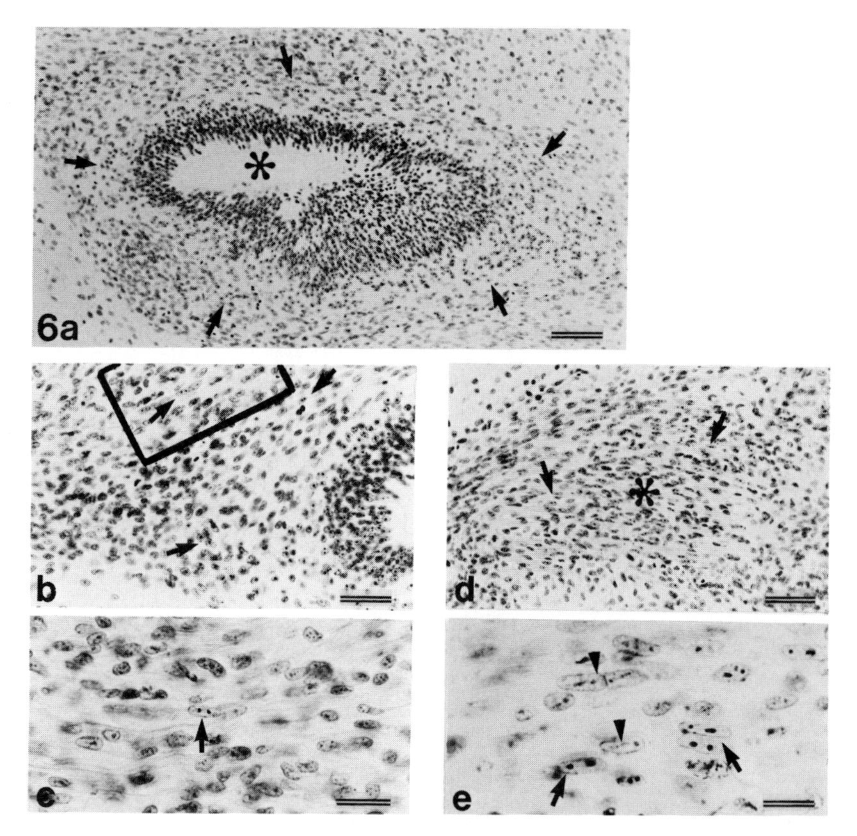


Fig.6. Results after grafting of prechordal mesoderm into a wing bud: a) The implantation site is characterized by ectopic development of graft-derived neural tissue (asterisk). Quail cells (arrows) can be found to have left the grafting site, bar = 90μm. b) At higher magnification the leaving quail cells can be followed over a short distance (arrows) migrating into the surrounding host mesenchyme, bar = 35μm. c) The detail marked in fig.b) shows a quail cell incorporated into a myotube (arrow), bar = 18μm. d) Within the chimeric wing bud of another embryo of the same series local musculature (asterisk) can be seen made up of graft-derived quail cells (arrows), bar = 80μm. e) Higher magnification of fig.d) shows myotubes in which chick nuclei (arrowheads) and quail nuclei (arrows) can be distinguished, bar = 10μm.

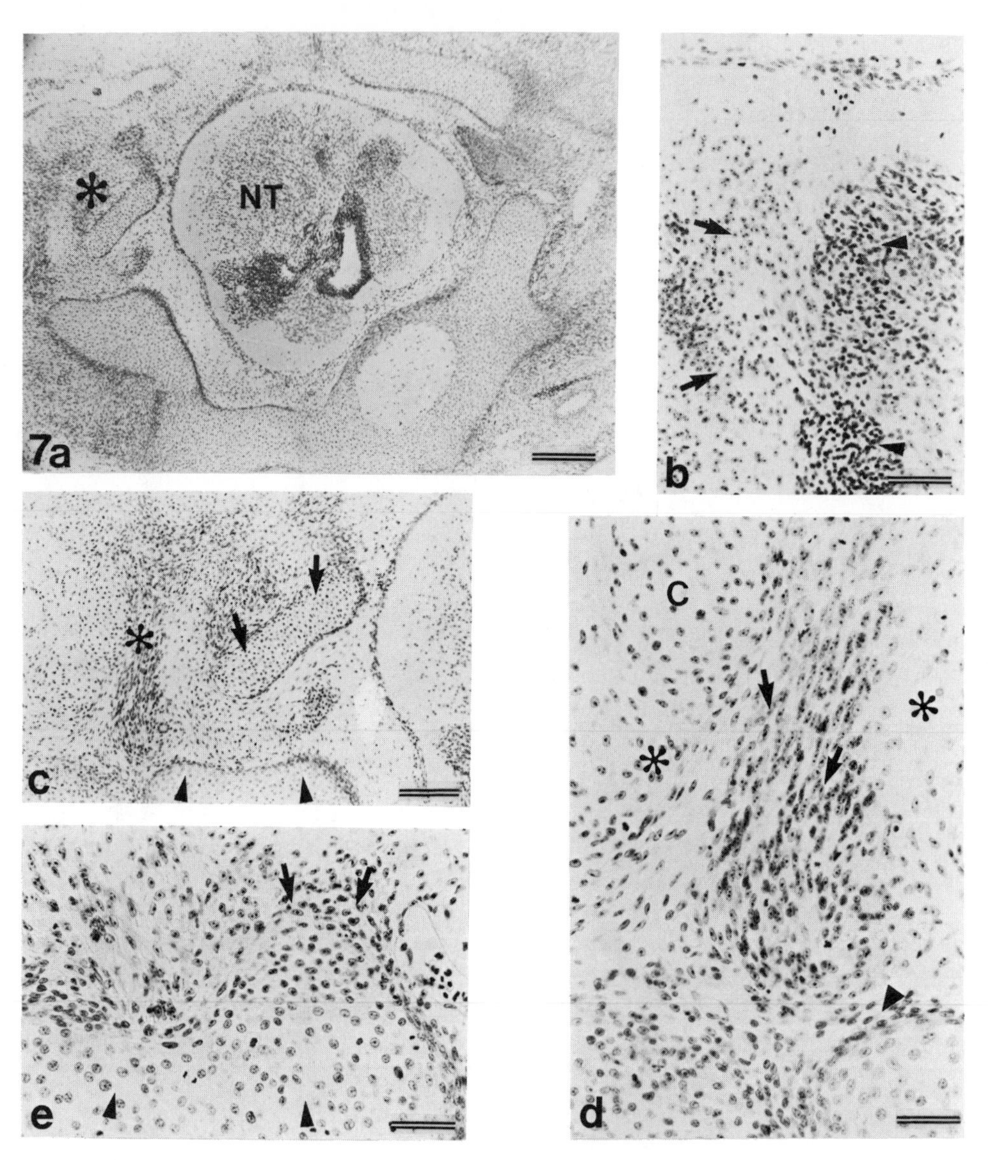


Fig.7. Results after replacement of thoracic paraxial mesoderm by prechordal head mesoderm: a) The operated side (asterisk) is changed and the neural tube NT appears to be enlarged, but the development is not very different from the normal side, bar = 150µm. b) A higher magnification of the neural tube of the operated side shows that the grafted neural quail cells (arrows) have fused with the host neural tube (arrowheads), bar = 45µm. c) At the implantation site a piece of cartilage made up of grafted quail cells (arrows) and another one of host origin (arrowheads) can be observed, asterisk - muscle blastema, bar = 95µm. d) The musculature of fig.c) in higher magnification is completely made up of graft-derived quail cells (arrows), and a piece of cartilage C and the surrounding connective tissue (asterisks) are also of graft origin. The muscle blastema remains separated from the adjacent host cartilage (arrowhead), bar = 30µm. e) Occasionally grafted chondrogenic cells of the quail (arrows) fuse with local cartilage of the chick (arrowheads), bar = 35µ

The ability of the grafted prechordal mesodermal cells to take part in the establishment of normally patterned limb muscles indicates a comparable state of differentiation of myogenic cells located within the prechordal head mesoderm and those located within the wing bud, which normally have immigrated from the dermomyotomes of the thoracic somites (Christ et al., 1977, Jacob et al., 1978, Wachtler et al., 1982).

Replacement of thoracic paraxial mesoderm by prechordal mesoderm

In the second series of experiments the prechordal mesoderm as described above was grafted in place of thoracic paraxial mesoderm of two-day-old chick embryos (Fig.7a). The microscopical analysis of the fate of the grafted tissue shows that the adjacent neural floor gives rise to neural tissue which is able to fuse with the neural tube of the trunk (Fig.7b). Moreover at the grafting site cartilage and muscle blastemas can be found surrounded by connective tissue as in normal development, but they are exclusively made up of graft-derived quail cells (Fig.7c-d). Although occasionally chondrogenic cells fuse with a local piece of host cartilage (Fig.7e), neither chimeric muscle blastema nor chimeric myotubes are observed. Therefore it might be that the appropriate local cells are missing, although this seems unlikely since additional implantation of prechordal mesoderm to paraxial trunk mesoderm gives comparable results.

In all experiments of this series the graft seems to be delimited from the surrounding host tissue, and mixing of graft and host tissue does not occur. Moreover no migration of myogenic quail cells can be observed, although the myogenic cells of this region in normal development do migrate colonizing the wing bud. Therefore the grafted cells of the prechordal mesoderm does not seem able to imitate this, although the findings of Seifert and Christ (1990) after orthotopical transplantation of the prechordal mesoderm clearly indicate a migration ability of prechordal head mesodermal cells as far as into the visceral arches.

Addition of paraxial trunk to axial head mesoderm

In the third experimental series the caudal half of a somite or a small piece of unsegmented paraxial trunk mesoderm of two-day-old quail embryos was added to the region of the prechordal head mesoderm of a chick embryo in st. 6-9 HH. This procedure always led to developmental malformations at the transplantation site which is located within the region of the prospective diencephalon between the eye anlagen (Fig.8).

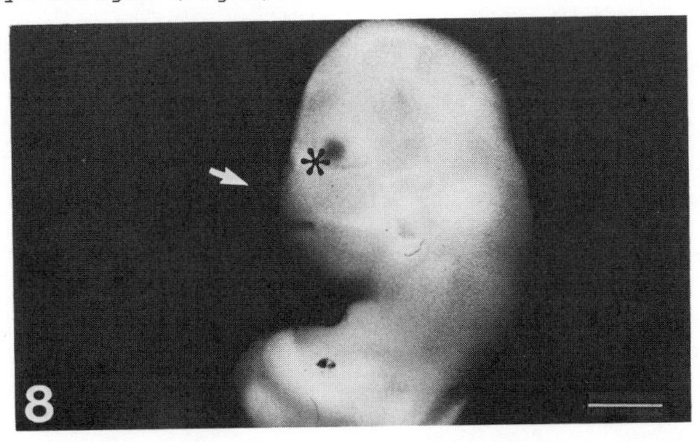


Fig.8. After implantation of paraxial trunk mesoderm into the axial prosencephalic mesoderm the chimeras show serious malformations such as absence of upper beak (arrow), rudimentary nasal process and eye anlagen (asterisk). The prosencephalon has not bent ventrally and lacks the typical telencephalic vesicles, bar = 2mm.

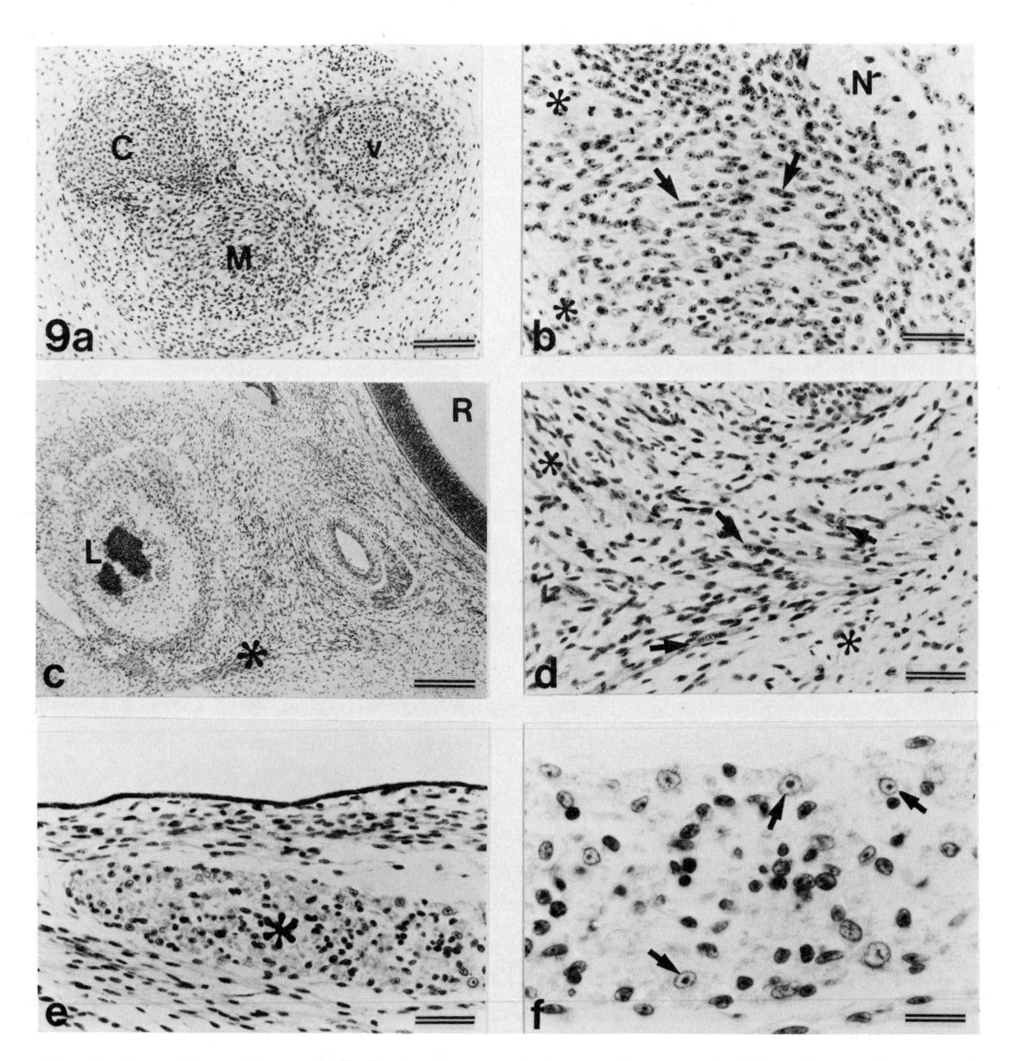


Fig.9. Results after addition of paraxial trunk to axial head mesoderm: a) After addition of the caudal half of a cervical somite a somite-like structure can be observed at the grafting site consisting of a chondrogenic C and a myogenic M region, completely made up of quail cells, V vessel, bar = 95µm. b) A detail of the myogenic region of fig.a) shows quail myotubes (arrows) and connective tissue (asterisks), whereas the nerve N is of chick origin, bar = 35µm. c) The right eye anlage R of the same embryo is normal, the left one L is only rudimentarily developed. Under the left eye anlage myotubes can be observed probably representing an extrinsic eye muscle (asterisk), bar = 200µm. d) In more detail the myotubes can be recognized containing quail (arrows) nuclei and surrounded by host-derived mesenchyme (asterisks), bar = 55µm. e) A normally developed extrinsic eye muscle (asterisk) of another embryo of this series after addition of the caudal half of a thoracic somite, bar = 35µm. f) This muscle also shows a participation of graft-derived quail myocytes (arrows), bar = 15µm.

The serious malformations concern the establishment of upper beak, egg tooth, eye anlagen and nasal cavity. All these structures are developed only rudimentarily or are missing. The development of the prosencephalon also has been disturbed. The telencephalon fails to bend ventrally, so that the eyes come to lie more medially and frontally.

According to the maps of Couly and Le Douarin (1987, 1988) it can be concluded that the implantation of paraxial trunk mesoderm probably has inhibited the migration of the neural crest cells into this part of the facial region.

In spite of the disturbed facial development the grafted tissue shows a self-differentiation of the somitic mesoderm. It forms somite-like structures containing a chondrogenic and a myogenic region and connective tissue , which are located ectopically within the prosencephalic mesenchyme (Fig.9a-b). Occasionally grafted cells can be made out within the extrinsic eye muscles, which once more indicates a participation of axially located mesoderm in their establishment (Fig.9c-f).

These results are comparable to those of Noden (1986). He also reported chimeric extrinsic eye muscles after replacement of chick paraxial head mesoderm of the mesencephalic and metencephalic level by quail paraxial trunk mesoderm. But he used chick embryos stages 9-11 HH. At these developmental stages, according to Wachtler et al.(1984) and Krenn et al.(1988), an immigration of myogenic precursor cells from axial into paraxial head mesoderm has probably taken place. Therefore it is possible that the axial mesodermal myogenic cells of the head take the pathway through the paraxial head mesenchyme, proliferating and migrating, to reach their final position.

However, Noden (1986) found many grafted myogenic cells taking part in the extrinsic eye muscles and moreover in jaw closing muscles. Compared with this in our experiments only a few grafted quail cells are observed within the extrinsic eye muscles and a participation of grafted myogenic cells in other head muscles is not certain. For these different results some reasons have to be considered: The described malformations or the low number of grafted myogenic cells in our experiments that perhaps does not allow a further distribution, and difficulties of the grafted cells in migrating out of the axially located grafting site.

## CONCLUDING REMARKS

The present morphological studies show

- the prechordal head mesoderm of quail embryos lies rostral to the notochord and on the dorsal wall of the foregut. In a 3-somite embryo it is already separated from the following head process.
- The quail prechordal mesoderm appears to be divided by a pit, probably made by a bulge of the dorsal foregut wall, into a posterior and an anterior part. From the anterior part cells seem to detach which are suggested to give rise to rostral prosencephalic mesoderm.
- A similar structure can also be found in the chick embryo.

The meaning of this pit, described here for the first time, remains still unknown. A possible comparability to previously described structures of ectodermal (Rathke, 1838) or endodermal origin (Seessel, 1877) needs further investigation.

The experimental procedures performed to improve knowledge about differentiation abilities and behaviour of the prechordal mesoderm show

- that the prechordal mesodermal cells possess myogenic potency. In this respect they are comparable to the paraxial trunk mesoderm.
- Moreover the prechordal head mesoderm of the quail has the capacity to

differentiate into cartilage and connective tissue in an appropriate environment, for example the paraxial trunk mesoderm. On the other hand a differentiation of cartilage and connective tissue does not occur if it is grafted into the wing bud. Therefore it might be that this ability is somehow suppressed by the mesenchymal environment or by preexisting host tissue.

- A migratory ability of prechordal mesodermal cells as provided by previous experimental results of Wachtler et al. (1984) and Seifert and Christ (1990) also seems to depend on the environmental situation. Grafted into the wing bud mesenchyme the prechordal myogenic cells leave the grafting site and migrate into the surrounding tissue, at least over short distances, whereas when grafted into thoracic paraxial mesoderm they do not migrate at all. This is probably due to the absence of an appropriate extracellular matrix as no host tissue can be found within or around the graft-derived musculature.
- The myogenic cells of the prechordal head mesoderm appear to be able to behave according to local cues of pattern formation within the wing bud mesenchyme. Transplanted in place of thoracic paraxial trunk mesoderm they differentiate into appropriate tissues, whose pattern in some cases resembles that in normal development, although no chimeric musculature can be observed. Vice versa, paraxial trunk mesoderm to a certain extent seems to be able to take part in normal muscle patterning of the head. If pattern formation is directed by mesenchymal properties, it must be remembered that the mesenchyme of the wing bud is of somatopleural origin, whereas that of the head consists of mesodermal and of mesectodermal neural crest-derived cells. Therefore in line with Noden (1988) we conclude that the mesenchymal environment generally influences the distribution and alignment of myogenic cells, irrespective of their origin.

ACKNOWLEDGEMENTS

The authors would like to thank H.Hake and U.Ritenberg for expert technical assistance and A. Jaeger and M. Köhn for excellent photographic work.

REFERENCES

Adelmann, H. B., 1922, The significance of the prechordal plate. An interpretative study, Amer. J. Anat., 31:55-101.

Adelmann, H. B., 1926, The development of the premandibular head cavities and the relations of the anterior end of the notochord in the chick and robin, J. Morph. Physiol., 42:371-439.

Christ, B., Brand-Saberi, B., Jacob, H. J., Jacob, M., and Seifert, R., 1990, Principles of early muscle development, in: The avian model in developmental biology: from organism to genes, editions du CNRS, pp.139-151.

Christ, B., Jacob, H. J., and Jacob, M., 1977, Experimental analysis of the origin of the wing musculature in avian embryos, Anat. Embryol., 150:171-186.

Christ, B., Jacob, M., Jacob, H. J., Brand, B., and Wachtler, F., 1986, Myogenesis: A problem of cell distribution and cell interactions, in: "Somites in Developing Embryos", R. Bellairs, D. A. Ede, J. W. Lash, eds., Plenum Press, New York, London, pp.261-275.

Couly, G. F., and Le Douarin, N. M., 1985, Mapping of the early neural primordium in quail-chick chimeras. I.Developmental relationships between placodes, facial ectoderm, and prosencephalon, Dev. Biol., 110:422-439.

Couly, G. F., and Le Douarin, N. M., 1987, Mapping of the early neural primordium im quail-chick chimeras. II. The prosencephalic neural plate and neural folds: Implications for the genesis of cephalic human congenital abnormalities, Dev. Biol., 120,198-214.

Couly, G. F., and Le Douarin, N. M., 1988, The fate map of the cephalic neural primordium at the presomitic to the 3-somite stage in the avian embryo, Development, 103,Suppl.:101-113.

Couly, G. F., and Le Douarin, N. M., 1990, Head morphogenesis in embryonic avian chimeras:evidence for a segmental pattern in the ectoderm corresponding to the neuromeres, Development, 108:543-558.

Fraser, S., Keynes, R., and Lumsden, A., 1990, Segmentation in the chick embryo hindbrain is defined by cell lineage restrictions, Nature, 344:431-435.

Guthrie, S., and Lumsden, A., 1991, Formation and regeneration of rhombomere boundaries in the developing chick hindbrain, Development, 112:221-229.

Hamburger, V., and Hamilton, H. L., 1951, A series of normal stage in the development of the chick embryo, J. Morphol., 88:49-92

Hamilton, H. L., 1965, Lillie´s Development of the chick, Holt, Rinehart and Winston, New York.

Hinrichsen, K. V., 1990, "Humanembryologie", Springer-Verlag, Berlin, Heidelberg, New York.

Holland, P. W. H., 1988, Homeobox genes and the vertebrate head, Development, 103,Suppl.:17-24.

Jacob, M., Christ, B., and Jacob, H. J., 1978, On the migration of myogenic stem cells into the prospective wing region of chick embryos, Anat. Embryol., 153:179-193.

Jacob, M., Jacob, H. J., Wachtler, F., and Christ, B., 1984, Ontogeny of avian extrinsic ocular muscles, Cell Tissue Res., 237:549-557.

Jacob, M., Wachtler, F., Jacob, H. J., and Christ, B., 1986, On the problem of metamerism in the head mesenchyme of chick embryos, in:"Somites in Developing Embryos", R. Bellairs, D. A. Ede, J. W. Lash, eds., Plenum Press, New York, London, pp.79-89.

Johnston, M. C., Noden, D. M., Hazelton, R. D., Coulombre, J. L., and Coulombre, A. J., 1979, Origin of avian ocular and periocular tissues, Exp. Eye Res., 29:27-43.

Krenn, V., Gorka, P, Wachtler, F., Christ, B., and Jacob, H. J., 1988, On the origin of cells determined to form skeletal muscle in avian embryos, Anat. Embryol. 179:49-54.

Le Douarin, N. M., 1982, "The Neural Crest", Cambridge Univ. Press, Cambridge, London, New York.

Le Lievre, C. S., and Le Douarin, N. M., 1975, Mesenchymal derivatives of the neural crest: analysis of chimaeric quail and chick embryos, J. Embryol. exp. Morph., 34:125-154.

McClearn, D., and Noden, D. M., 1988, Ontogeny of architectural complexity in embryonic quail visceral arch muscles, Amer. J. Anat., 183:277-293.

Meier, S., 1979, Development of the chick embryo mesoblast. Formation of the embryonic axis and establishment of the metameric patten, Dev. Biol., 73:24-45.

Meier, S., 1981, Development of the chick embryo mesoblast: Morphogenesis of the prechordal plate and cranial segments, Dev. Biol., 83:49-61.

Nicolet, G., 1970, Analyse autoradiographique localisation des differentes ebauches presomtives dans la ligne primitive de l´embryon de Poulet, J. Embryol. exp. Morph., 23:79-108.

Noden, D. M., 1978a, The control of avian cephalic neural crest cytodifferentiation. I. Skeletal and connective tissues, Dev. Biol., 67:296-312.

Noden, D. M., 1978b, The control of avian cephalic neural crest cytodifferentiation. II. Neural tissues, Dev. Biol., 67:313-329.

Noden, D. M., 1983, The embryonic origins of avian cephalic and cervical muscles, and associated connective tissues, Amer. J. Anat., 168:257-276.

Noden, D. M., 1984, Craniofacial development: New views on old problems, Anat. Rec., 208:1-13.

Noden, D. M., 1986, Patterning of avian craniofacial muscles, Dev. Biol., 116:347-356.

Noden, D. M., 1988, Interactions and fates of avian craniofacial mesenchyme, Development, 103,Suppl.:121-140.

Rathke, H., 1838, Über die Entwicklung der Glandula pituitaria, Arch. Anat. Physiol. Wiss. Med., 4:482-485.

Seessel, A., 1877, Zur Entwicklungsgeschichte des Vorderdarmes, Arch. Anat. Entw., 1:449-467.

Seifert, R., and Christ, B., 1990, On the differentiation and origin of myoid cells in the avian thymus, Anat. Embryol., 181:287-298.

Wachtler, F., Christ, B., and Jacob, H. J., 1982, Grafting experiments on determination and migratory behaviour of presomitic, somitic, and somatopleural cells in avian embryos, Anat. Embryol., 164:369-378.

Wachtler, F., Jacob, H. J., Jacob, M., and Christ, B., 1984, The extrinsic ocular muscles in birds are derived from the prechordal plate, Naturwissenschaften, 71:379.

Wachtler, F., and Jacob, M., 1986, Origin and development of the cranial skeletal muscles, Biblthca anat., 29:24-46.

# ON THE DIFFERENTIATION AND MIGRATION OF THE WOLFFIAN DUCT IN AVIAN EMBRYOS *

Monika Jacob, Heinz Jürgen Jacob, and Roswitha Seifert

Ruhr-Universität Bochum
Abteilung für Anatomie und Embryologie
Universitätsstr. 150
D-4630 Bochum 1, Germany (FRG)

## INTRODUCTION

It has long been believed that the Wolffian (pro- or mesonephric) duct is formed by the fusion of the pronephric tubules, although Gasser (1877) had described its origin from an unsegmented ridge at the dorsal side of the intermediate mesoderm. This observation was confirmed by Torrey (1954) on human embryos, and finally, by SEM investigations (Jacob et al., 1979; Jarzem and Meier, 1987) on chick pronephros.

Whereas the cranial part of the pronephric duct primordium develops into an epithelial, canalized duct, the distal part remains mesenchymal and grows in a caudal direction to join the cloaca. Since the growth of the Wolffian duct is a prerequisite for the development of the mesonephros as well as of the metanephros, we have studied the modalities of this growth in some previous investigations (Jacob and Christ, 1978; Jacob et al., 1986; Jacob et al. 1990; Jacob et al., 1991).

This article reviews the most important data from our former investigations on chick and quail embryos and introduces at the same time some new data determining reasons and conditions for the directed growth of the duct. In particular, the following questions are answered:

1. Is the duct shifted caudad by the growth of the adjacent structures, particularly the paraxial mesoderm?
2. Is the elongation of the duct caused by cells in the distal tip dividing faster than in the adjacent regions?
3. What is the role of the extracellular matrix, especially fibronectin, during the movement of the duct?
4. Is the HNK-1 epitope involved in the adhesion between duct cells and their matrix?

## POSITION AND SHAPE CHARACTERISTICS OF THE CAUDAL END OF THE WOLFFIAN DUCT PRIMORDIUM

To analyze whether the Wolffian duct elongates in synchronization with somitogenesis and whether there is a causal relationship between the growth of

* Dedicated to Prof. Dr. med. Klaus V. Hinrichsen on the Occasion of his $65^{th}$ Birthday.

the paraxial mesoderm and the duct as seen in Ambystoma (Poole and Steinberg, 1982), we investigated the position of the pronephric duct tip relative to the new forming somites as shown in Fig.1. The drawings are based on SEM micrographs of identical magnifications. In embryos at stage 10 HH, the tip of the Wolffian duct is located about two somite lengths posterior to the last formed somite. In St 11 HH embryos, the tip of the duct has extended about 3 somite lengths caudal the last somite and by St 13 HH about 6 somite lengths. Thus, in chick embryos no synchronization between somitogenesis and caudad movement of the Wolffian duct can be found.

In Ambystoma, unlike in chick embryos, somite segmentation precedes pronephric duct tip advance always by two somite lengths (Poole and Steinberg, 1982) and both processes progress by tandem. Although the general pattern of pronephric duct formation is the same in various amphibians, morphometric data (Lynch and Fraser, 1990) have revealed some differences in duct elongation between urodeles and anurans: in Xenopus, at first the pronephric duct lengthens synchronously with the surrounding structures, but at later stages, the duct actively elongates along the caudal myotomes.

Our results suggest that the growth of the Wolffian duct in avian embryos is not due to a passive shifting caused by the paraxial mesoderm, but that it rather seems to be an active migration. If this migration is guided by a cranio-caudally travelling adhesion gradient, as postulated for Ambystoma (Poole and Steinberg, 1982; Gillespie et al., 1985), it could be, at best, indirectly correlated with somitogenesis.

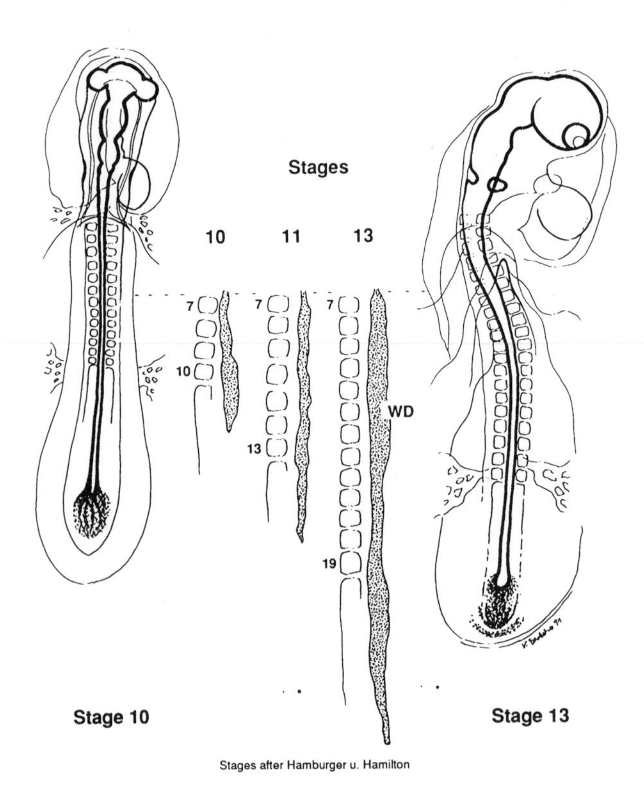


Fig. 1. Scheme showing the position of the Wolffian duct relative to the new forming somites and the unsegmented paraxial mesoderm during St 10, 11 and 13 HH. For better understanding, the first and last stages are shown in their complete form at identical magnification on either side.

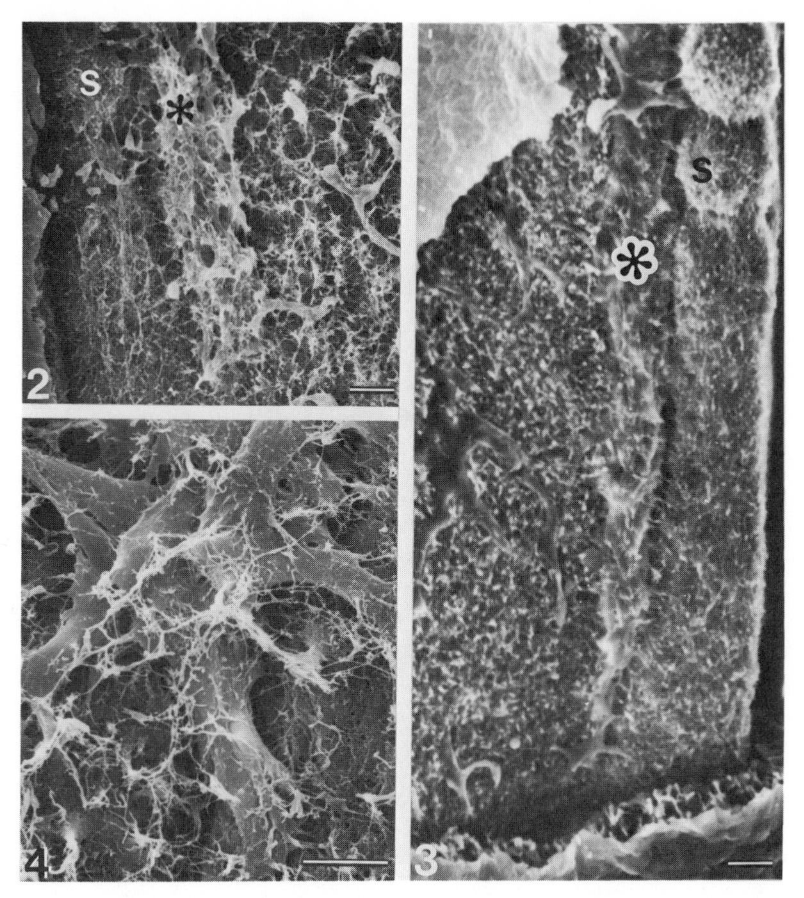


Fig. 2. Scanning electron micrograph of an embryo at stage 10 HH, from which the ectoderm has been removed. S, somite; asterisk, Wolffian duct. Bar = 25 µm

Fig. 3. Scanning electron micrograph of an embryo at stage 13 HH. S, last somite; asterisk, Wolffian duct. Bar = 25 µm

Fig. 4. Greater magnification of the tip of the Wolffian duct. Cell processes in contact with extracellular matrix. Bar = 5 µm

The assumption that the tip of the Wolffian duct migrates actively on the intermediate mesoderm is also supported by the morphological features of the distal cells. In St 10 HH embryos, the Wolffian duct is a uniform ridge with a broad distal end (Fig.2). By St 13 HH the tip becomes narrower towards the caudal end and the cells are elongated in cranio-caudal direction (Fig.3). The most distal cells are not identically oriented and extend their processes in various directions as if searching for the right way (Fig.4). Filopodia are in close contact with a dense network of extracellular matrix. Thus their morphology corresponds to those of other migrating cells.

MITOTIC RATE OF THE WOLFFIAN DUCT

In order to test whether the migration of the Wolffian duct is the result of a "growth energy" (Grünwald, 1942) of cells in the duct tip dividing faster, we estimated their mitotic rate using the BrdU-antiBrdU technique (Gratzner, 1982). Since SEM micrographs have revealed considerable elongation of the duct primordia during St 12 and 13 HH, we chose these stages for our study.

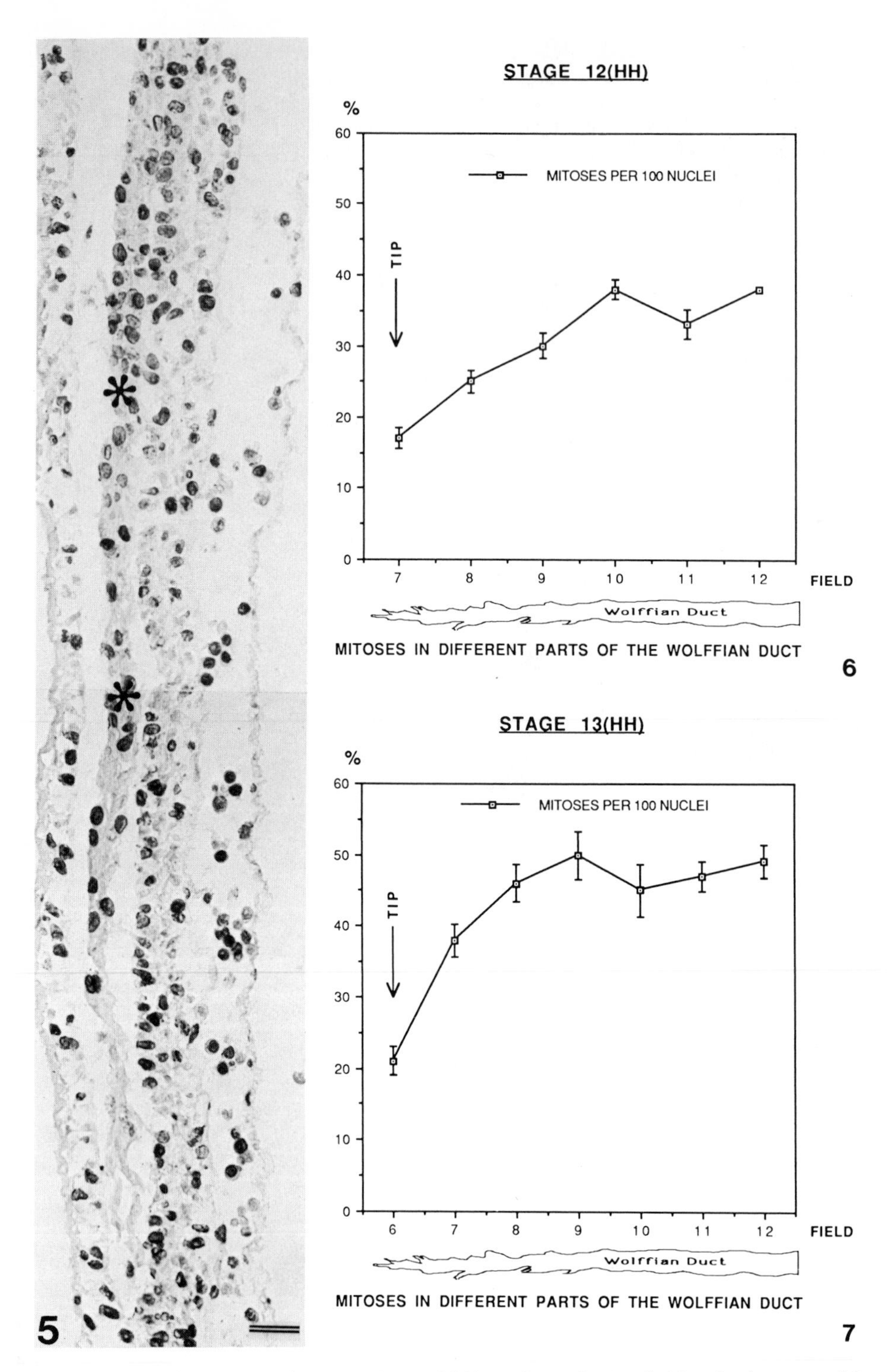


Fig. 5. Sagittal section of the Wolffian duct (asterisk). S-phase cells are stained using the BrdU-anti BrdU technique. For the micrograph only weak counterstaining was used. Bar = 25 µm

Fig. 6.–7. Mitotic rate of the Wolffian duct. Values are means ±S.E. of three to nine counts.

After incubation for 1 h with 25 µl of 0.4 M BrdU in PBS, specimens were fixed, embedded, and cut at 4 µm. Sections were deparaffinized and rehydrated, and the standard indirect immunoperoxidase technique was performed with a monoclonal anti-BrdU (Dakopatts). To allow greater accessibility to the antibodies, the sections had been treated with HCl to denature the DNA. Slides were counterstained with hematoxylin. Using oil-immersion, counts were made in fields of 180 µm length in every second sagittal section (Fig.5). Thus, the tip of the Wolffian duct is found in field 6 or 7 respectively, starting at the tail bud.In stage 12 HH embryos (Fig.6), the mitotic rate is significantly lower in the tip than in the more cranial field ($p<0.05$ by Student t test).

In stage 13 HH embryos (Fig.7), the mitotic rate is also high in the most cranial part of the duct and decreases rapidly towards the tip which shows a significantly lower rate than the immediately adjacent field of the duct ($p<0.001$, t test). The high mitotic rate in the cranial part of the Wolffian duct corresponds to that in the paraxial mesoderm at the same level. This is in line with the results of Overton (1959) on colchicine-treated stage 14 HH embryos. But, in contrast to us, she found the mitotic frequency to be greatest towards the posterior tip of the elongating pronephric duct.Our results as well as the morphological data lead us to conclude that the elongation of the Wolffian duct is due to an active migration of the cells at the tip of the duct.

## THE ROLE OF THE EXTRACELLULAR MATRIX IN THE ACTIVE MIGRATION OF THE WOLFFIAN DUCT

According to experimental studies of Jacob and Christ (1978) on chick-quail chimeras, even cranial parts of the Wolffian duct are able to migrate caudally and to join the cloaca if grafted in place of the duct tip. Similar results were obtained by Poole and Steinberg (1982) who, in a series of grafting experiments, showed that in Ambystoma only a particular region of the flank mesoderm promotes duct migration. They proposed that the duct-guiding information comes from a craniocaudally travelling adhesion gradient. Therefore, it seemed likely that cell-substratum adhesion is involved in the directional migration of the duct mesoderm.

As fibronectin has been shown to be the substratum for many migrating cells, e.g. neural crest cells (Newgreen and Thiery, 1980) and precardiac mesoderm (Linask and Lash, 1986), we considered this glycoprotein the main candidate for guiding duct migration in a haptotactic manner. Recently we have confirmed the importance of fibronectin in duct migration by using synthetic peptides which mimic the cell binding domain of this matrix molecule (Jacob et al., 1990; Jacob et al., 1991): After competitive inhibition of the fibronectin receptor by the pentapeptide GRGDS, the cells at the tip of the Wolffian duct are rounded and their migration is inhibited. Thus, the substance for a craniocaudally travelling adhesion gradient seems to have been found. Unfortunately these findings are not consistent with the results of Ostrovsky et al. (1983) concerning fibronectin distribution during somitogenesis in chick embryos. Adjacent to the migrating Wolffian duct, fibronectin increases towards the anterior end of the segmental plate which will form the next somite, and is localized mainly in the basal lamina of the newly formed somites. Therefore, the hypothesis that the migration of the mesodermal cells of the Wolffian duct is possibly guided in a haptotactic manner by fibronectin needs to be re-examined.

Consequently, we investigated the distribution of fibronectin along the migration pathway of the Wolffian duct at first at the light microscopic level using the indirect immunoperoxidase reaction. Paraffin sections were treated with a polyclonal anti-fibronectin (gift from Dr.Poelmann, Leiden). Fibronectin is found at the surface of the paraxial mesoderm, the somatopleure and at the cells of the duct tip (Fig. 8). More cranially (Fig. 9), the basal laminae of the somites are strongly stained for fibronectin, and an immunoreaction is also found predominantly at the basal surface of the Wolffian duct, indicating the onset of polarization and epithelialization within this part of the duct.

The ultrastructural localization of fibronectin was investigated using the peroxidase-antiperoxidase (PAP) technique ( according to Mayer et al., 1981 and Sanders, 1982). The mesenchymal cells at the tip of the Wolffian duct are coated with a thin layer of fibronectin (Fig. 10). The staining is not as heavy as in the basal laminae of the somites (Fig. 11) but is found all over the cell surface (Fig. 12). A strong staining is exhibited on the filopodia of duct cells, especially where they are in contact with fibrils (Fig.13).

Furthermore, the distribution of laminin was examined using the indirect immunoperoxidase technique with a polyclonal anti-laminin (Eurodiagnostics). Laminin is found on the basal surfaces of the unsegmented paraxial mesoderm and the somatopleure and especially in the basal laminae of somites, but it was nearly undetectable on the tip of the Wolffian duct ( Figs. 14 and 15).

Our results establish that fibronectin is a conditio sine qua non for the directed migration of the Wolffian duct tip, but not in a strong gradient depending haptotactic manner. Fibronectin is found as a thin but uniform coat on the migrating cells and at the surfaces of the adjacent mesenchymal tissues. The strongest staining, however, was seen more cranially in the basal laminae of the somites as well as in the epithelial part of the Wolffian duct. Thus, a very high amount of fibronectin is correlated with the polarization and epithelialization of mesenchymal cells. Obviously, a reversible adhesion and consequently a migration of mesenchymal cells needs fibronectin in a medium concentration. A hypothetical model of fibronectin mediated cell-substratum adhesion is shown in Fig. 16.

Our model does not have to be inconsistent with previous findings of Linask and Lash (1986) on migrating precardiac mesoderm or with the work of Zackson and Steinberg (1987) on Ambystoma pronephric duct. We agree with these authors that the information for a guided migration is local to the substratum upon which cells migrate. Therefore, this guidance may be named haptotaxis, in contrast to chemotaxis which implies that the guidance information originates from a distant source. But we think that at least in the case of the migrating Wolffian duct this haptotaxis does not mean movement towards surfaces offering greater adhesion, but movement towards surfaces offering the right degree of adhesion.

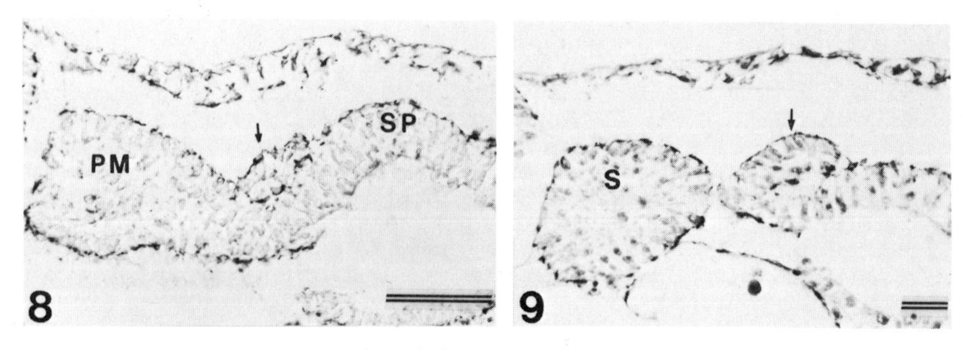


Fig. 8. Localization of fibronectin on a stage 11 HH embryo. Indirect immunoperoxidase reaction and counterstaining with haematoxylin. PM, paraxial mesoderm; SP, somatopleure; arrow, Wolffian duct tip.

Fig. 9. Fibronectin at the somitic level of a stage 12 HH embryo. S, somite; arrow, Wolffian duct in continuity with the intermediate mesoderm. Bars = 50 μm

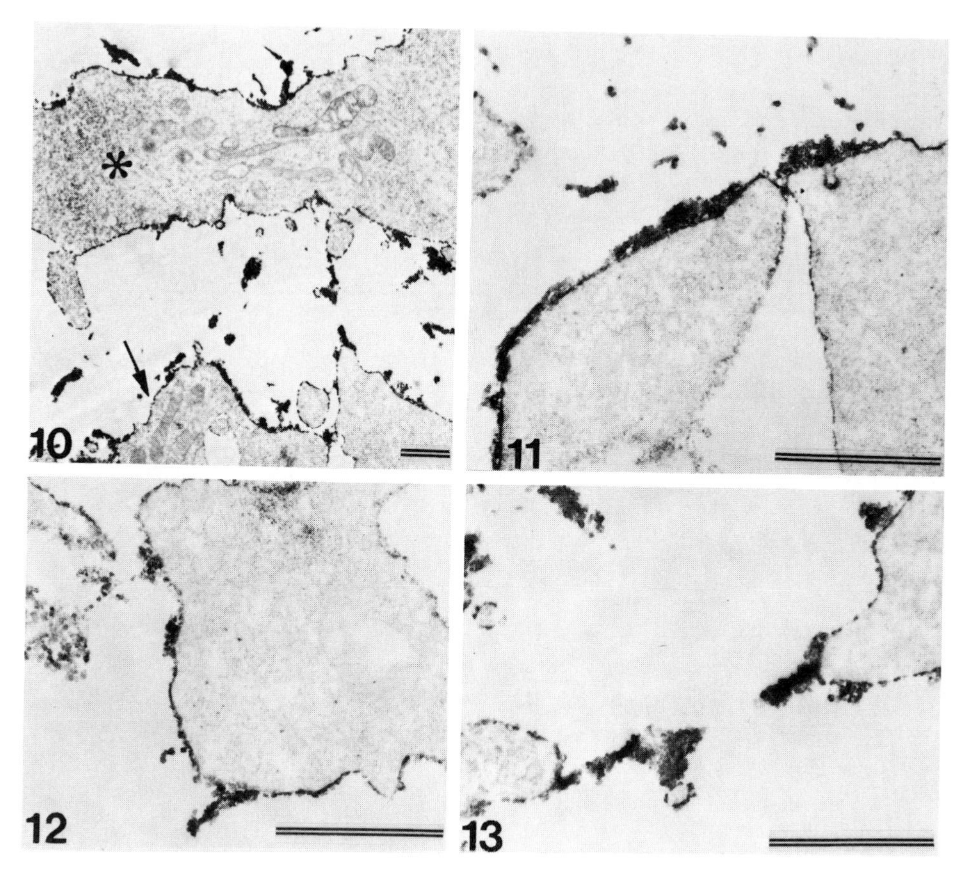


Fig. 10. Mesenchymal cell of the Wolffian duct tip (asterisk) coated with fibronectin. Arrow, intermediate mesoderm. PAP technique.

Fig. 11. Basal lamina of a somite containing fibronectin.

Fig. 12. Thin, but uniform layer of fibronectin at the surface of migrating cells of the duct tip

Fig. 13. Fibronectin at the filopodia mainly in contact with fibrils. Bars = 1 µm

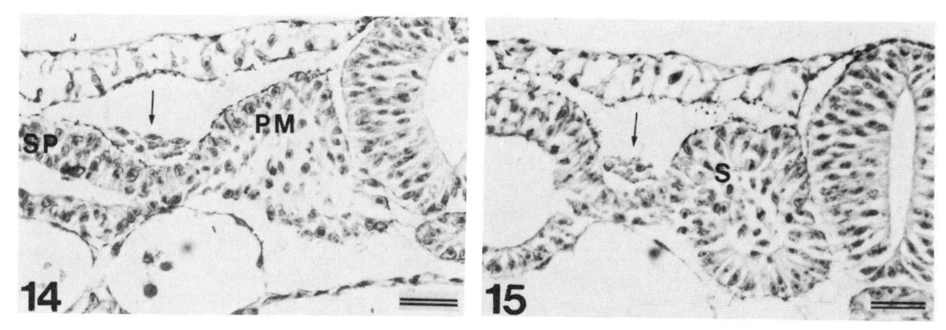


Fig. 14. Laminin at the surface of the unsegmented paraxial mesoderm (PM) and of the somatopleure (SP). Arrow, Wolffian duct. Indirect immunoperoxidase reaction and counterstaining with haematoxylin.

Fig. 15. Laminin in the basal lamina of a somite (S). Arrow, Wolffian duct. Bars = 25 µm

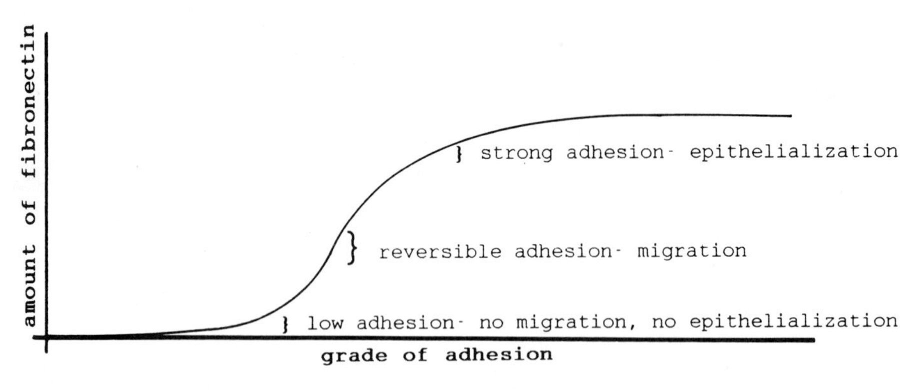


Fig. 16. Hypothetical mode of adhesion during differentiation of the Wolffian duct.

Since laminin is found on the basal surface of the paraxial mesoderm and the somatopleure but scarcely at the migrating duct cells, it does not seem to be necessary for duct migration. This is in line with our results using the synthetic peptide YIGSR (Jacob et al., 1991) which is thought to mimic a cell binding domain of laminin. These experiments led to the conclusion that laminin may be involved in the directional migration of the Wolffian duct by means of contact inhibition.

## EXPRESSION OF THE HNK-1 EPITOPE IN THE EPITHELIAL PART OF THE WOLFFIAN DUCT

The HNK-1 epitope, a tetrasaccharide, is expressed on many glycoproteins and may play a major role in modulating cell adhesion during development (Kruse et al., 1984). The monoclonal HNK-1 antibody is primarily used to stain migrating neural crest cells. According to Pesheva et al.(1987), integrins, the cell surface receptors for fibronectin and laminin also express the HNK-1 structure. Therefore, we enquired whether the HNK-1 epitope is involved in the adhesion between duct cells and their matrix.

In our investigations we applied the HNK-1 antibody (method of Lipinski et al., 1983) on paraplast sections. We did not find HNK-1 in the migrating mesenchymal part of the Wolffian duct, but in the already epithelial part at stages preceding the condensation of the mesonephric blastema (Fig.17). The immunoreaction was found intracellularly (Fig. 18) perhaps at the site of glycosylation of the HNK-1 associated molecules.

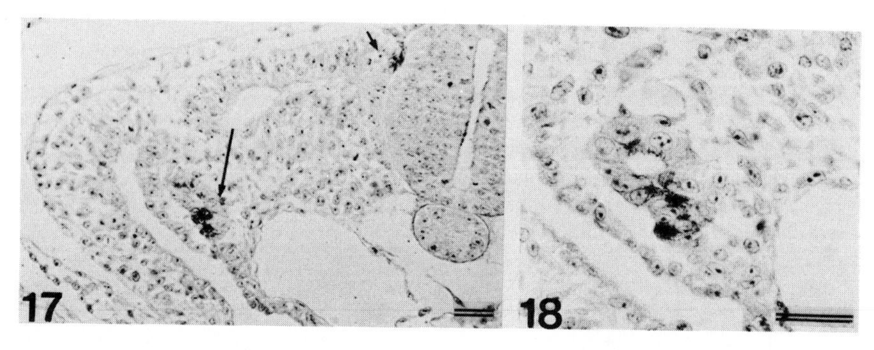


Fig. 17. Indirect immunoperoxidase reaction with HNK-1 mouse IgM at a stage 15 HH embryo. Counterstaining with haematoxylin. Short arrow, neural crest cells; long arrows, Wolffian duct.

Fig. 18. Intracellular reaction of the Wolffian duct. Bars = 25 μm

This may support the hypothesis of Canning and Stern (1988), that this antibody reveals changes associated with mesoderm induction in early chick embryos. Possibly, cell-cell adhesion molecules may be synthesized by the epithelial Wolffian duct which enable the condensation of the mesonephric anlage.

## CONCLUSIONS

1. The caudad directed growth of the Wolffian duct does not happen synchronously with somitogenesis, but precedes somitogenesis.

2. The mitotic rate decreases towards the tip of the Wolffian duct. There it is significantly lower than in the paraxial mesoderm and in the cranial part of the duct. Thus we suggest that the growth of the duct is not due to a passive shifting by adjacent tissues or by cells dividing faster, but that it is due to an active migration of the cells at the duct tip.

3. It can be deduced from morphological and experimental data that the extracellular matrix plays a major role in the directed migration of the Wolffian duct.

4. Fibronectin is a conditio sine qua non for the migration of the cells at the tip of the duct as revealed by studies with synthetic peptides. However, the cells do not follow a craniocaudal gradient. Since a thin and uniform coat of fibronectin is adequate for cell migration and a strong immunoreaction is correlated with a polarisation and immobilisation of mesodermal cells, we suggest that the Wolffian duct migration is supported by a medium amount of fibronectin. The duct tip follows the zone with the adequate amount of fibronectin to promote migration which travels caudally with the differentiation of the paraxial mesoderm.

5. Laminin does not seem to be directly involved in the migration of the duct. According to the investigations of Jacob and al.(1991) with synthetic peptides it might influence the direction of migration by a kind of contact inhibition.

6. The expression of the HNK-1 epitope can perhaps be correlated with the induction of the mesonephric blastema.

## ACKNOWLEDGEMENTS

We thank H. Hake, M. Köhn, U. Ritenberg, and I. Schaeben-Hamm for excellent technical assistance and K. Barteczko for drawings.

## REFERENCES

Canning, D. R., and Stern, C. D., 1988, Changes in the expression of the carbohydrate epitope HNK-1 associated with mesoderm induction in the chick embryo, Development 104: 643-655

Gasser, E., 1877, Beobachtungen über die Entstehung des Wolff'schen Ganges bei Embryonen von Hühnern und Gänsen, Arch. f. mikr. Anat. 14: 442-459

Gillespie, L. L., Armstrong, J. B., Steinberg, M. S., 1985, Experimental evidence for a proteinaceous presegmental wave required for morphogenesis of axolotl mesoderm, Develop. Biol. 107: 220-226

Gratzner, H. G., 1982, Monoclonal antibody to 5-Bromo- and 5-Jodo-deoxyuridine: A new reagent for detection of DNA replication, Science, 218:474-475

Grünwald, P., 1942, Experiments on distribution and activation of the nephrogenic potency in the embryonic mesenchyme, Physiol. Zool. 15: 396-409

Jacob, H. J., and Christ, B.,1978, Experimentelle Untersuchungen am Exkretionsapparat junger Hühnerembryonen. XIXth Morphol. Congress Symposia Charles University Prague, pp 219-225

Jacob, H. J., Jacob, M., Christ, B., 1979, Feinstrukturelle Untersuchungen zur Entwicklung der Vorniere von Hühnerembryonen, Verh. Anat. Ges. 73 :547-554

Jacob, H. J., Jacob, M., Christ, B., 1986, The early development of the intermediate mesoderm in the chick, in:" Somites in Developing Embryos," NATO ASI Series, Series A: Life Sciences Vol. 118, R. Bellairs, D. A. Ede, J.W.Lash, eds., Plenum Press, New York pp 61-68

Jacob, M., Christ, B., Poelmann, R. E., Jacob, H. J., Verbout, A. G., Geiger, R., 1990, Die Bedeutung von Zell-Fibronektin-Interaktionen für die Somitenbildung und das Wachstum des Wolffschen Ganges bei Hühnerembryonen. Verh. Anat. Ges. 83: 91-92

Jacob, M., Christ, B., Jacob, H. J., Poelmann, R. E., 1991, The role of fibronectin and laminin in development and migration of the avian Wolffian duct with reference to somitogenesis, Anat. Embryol. 138: 385-395

Jarzem, J., and Meier, S.P., 1987, A scanning electron microscope survey of the origin of the primordial duct cells in the avian embryo, Anat. Rec. 218: 175-181

Kruse, J., Mahlhammer, R., Wernecke, H., Faissner, A., Sommer, I., Goridis, C., Schachner, M., 1984, Neural cell adhesion molecules and myelin-associated glycoprotein share a carbohydrate moiety recognized by monoclonal antibodies L2 and HNK-1, Nature 311: 153-155

Linask, K. K., and Lash, J.W., 1986, Precardiac cell migrations: Fibronectin localization at mesoderm-endoderm interface during directional movement, Develop. Biol. 114: 87-101

Lipinski, M., Braham, K., Cailland, J.-M-, Carhi, C., Tursz, T., 1983, HNK-1 antibody detects an antigen expressed on neuroectodermal cells, J.Exp. Med. 158: 1775-1780

Lynch, K., and Fraser, S. E., 1990, Cell migration in the formation of the pronephric duct in Xenopus laevis, Develop. Biol. 142: 283-292

Mayer, B. W., Hay, E. D., Hynes, R. O., 1981, Immunocytochemical localization of fibronectin in embryonic chick trunk and area vasculosa, Develop. Biol. 82: 267-286

Newgreen, D., and Thiery, J.-P., 1980, Fibronectin in the early avian embryo: Synthesis and distribution along the migration pathways of neural crest cells, Cell Tiss. Res. 211: 269-291

Ostrovsky, D., Cheney, C. M., Seitz, A. W., Lash, J. W., 1983, Fibronectin distribution during somitogenesis in the chick embryo, Cell Differ. 13: 217-223

Overton, J., 1959, Mitotic pattern in the chick pronephric duct, J. Embryol. exp. Morph. 7: 275-280

Poole, T. J., and Steinberg, M. S., 1982, Evidence for the guidance of pronephric duct migration by a craniocaudally traveling adhesion gradient, Develop. Biol. 92: 144-158

Pesheva, P., Horwitz, A. F., Schachner, M., 1987, Integrin, the cell surface receptor for fibronectin and laminin, expresses the L2/HNK-1 and 13 carbohydrate structures shared by adhesion molecules. Neurosci. Lett. 83: 303-306

Sanders, E. J., 1982, Ultrastructural immunocytochemical localization of fibronectin in the early chick embryo, J. Embryol. exp. Morph. 71: 155-170

Torrey, T. W., 1954, The early development of the human nephros, Contrib. Embryol. 239: 175-197

Zackson, S. L., Steinberg, M.S., 1987, Chemotaxis or adhesion gradient? Pronephric duct elongation does not depend on distant sources of guidance information, Develop. Biol. 124: 418-422

# THE NEURAL CREST AS A MODEL SYSTEM TO STUDY MODULATIONS OF CELLULAR ADHESIVENESS DURING AVIAN EMBRYONIC DEVELOPMENT

Jean-Loup Duband and Muriel Delannet

Institut Jacques Monod, Université Paris 7
2, place Jussieu
75251 Paris Cedex 05, France

In Vertebrates, the ability of cells to detach from a cohesive structure and to migrate and contribute to the formation of new tissues is a basic event during morphogenesis. A shift from an intercellular mode of adhesion towards a preferential adhesion to extracellular material is therefore necessary if active cell migration occurs. The dynamic modulation of the expression of cell-to-cell or cell-to-substratum adhesion systems will dictate whether a cell will remained associated with its neighbors or undergo migration. Insights into this issue can be provided by understanding the molecular basis of cell adhesion. A search for promoters of adhesion led to the isolation of two major families of molecules, the cell adhesion molecules (CAM) and the receptors to the components of the extracellular matrix (integrins), which play a major role in the control of cell interactions.

The ontogeny of the neural crest is a remarkable example of rapid changes in the mode of adhesion of cells. It starts with the loss of the epithelial arrangement of presumptive neural crest cells in the dorsal side of the closing neural tube and is followed by extensive migration along defined pathways through adjacent structures. After reaching their sites of arrest in various areas of the embryo, neural crest cells often coalesce into compact cell cohesive before undergoing differentiation (Le Douarin, 1982; Newgreen and Erickson, 1986; Levi et al., 1990). The development of the neural crest shares a number of common features with the formation of the early mesoderm during gastrulation and, in this respect, can be considered as a secondary gastrulation. Thus, both mesodermal cells and neural crest cells arise from a local delamination of the epiblast during early morphogenesis, and both cell populations show a strong ability to venture into extracellular matrix material to reach their target sites. Therefore, studying the formation of the neural crest may shed light on some aspects of gastrulation that are not readily accessible to experimentation. Among the reasons that make the neural crest attractive are (i) the ability to culture crest cells and somewhat mimic their in vivo translocations in an in vitro system and (ii) the possiblity of perturbing directly their migration in vivo. In the present review, we shall address two questions regarding cellular adhesiveness during the ontogeny of the neural crest: (i) How cell-substratum adhesions are modulated during migration of crest cells, and (ii) what are the major changes in cell adhesion at the onset of crest cell migration.

## MODULATIONS OF CELLULAR ADHESIVENESS DURING LOCOMOTION OF NEURAL CREST CELLS

### Requirement of Fibronectin for Neural Crest Cell Migration

Neural crest cells are intrinsically highly motile. When neural tube explants are cultured in vitro, crest cells emigrate from the neural tube and migrate over considerable distances, if they are provided with an adequate substrate. Ultrastructural and immunohistological studies revealed the presence of extracellular matrix material in the spaces

*Formation and Differentiation of Early Embryonic Mesoderm*
Edited by R. Bellairs *et al.*, Plenum Press, New York, 1992

colonized by crest cells (Bancroft and Bellairs, 1976; Tosney, 1978; Newgreen et al., 1982). This matrix has been found to contain fibronectin, collagens, laminin, cytotactin, and a variety of proteoglycans (Derby, 1978; Thiery et al., 1982; Krotoski et al., 1986; Sternberg and Kimber, 1986; Duband and Thiery, 1987; Tan et al., 1987; Mackie et al., 1988). Some of these components (like laminin and collagen IV) have been detected exclusively in the basal lamina surrounding the tissues lining crest cell migratory pathways.

The role of the various matrix molecules during neural crest migration has been analyzed extensively in vitro. Thus, fibronectin appeared as the most efficient substrate molecule to promote attachment, spreading, and migration of neural crest cells (Newgreen, et al., 1982; Rovasio et al., 1983; Tucker and Erickson, 1984). In addition, a number of perturbation experiments performed with synthetic peptides containing the RGDS cell-binding domain of fibronectin and with antibodies directed either to fibronectin or to the ß1 subunit of its corresponding integrin receptor have demonstrated that direct interaction between fibronectin and the surface of neural crest cells is required for their adhesion and motility both in vivo and in vitro (Rovasio, et al., 1983; Boucaut et al., 1984; Bronner-Fraser, 1985; Duband et al., 1986). Laminin has also been found to support neural crest migration in vitro, but so far there are no evidence for its direct implication in crest cell migration in vivo. In contrast to fibronectin and laminin, collagens, proteoglycans, and cytotactin, cannot support crest cell migration.

Mode of Interaction of Neural Crest Cells with Fibronectin

If neural crest cells require fibronectin for their migration, it should be recalled that non-motile cells also depend on this molecule for substratum anchorage. In addition, fibronectin-binding integrins are expressed by both motile and nonmotile cells and, therefore, the sole fact of expressing this molecule is not sufficient to account for locomotion of cells. How can a single molecule promote both the permanent substrate anchorage of an immobile cell and the labile adhesion of a motile cell? This apparent dual function of fibronectin may result in part from differences in the modes of interaction between locomoting and stationary cells and fibronectin molecules.

The primary condition required for cell locomotion is that the cell should not adhere too firmly to the substratum to avoid paralysis. This has been demonstrated using silicone-rubber sheets as a substrate for migration (Tucker et al., 1985). There appears to be an inverse correlation between the degree of motility of a cell and its ability to induce distortions in the rubber sheet, suggesting that motile cells are not attached firmly to the substratum or that they do not generate sufficient forces for wrinkling the rubber sheet. In addition, treatment of slowly migrating cells with certain anti-integrin antibodies can stimulate migration (Akiyama et al., 1989). The importance of weak transient adhesion to the substratum in cell locomotion has been further demonstrated and quantitated by culturing neural crest cells onto antibodies of differing affinities to the fibronectin receptor (Duband et al., 1991). On low affinity antibodies, neural crest cells are able to locomote at rates very similar to those obtained on fibronectin. In contrast, on high-affinity antibodies to integrins, the rate of migration of neural crest cells is considerably reduced. A detailed analysis of the locomotion of crest cells onto high-affinity substrates showed that fibronectin receptors bind almost irreversibly to the antibodies thus, preventing cell detachment from the substratum.

Differences in the strength of interaction with the substratum are presumably reflected in the organization of both the substratum-contact sites and the cytoskeleton. The primary mode of anchorage to the substratum of a stationary cell is via restricted sites called focal and close contacts where the ventral membrane is in close proximity to the substratum. Much of the rest of the ventral plasma membrane is further away from the substratum (Chen and Singer, 1982). In contrast, locomotory cells interact more uniformly with the substratum at broad close contacts, and develop only a limited number of focal contacts (Couchman and Rees, 1979; Duband, et al., 1986). In stationary cells, actin microfilaments are bundled into stress fibers that terminate in focal contacts. In these regions, actin bundles are attached to the membrane via the cytoskeletal proteins talin, vinculin, and α-actinin (Burridge, 1986). In motile cells, actin microfilaments are not extensively bundled and are mainly distributed as a network in the cellular cortex (Tucker, et al., 1985; Duband, et al., 1986), while talin, vinculin, and α-actinin remain primarily diffuse throughout the cytoplasm. Integrin receptors are distributed as a nearly homogeneous pattern over the entire surface of motile cells, in striking contrast to stationary cells where they are concentrated in clusters around focal contacts and in fibrillar streaks that align with fibronectin fibrils externally and actin microfilament bundles internally (Chen et al., 1985; Damsky et al., 1985).

Thus, the immobilization and concentration of integrin receptors at focal contacts and fibrillar streaks in coincidence with actin bundles may increase locally the binding strength between the receptor, fibronectin, and the cytoskeleton, providing a strong and stable attachment to the substratum and inducing complete immobilization of the cell. In contrast, the absence of concentration of receptors at specific sites on the cell surface together with a poorly organized cytoskeleton would enable the cell to change shape rapidly, and allow labile adhesions to the substratum required for motility.

Aside from differences in the subcellular distribution of molecules involved in substratum adhesion, the specificity of the locomotory behavior of crest cells may also reside in the mode of recognition of fibronectin molecules by integrin receptors. Indeed, advances in our understanding of the cell-binding domains of fibronectins have revealed a far more intricate way of interaction of a cell with fibronectin than was previously thought. Recent experiments have suggested that the various cell-binding sites along fibronectin show specificity not only for the cells they bind, but also for the function they mediate (Humphries et al., 1986; Humphries et al., 1987). Thus, it was found that neural crest cells recognize the major cell-binding domain of fibronectin containing the RGDS sequence, but also the CS1-adhesion site located in the IIICS region in the C-terminal portion of the fibronectin molecule. Interestingly, the RGDS domain is able to promote both attachment and spreading of crest cells, while the CS1 site only permits crest cell attachment. In addition, neither the RGDS site nor the CS1 site alone are able to promote the migration of neural crest cells at rates comparable to fibronectin itself. This process requires at least the presence of both sites, but it cannot be excluded that other sequences within fibronectin may also be involved. This study clearly indicates that the various adhesion sites of fibronectin possess functional specificities and can act synergistically, resulting in the acquisition by cells of complex, coherent behaviors (Dufour et al., 1988). Integrin receptors for the RGDS and the CS1 adhesion sites of fibronectin have not been determined yet on crest cells. Because of their binding specificities, it can be postulated that the $\alpha3\beta1$ and $\alpha5\beta1$ integrins interact with the RGDS site and $\alpha4\beta1$ integrin recognizes the CS1 site (Humphries, 1990). The precise function of these integrin receptors in the process of locomotion is unclear, but can be postulated through the examination of changes induced in cell motility by probes that specifically inhibit each fibronectin cell-binding site. It thus seems that the $\alpha4\beta1$ receptor is not connected with the cytoskeleton and is mostly involved in the elongation of lobopodia while the $\alpha3\beta1$ or $\alpha5\beta1$ receptor is responsible for the formation of transient firm anchorages to the matrix and association with the cytoskeleton to permit the retraction of the remainder of the cell (figure 1).

## Regulation of Interactions between Fibronectin Receptors and their Ligands

Direct binding of the integrin receptors with fibronectin, on one hand, and talin and α-actinin, on the other hand, has been described with the isolated molecules (Horwitz et al., 1986; Otey et al., 1990), thus confirming in situ observations that integrins provide a transmembrane linkage between the extracellular matrix and the cytoskeleton. Differences in the distribution of integrin receptors on the surface of motile and stationary cells then suggest that the links between the receptor and its ligands would differ in the two states. The use of fluorescence recovery after photobleaching technique appeared particularly suitable to examine the dynamics of the interaction between the receptor and its ligands. Indeed, the lateral mobility of integral membrane proteins within the plane of the plasma membrane is indicative of the degree of association of the molecule with peripheral structures, including the extracellular matrix and the cytoskeleton (Jacobson et al., 1987). It was found that the population of fibronectin receptors is mobile in locomotory cells whereas it is almost totally immobile when it is concentrated in focal contacts and fibrillar streaks in stationary cells (Duband et al., 1988). Thus, the high lateral mobility of the receptor in motile cells suggests that a large majority of receptors are bound to fibronectin and cytoskeletal components, but these interactions are very transient; the receptors rapidly dissociate from their ligands to extablish new associations with other peripheral molecules. Alternatively, only a limited proportion of the receptors is bound to their ligands, the other receptors would be free and constitute a reservoir of receptors for the establishment of new substratum anchorage. Increase in the number of peripheral ligands simultaneously bound to the receptor would increase the immobile fraction of receptors. This multivalent, immobilized receptor state would provide stable anchorage to the substratum. It will be important in the future to examine how receptor is transported and localized to sites appropriate for locomotion, since the numbers and types of these sites presumably regulate the locomotory state of a cell.

What could be the regulatory mechanism of the interaction between the receptor and its ligands? A diffuse distribution of fibronectin receptors has been reported in virally-transformed cells (Chen et al., 1986). In such cells, it has also been shown that the receptor is phosphorylated on tyrosine residues (Hirst et al., 1986), which could possibly modify its binding properties. However, no receptor phosphorylation could be detected in embryonic motile cells, indicating that the process of regulation of receptor binding activity differs considerably in these cells and in transformed cells (Duband et al., 1988). Alternatively, regulation of receptor

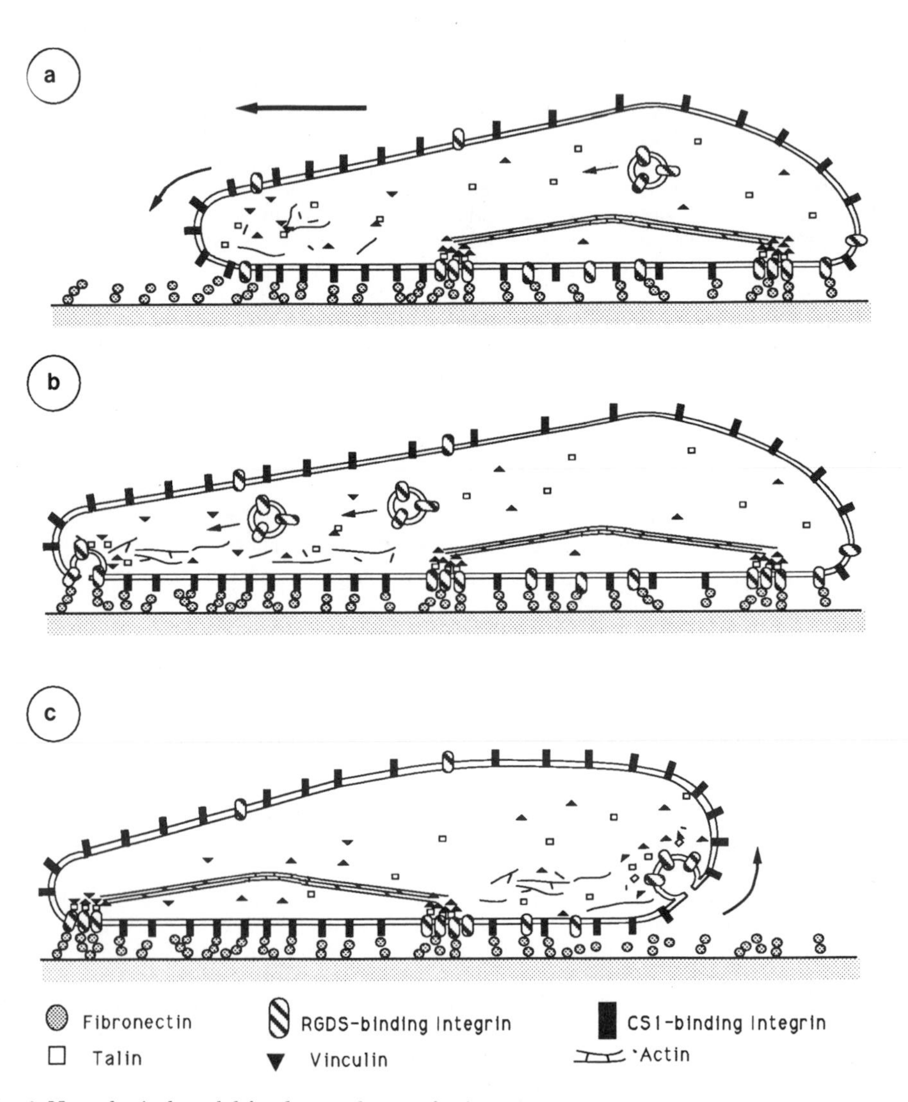


Fig. 1. Hypothetical model for the regulation of substratum interactions during crest cell motility. (a) Elongation of lobopodium resulting from the rapid binding of CS1-binding integrins to fibronectin molecules. These integrins are recruited from a diffuse pool of molecules that can navigate freely in the membrane. (b) Formation of transient firm anchorage sites involving the polarized intracellular transport and membrane incorporation of RGDS-binding integrins at the front of the cell. Actin microfilaments in the tailing edge become contracted. (c) Detachment of the trailing edge of the cell resulting from internalisation of integrins and contraction of actin microfilaments and formation of new actin microfilaments in the leading edge of the cell.

binding would preferentially involve rapid internalization of the receptor. When living cells are labeled with a rhodamine-conjugated monoclonal antibody directed to the receptor, fluorescence is rapidly seen in vesicles in the cytoplasm of motile cells but not in stationary cells. However, whether the receptor is internalized with fibronectin molecules and cytoskeletal elements remains to be determined (Duband et al., 1988).

Factors Involved in the Directionality of Migration

The determination of the pathways followed by neural crest cell results from specific interations between the cells and their surroundings. When crest cells are transplanted from one axial level to another, they follow the correct pathway for the level to which they have been transplanted, demonstrating that directional cues defining the pathways are present at each given level and that crest cells are not pre-programmed to follow certains route of migration (Le Douarin and Teillet, 1974; Noden, 1978). However, the capability to read correctly the directional cues present in the extracellular space resides largely in intrinsic properties of the migratory population (Erickson et al., 1980). The environment in which crest cells migrate is, therefore, to a large extent responsible for the determination of the migratory pathways. The factors in the extracellular environment that are believed to contribute to the specification of pathways are of two different types: molecular cues constituted by matrix molecules and physical factors that permit or prevent cell motion.

As mentionned above, a number of matrix components can support the migratory behavior of crest cells in vitro. Several attempts have been made to understand the specific role of these molecules, and more particularly of fibronectin, in the control of cell migration. Fibronectin, however, beside being ubiquitously present in the pathways of migration of neural crest cells is also present in several locations in the embryo that are not occupied by crest cells. The presence of fibronectin is therefore not sufficient to determine the pathway of neural crest cell migration. It is conceivable that a pathway might be determined by the simultaneous local modulations in the expression of several matrix molecules or of several forms of the same matrix molecule, leading to subtle changes in the local capacity of the matrix to support cell migration. Tenascin is one of the leading candidates to play a regulatory role in the migratory capabilities of neural crest cells, because it is the only component known so far to be restricted to regions occupied by crest cells. However, the exact function of tenascin in crest migration is still unclear and controversial (Tucker and McKay, 1991).

The directional migration of neural crest cells might also result from a differential distribution of some forms of fibronectin molecules within the embryo. Indeed, there are several variants of fibronectin which differ by three domains, namely EDI, EDII and IIICS, that undergo alternative splicings. Although the expression pattern of the fibronectin variants was apparently found to be uniform in the early embryo (ffrench-Constant and Hynes, 1988), the distribution of at least one of these variants (EDI) appears restricted to some specific sites that are colonized by certain neural crest subpopulations.(our unpublished results). However, a direct role of this fibronectin variant in the migration process of neural crest cells remains to be established.

Several different physical factors may contribute to the control of the migratory capacity of crest cells; these are the orientation of fibrils in the extracellular matrix, the size of the microspaces within the matrix and the presence of barriers blocking migration. Orientation of fibrils in the extracellular matrix have been detected along the migratory pathways of neural crest cells (Löfberg et al., 1980; Newgreen, 1989). This hypothesis has been further supported by the observation that crest cells allowed to migrate in vitro on extracellular matrix fibrillar meshworks of different degrees of orientation align with the predominant orientation of the fibrils (Newgreen, 1989). However, fibril orientation cannot account solely for the precise directionality of movement of crest cells at all embryonic levels.

Onother elements controlling neural crest cell migration could be the appearence of cell-free spaces sufficiently large to permit the passage of the migrating cells. This role has been assigned to hyaluronate which is known to expand acellular spaces through its hydrophilic properties and is synthesized by migrating neural crest cells (Pratt et al., 1975; Pintar, 1978). Recently, however, it has been directly determined that the minimal diameter of a hole permitting the passage of neural crest cells is less than 1 mm, i.e. less than most cell-free spaces in the embryo (Newgreen, 1989). It appears, therefore, that, more than the presence of a physical space between cells, spatial restrictions will be imposed on migrating neural crest cells by the environment and more specifically by the presence of basal laminae. Indeed, although crest cells possess the capacity to degrade extracellular matrix molecules (Valinsky and Le

Douarin, 1985), they lack the ability to cross basal laminae of epithelial sheets (Erickson, 1987). Interestingly, epithelia is the predominent tissue organization found in the embryo at the time of crest cell migration, therefore providing spatially-restricted pathways of migration.

## MODULATIONS OF CELLULAR ADHESIVENESS AT THE ONSET OF MIGRATION OF NEURAL CREST CELLS

In amphibians, neural crest cells are anatomically separated from the neural tube prior to migration, and can thus be identified precociously. In amniotes, in contrast, both segregation from the neural tube and dispersion occur simultaneously and, consequently, the unequivocal designation of neural crest cells resides only in cell dispersion. The process of crest cell separation from the neural epithelium is precisely regulated both spatially and temporally. Neural crest cells first undergo migration from the most rostral levels of the neural tube in the forebrain and continue to separate in a rostrocaudal wave approximately parallel to the wave of segmentation of the axial mesoderm (for a review, see Newgreen and Erickson, 1986). Morphological studies at both the light and electron microscopic levels have permitted a detailed description of the main cellular events that accompany the onset of neural crest cell migration in the trunk (reviewed in Martins-Green and Erickson, 1986; Newgreen and Erickson, 1986; Martins-Green and Erickson, 1987; Levi, et al., 1990). These include the complete disruption of the basal lamina that partially covers the dorsal part of the neural tube, the disappearance of cell-cell adhesion molecules from the surface of the emigrating neural crest cells, the ability of cells to adhere to extracellular matrix material such as fibronectin, and the acquisition of motile properties. However, these studies did not allow the exact determination of the sequence and the relative importance of the events that are responsible for crest cell dispersion, because they occur almost simultaneously and in a restricted portion of the neural tube.

It is possible to reproduce to the some extent in an in vitro system the major events that accompany neural crest emigration in avians, taking advantage of the fact that it occurs as a gradient from the rostral to the caudal part of the embryo (Newgreen and Gibbins, 1982; Newgreen and Gooday, 1985). Indeed, if the portion of the neural tube corresponding to the last somites of a 2.5 day-old embryo is explanted in culture onto an appropriate substratum, neural crest cells readily emigrate from the explant. In contrast, if the neural tube corresponding to the presomitic region is explanted in culture, neural crest cells are seen migrating only after a 6 to 7 hr-delay (figure 2). This observation would then indicate that neural crest cells originating from the neural tube corresponding to the pre-somitic region have not entirely achieved the whole sequence of events accompanying initiation of emigration. Neural crest cells from the somitic region, in contrast, have already been induced to emigrate and are undergoing migration prior to neural tube explantation. Accordingly, these axial levels have been termed migratory and premigratory (Newgreen and Gibbins, 1982; Newgreen and Gooday, 1985). Hence, this in vitro system is suitable for analyzing sequentially the modulations of adhesive properties of neural crest cells originating from the two axial levels and for probing various factors for their ability to control emigration of cells.

### Modulations of Substratum Adhesion during Crest Cell Emigration

Previous studies on the spatio-temporal distribution of extracellular matrix components during early embryonic development did not reveal extensive changes in the macromolecular composition of the matrix at the onset of crest cell migration, thus indicating that the direct extracellular environment of presumptive crest cells is permissive for emigration long before its occurence (Thiery, et al., 1982; Sternberg and Kimber, 1986). In addition, a complete basal lamina which would potentially constitute a physical barrier preventing the release of cells is never observed along the dorsal side of the neural tube prior to crest cell emigration (Tosney, 1978; Newgreen and Gibbins, 1982; Martins-Green and Erickson, 1986; Duband and Thiery, 1987; Martins-Green and Erickson, 1987). In vitro, none of the extracellular matrix components known to promote cell locomotion are able to trigger individually or in combination premature emigration of neural crest cells from premigratory axial levels (Newgreen and Gibbins, 1982; Delannet and Duband, 1991). It thus seems that, although a favorable extracellular matrix must be a prerequisite, changes in the extracellular matrix composition probably do not induce initiation of migration. The inability of neural crest cells to separate rapidly from the neural tube and undergo migration would then possibly reside in the inability of cells to adhere to the matrix. Using an in vitro adhesion assay, we found that, in contrast to migratory-level neural crest cells, premigratory-level crest cells lack the capacity of spreading onto extracellular matrix components as they are integrated in the neural tube, and gradually acquire substratum adhesion only a few hours preceding the onset of migration (Delannet and Duband, 1991). These results strongly suggest that the lack of appropriate receptors for extracellular matrix material can constitute a major restrain for crest cell emigration.

Role of Intercellular Adhesions during Neural Crest Cell Emigration

During neurulation, presumptive crest cells initially extend like neural tube cells to the apical surface of the neural epithelium where they exhibit adherens junctions visible with the electron microscope. Several hours before migration, in contrast, presumptive crest cells are seen without any morphologically-defined junctions, including adherens, tight and gap junctions. Coincidently, a number of cells lose polarity and no longer extend to the apical surface of the neural epithelium. Finally, calcium-dependent cell adhesion molecules disappear from the surface of cells upon initiation of migration (Revel and Brown, 1975; Bancroft and Bellairs, 1976; Tosney, 1978; Newgreen and Gibbins, 1982; Duband et al., 1988). In vitro, premature emigration of premigratory-level crest cells can be obtained upon exposure to calcium-free medium or calcium-channel antagonists. This inductive effect of calcium depletion has been interpreted as inactivation of calcium-dependent cell-cell adhesion systems, thus allowing the release of crest cells from the neural epithelium (Newgreen and Gooday, 1985).

However, we found that, if A-CAM-mediated cell-cell interactions are specifically perturbed in vitro, the entire neural tube disaggregates, yet emigration of neural crest cells from PM explants is not anticipated, but even retarded (Delannet and Duband, 1991). This observation

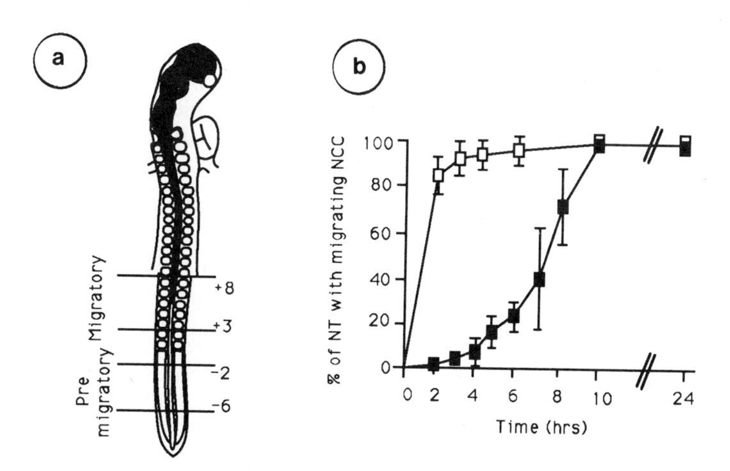


Fig. 2. (a) Schematic representation of a 20-somite embryo showing the axial levels at which crest cell emigration has started (the corresponding neural tube is filled) and the positions of the migratory and premigratory levels. (b) Timetable of in vitro emigration of crest cells obtained from migratory (open squares) and premigratory (plain squares) axial levels. Results are expressed as the percentage of explants that produced crest cells as a function of the duration of culture.

would indicate that, in contrast to what was originally proposed, the disappearance of A-CAM-dependent adhesions is not sufficient to trigger initiation of migration. Morever, we could show that the presence of intercellular adhesions among the presumptive crest cell population is necessary for the transition from the premigratory to the migratory state. Indeed, the development of substratum-adhesion capabilities of neural crest cells is severely prevented, if cell cohesion is abolished using EDTA treatment or anti-A-CAM antibodies. Consistent with these data, morphological studies have revealed an increase in cell density in the dorsal aspect of the neural tube a few hours before crest cell migration associated with enhancement in A-CAM expression throughout the neural tube (Bancroft and Bellairs, 1976; Tosney, 1978; Duband, et al., 1988). A similar critical role of cellular cohesion in the capacity of a cell population to respond to external signals has been described during mesodermal induction in amphibians and has been termed the community effect (Gurdon, 1988).

### Involvement of Transforming Growth Factor-ß in Crest Cell Emigration

The mechanisms regulating the onset of neural crest cell migration are unknown. It has been suggested that emigration is under the control of external signals released in the extracellular milieu by neighboring tissues. Using tissue grafting or transplantation of extracellular matrix material adsorbed onto microcarriers, Löfberg et al. (1985) found that amphibian neural crest cells are triggered to initiate rapid emigration by extracellular material originating from the dorsal epidermis of the rostral half of the embryo. Since extracellular matrix components either isolated or in combination cannot trigger crest cell emigration, it is likely that soluble factors that are trapped in the extracellular matrix are responsible for the inducing activity.

Because of its biological effects and of its tissue distribution, transforming growth factor-ß (TGF-ß) appears as a leading candidate to play an important role in the control of crest cell emigration. TGF-ß has been implicated in a variety of biological events involving epithelial-mesenchymal transformations and cell dispersion. For example, it stimulates invasion and metastatic potential of certain cells (Welch et al., 1990) and, during embryonic development, it has been found to promote in vitro phenotypic change of epiblastic cells into mesoderm during gastrulation and migration of cardiac cushion cells from the endocardium (Potts and Runyan, 1989; Potts et al., 1991; Sanders and Prasad, 1991). TGF-ß also promotes the synthesis of fibronectin and collagens and causes a marked alteration in the repertoire of integrins (for a review, see Massagué, 1990). In particular, it modifies the adhesive behavior of MG-63 human osteosarcoma cells by switching integrin pattern from laminin receptor to fibronectin and collagen receptors (Heino and Massagué, 1989).

We found that TGF-ß is able to induce premature emigration of premigratory-level neural crest cells and to enhance fibronectin adhesion of these cells (figure 3; Delannet and Duband, 1991). In addition, neural crest cells either before or during migration are able to bind TGF-ß. We propose that TGF-ß may stimulate epithelio-mesenchymal transformation of neural crest cells by acting primarily on the expression pattern of members of the integrin family on the surface of these cells.

Using radioimmunoassay analyses, we found that TGF-ß-like molecules are present in the area of crest cell emergence. However, the exact tissue origin of TGF-ß and which of the various members of the TGF-ß family is responsible for the inducing activity remain to be determined. The patterns of distribution and expression of members of the TGF-ß family during embryonic development have been analyzed using immunofluorescence and in situ hybridization techniques. TGF-ß1, ß2, and ß3 were all evidenced in multiple sites in the embryo predominantly in mesodermal tissues and at sites where epithelial-mesenchymal transformations are likely to occur (Heine et al., 1987; Lehnert and Akhurst, 1988; Pelton et al., 1989; Akhurst et al., 1990; Pelton et al., 1990; Millan et al., 1991; Potts, et al., 1991; Sanders and Prasad, 1991; Schmid et al., 1991). Activin and TGF-ß1 have been described respectively in the hypoblast and in the mesoderm of early gastrulating avian embryos (Mitrani et al., 1990; Sanders and Prasad, 1991). Several reports have described the presence of TGF-ß1 transcripts in migrating neural crest cells lining the neural tube and somites (Schmid, et al., 1991), and of TGF-ß2 mRNA in the ventral portion of the neural tube at the time of neural crest cell migration (Millan et al., 1991). In addition, using reduced hybridization stringency conditions, Lehnert and Akhurst (1988) detected staining in the notochord, spinal cord and dorsal root ganglia, suggesting that TGF-ß1-related molecules are expressed in these tissues. Very recently, TGF-ß1 and TGF-ß2 have been found in the nervous system in the mouse embryo (Flanders et al., 1991). Finally, TGF-ß has been detected in the somitic epithelium and not in the segmental plate (Heine, et al., 1987), correlating with the spatio-temporal pattern of crest cell emigration. It is therefore possible that a coincidental synthesis of TGF-ß by the somitic mesoderm, the neural tube, and crest cells themselves, results in the local increase of TGF-ß necessary for its action.

## CONCLUDING REMARKS

Formation and subsequent migration of the neural crest in avians is a complex process that involves a sequence of timely and spatially regulated modulations of cellular adhesiveness. Among the various events implicated, acquisition of substratum adhesion by presumptive crest cells is a major limiting step in the development of this structure. Intercellular adhesions are also of a great importance because they allow crest cells to respond to morphogens under certain circumstances. Growth factors, including TGF-ß, may be responsible for the coordinated occurence of the various events that lead to the complete segregation of the neural crest from the rest of the neural epithelium and its subsequent migration and differentiation.

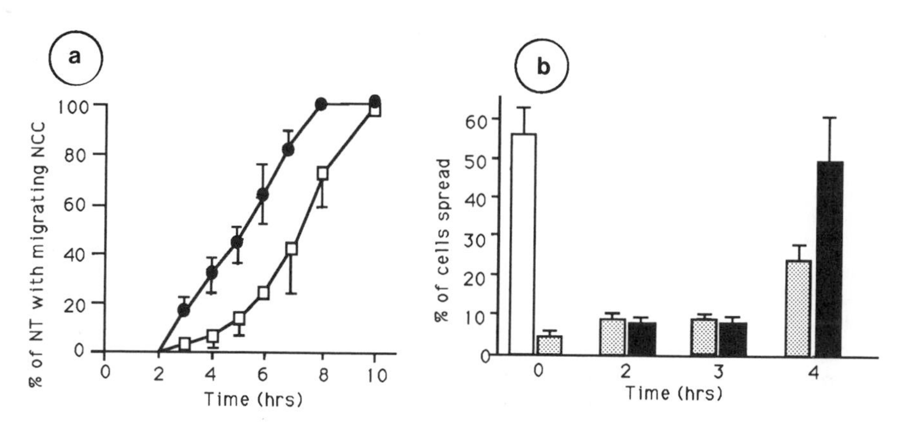


Fig. 3. Effect of TGF-ß on the timing of crest cell emigration in vitro (a) and on the substratum adhesion properties (b) of premigratory neural crest cells. (a) TGF-ß1 at 20 ng/ml reduces the delay of crest emigration by about two hrs. Open squares, control cultures; plain circles, cultures with TGF-ß. (b) TGF-ß increases fibronectin adhesion of crest cells within 4 hrs. White column, control migrating neural crest cells; dotted columns, without TGF-ß; plain columns, TGF-ß at 20 ng/ml.

## REFERENCES

1. Akhurst, R. J., Lehnert, S. A., Faissner, A., and Duffie, E., 1990, TGF-β in murine morphogenetic processes: The early embryo and cardiogenesis, Development 108:645.
2. Akiyama, S. K., Yamada, S. S., Chen, W.-T., and Yamada, K. M., 1989, Analysis of fibronectin receptor function with monoclonal antibodies: roles in cell adhesion, migration, matrix assembly, and cytoskeletal organization, J. Cell Biol. 109:863.

3. Bancroft, M., and Bellairs, R., 1976, The neural crest cells of the trunk region of the chick embryo studied by SEM and TEM, Zoon 4:73.
4. Boucaut, J.-C., Darribère, T., Poole, T. J., Aoyama, H., Yamada, K. M., and Thiery, J. P., 1984, Biological active synthetic peptides as probes of embryonic development: A competitive peptide inhibitor of fibronectin function inhibits gastrulation in amphibian embryos and neural crest cell migration in avian embryos, J. Cell Biol. 99:1822.
5. Bronner-Fraser, M., 1985, Alteration in neural crest migration by a monoclonal antibody that affects cell adhesion, J. Cell Biol. 101:610.
6. Burridge, K., 1986, Substrate adhesions in normal and transformed fibroblasts: Organization and regulation of cytoskeletal, membrane and extracellular matrix components at focal contacts, Cancer Rev. 4:18.
7. Chen, W.-T., and Singer, S. J., 1982, Immunoelectron microscopic studies of the sites of cell-substratum and cell-cell contacts in cultured fibroblasts, J. Cell Biol. 95:205.
8. Chen, W.-T., Hasegawa, E., Hasegawa, T., Weinstock, C., and Yamada, K. M., 1985, Development of cell surface linkage complexes in cultured fibroblasts, J. Cell Biol. 100:1103.
9. Chen, W.-T., Wang, J., Hasegawa, T., Yamada, S. S., and Yamada, K. M., 1986, Regulation of fibronectin receptor distribution by transformation, exogenous fibronectin, and synthetic peptides, J. Cell Biol. 103:1649.
10. Couchman, J. R., and Rees, D. A., 1979, The behaviour of fibroblasts migrating from chick heart explants: Changes in adhesion, locomotion and growth, and in the distribution of actomyosin and fibronectin, J. Cell Sci. 39:149.
11. Damsky, C. H., Knudsen, K. A., Bradley, D., Buck, C. A., and Horwitz, A. F., 1985, Distribution of the cell substratum attachment (CSAT) antigen on myogenic and fibroblastic cells in culture, J. Cell Biol. 100:1528.
12. Delannet, M., and Duband, J.-L., 1991, Transforming growth factor-β control of modulations of cellular adhesiveness during avian neural crest cell emigration in vitro, Submitted.
13. Derby, M. A., 1978, Analysis of glycosaminoglycans within the extracellular environment encountered by migrating neural crest cells, Dev. Biol. 66:321.
14. Duband, J.-L., Rocher, S., Chen, W.-T., Yamada, K. M., and Thiery, J. P., 1986, Cell adhesion and migration in the early vertebrate embryo: Location and possible role of the putative fibronectin-receptor complex, J. Cell Biol. 102:160.
15. Duband, J.-L., and Thiery, J. P., 1987, Distribution of laminin and collagens during avian neural crest development., Development 101:461.
16. Duband, J.-L., Dufour, S., and Thiery, J. P., 1988, Extracellular matrix-cytoskeleton interactions in locomoting embryonic cells, Protoplasma 145:112.
17. Duband, J.-L., Dufour, S., Yamada, K. M., and Thiery, J. P., 1988, Phosphorylation of the fibronectin-receptor complex occurs during the acquisition of the stationary state by avian embryonic locomoting cells, FEBS Lett. 230:181.
18. Duband, J.-L., Nuckolls, G. H., Ishihara, A., Hasegawa, T., Yamada, K. M., Thiery, J. P., and Jacobson, K., 1988, The fibronectin receptor exhibits high lateral mobility in embryonic locomoting cells but is immobile in focal contacts and fibrillar streaks in stationary cells, J. Cell Biol. 107:1385.
19. Duband, J.-L., Volberg, T., Sabanay, I., Thiery, J. P., and Geiger, B., 1988, Spatial and temporal distribution of adherens-junction associated adhesion molecule A-CAM during avian embryogenesis, Development 103:325.
20. Duband, J.-L., Dufour, S., Yamada, S. S., Yamada, K. M., and Thiery, J. P., 1991, Neural crest cell locomotion induced by antibodies to β1 integrins: A tool for studying the roles of substratum molecular avidity and density in migration, J. Cell Sci. 98:517.
21. Dufour, S., Duband, J.-L., Humphries, M. J., Obara, M., Yamada, K. M., and Thiery, J. P., 1988, Attachement, spreading, and locomotion of avian neural crest cells are mediated by multiple adhesion sites on fibronectin molecules, EMBO J. 7:2661.
22. Erickson, C. A., 1987, Behavior of neural crest cells on embryonic basal laminae, Dev. Biol. 120:38.
23. Erickson, C. A., Tosney, K. W., and Weston, J. A., 1980, Analysis of migratory behavior of neural crest and fibroblastic cells in embryonic tissues, Dev. Biol. 77:142.
24. ffrench-Constant, C., and Hynes, R. O., 1988, Pattern of fibronectin gene expression and splicing during cell migration in chicken embryo, Development 104:369.
25. Flanders, K. C., Lüdecke, G., Engels, S., Cissel, D. S., Roberts, A. B., Kondaiah, P., Lafyatis, R., Sporn, M. B., and Unsicker, K., 1991, Localization and actions of transforming growth factor-βs in the embryonic nervous system, Development 113:183.
26. Gurdon, J. B., 1988, A community effect in animal development, Nature 336:772.

27. Heine, U. I., Munoz, E. F., Flanders, K. C., Ellingsworth, L. R., Lam, H.-Y. P., Thompson, N. L., Roberts, A. B., and Sporn, M. B., 1987, Role of transforming growth factor-β in the development of the mouse embryo, J. Cell Biol. 105:2861.
28. Heino, J., and Massagué, J., 1989, Transforming growth factor-β switches the pattern of integrins expressed in MG-63 human osteosarcoma cells and causes a selective loss of cell adhesion to laminin, J. Biol. Chem 264:21806.
29. Hirst, R., Horwitz, A., Buck, C., and Rohrschneider, L., 1986, Phosphorylation of the fibronectin receptor complex in cells transformed by oncogenes that encode tyrosine kinases, Proc. Natl. Acad. Sci. USA 83:6470.
30. Horwitz, A., Duggan, K., Buck, C., Beckerle, M. C., and Burridge, K., 1986, Interaction of plasma membrane fibronectin receptor with talin. A transmembrane linkage, Nature 320:531.
31. Humphries, M. J., 1990, The molecular basis and specificity of integrin-ligand interactions, J. Cell Sci. 97:585.
32. Humphries, M. J., Akiyama, S. K., Komoriya, A., Olden, K., and Yamada, K. M., 1986, Identification of an alternatively-spliced site in human plasma fibronectin that possesses cell-type specificity , J. Cell Biol. 103:2637.
33. Humphries, M., Komoriya, A., Akiyama, S. K., Olden, K., and Yamada, K. M., 1987, Identification of two distinct regions of the type III connecting segment of human plasma fibronectin that promote cell type-specific adhesion, J. Biol. Chem. 262:6886.
34. Jacobson, K., Ishihara, A., and Inman, R., 1987, Lateral diffusion of proteins in membranes, Ann. Rev. Physiol. 49:163.
35. Krotoski, D. M., Domingo, C., and Bronner-Fraser, M., 1986, Distribution of a putative cell surface receptor for fibronectin and laminin in the avian embryo, J. Cell Biol. 103:1061.
36. Le Douarin, N. M., 1982, "The Neural Crest", Cambridge University Press, Cambridge.
37. Le Douarin, N. M., and Teillet, M.-A., 1974, Experimental analysis of the migration and differentiation of neuroblasts of the autonomic nervous system and of neurectodermal mesenchymal derivatives, using a biological cell marking technique, Dev. Biol. 41:162.
38. Lehnert, S. A., and Akhurst, R. J., 1988, Embryonic expression pattern of TGF-β type 1 RNA suggests both paracrine and autocrine mechansims of action, Development 104:263.
39. Levi, G., Duband, J.-L., and Thiery, J. P., 1990, Modes of cell migration in the vertebrate embryo, Int. Rev. Cytol. 123:201.
40. Löfberg, J., Ahlfors, K., and Fällström, C., 1980, Neural crest cell migration in relation to extracellular matrix organization in the embryonic Axolotl trunk, Dev. Biol. 75:148.
41. Löfberg, J., Nynäs-McCoy, A., Olsson, C., Jonsson, L., and Perris, R., 1985, Stimulation of initial neural crest cell migration in the Axolotl embryo by tissue grafts and extracellular matrix transplanted on microcarriers, Dev. Biol. 107:442.
42. Mackie, E. J., Tucker, R. P., Halfter, W., Chiquet-Ehrismann, R., and Epperlein, H. H., 1988, The distribution of tenascin coincides with pathways of neural crest cell migration, Development 102:237.
43. Martins-Green, M., and Erickson, C. A., 1986, Development of neural tube basal lamina during neurulation and neural crest cell emigration in the trunk of the mouse embryo, J. Embryol. Exp. Morph. 98:219.
44. Martins-Green, M., and Erickson, C. A., 1987, Basal lamina is not a barrier to neural crest cell emigration: Documentation by TEM and by immunofluorescent and immunogold labelling, Development 101:
45. Massagué, J., 1990, The transforming growth factor-β family, Annu. Rev. Cell Biol. 6:597.
46. Millan, F. A., Denhez, F., Kondaiah, P., and Akhurst, R. J., 1991, Embryonic gene expression patterns of TGF β1, β2, and β3 suggest different developmental functions in vivo, Development 111:131.
47. Mitrani, E., Zic, T., Thomsen, G., Shimoni, Y., Melton, D. A., and Bril, A., 1990, Activin can induce the formation of axial structures and is expressed in the hypoblast of the chick, Cell 63:495.
48. Newgreen, D. F., 1989, Physical influences on neural crest cell migration in avian embryos: Contact guidance and spatial restrictions, Dev. Biol. 131:136.
49. Newgreen, D. F., and Erickson, C. A., 1986, The migration of neural crest cells, Int. Rev. Cytol. 103:89.
50. Newgreen, D. F., and Gibbins, I. L., 1982, Factors controlling the time of onset of the migration of neural crest cells in the fowl embryo, Cell Tiss. Res. 224:145.
51. Newgreen, D. F., and Gooday, D., 1985, Control of the onset of migration of neural crest cells in avian embryos: Role of $Ca^{++}$-dependent cell adhesions, Cell Tiss. Res. 239:329.

52. Newgreen, D. F., Gibbins, I. L., Sauter, J., Wallenfels, B., and Wütz, R., 1982, Ultrastructural and tissue-culture studies on the role of fibronectin, collagen and glycosaminoglycans in the migration of neural crest cells in the fowl embryo, Cell Tiss. Res. 221:521.
53. Noden, D. M., 1978, The control of avian cephalic neural crest cytodifferentiation. II. Neural tissues, Dev. Biol. 67:296.
54. Otey, C. A., Pavalko, F. M., and Burridge, K., 1990, An interaction between α-actinin and the β1 integrin subunit in vitro, J. Cell Biol. 111:721.
55. Pelton, R. W., Nomura, S., Moses, H. L., and Hogan, B. L. M., 1989, Expression of transforming growth factor β2 RNA during murine embryogenesis, Development 106:759.
56. Pelton, R. W., Dickinson, M. E., Moses, h., and Hogan, B. L. M., 1990, In situ hybridization analysis of TGFß3 RNA expression during mouse development: comparative studies with TGFß1 and ß2, Development 110:609.
57. Pintar, J. E., 1978, Distribution and synthesis of glycosaminoglycans during quail neural crest morphogenesis, Dev. Biol. 67:444.
58. Potts, J. D., and Runyan, R. B., 1989, Epithelial-mesenchymal cell transformation in the embryonic heart can be mediated, in part, by transforming growth factor ß, Dev. Biol. 134:392.
59. Potts, J. D., Dagle, J. M., Walder, J. A., Weeks, D. L., and Runyan, R. B., 1991, Epithelial-mesenchymal transformation of embryonic cardiac endothelial cells is inhibited by a modified antisense oligodeoxynucleotide to transforming growth factor ß3, Proc. Natl. Acad. Sci. USA 88:1516.
60. Pratt, R. M., Larsen, M. A., and Johnston, M. C., 1975, Migration of cranial neural crest cells in a cell-free hyaluronate-rich matrix, Dev. Biol. 44:298.
61. Revel, J. P., and Brown, S. S., 1975, Cell junctions in development with particular reference to the neural tube, Cold Spring Harbor Symp. Quant. Biol. 40:433.
62. Rovasio, R. A., Delouvée, A., Yamada, K. M., Timpl, R., and Thiery, J. P., 1983, Neural crest cell migration: Requirement for exogenous fibronectin and high cell density, J. Cell Biol. 96:462.
63. Sanders, E. J., and Prasad, S., 1991, Possible roles for TGF-β1 in the gastrulating chick embryo, J. Cell Sci. 99:617.
64. Schmid, P., Cox, D., Bilbe, G., Maier, R., and McMaster, G. K., 1991, Differential expression of TGF β1, β2, β3 genes during mouse embryogenesis, Development 111:117.
65. Sternberg, J., and Kimber, S. J., 1986, Distribution of fibronectin, laminin and entactin in the environment of migrating neural crest cells in early mouse embryos, J. Embryol. Exp. Morph. 91:267.
66. Tan, S. S., Crossin, K. L., Hoffman, S., and Edelman, G. M., 1987, Asymmetric expression in somites of cytotactin and its proteoglycan ligand is correlated with neural crest cell distribution, Proc. Natl. Acad. Sci. USA 84:7977.
67. Thiery, J. P., Duband, J.-L., and Delouvée, A., 1982, Pathways and mechanism of avian trunk neural crest cell migration and localization, Dev. Biol. 93:324.
68. Tosney, K. W., 1978, The early migration of neural crest cells in the trunk region of the avian embryo: An electron microscopic study, Dev. Biol. 62:317.
69. Tucker, R. P., and Erickson, C. A., 1984, Morphology and behavior of quail neural crest cells in artificial three-dimensional extracellular matrices, Dev. Biol. 104:390.
70. Tucker, R. P., and McKay, S. E., 1991, The expression of tenascin by neural crest cells and glia, Development 112:1031.
71. Tucker, R. P., Edwards, B. F., and Erickson, C. A., 1985, Tension in the culture dish: Microfilament organization and migratory behavior of quail neural crest cells, Cell Motility 5:225.
72. Valinsky, J. E., and Le Douarin, N. M., 1985, Production of plasminogen activator by migrating cephalic neural crest cells, EMBO J. 4:1403.
73. Welch, D. R., Fabra, A., and Nakajima, M., 1990, Transforming growth factor ß stimulates mammary adenocarcinoma cell invasion and metastatic potential., Proc. Natl. Acad. Sci. U.S.A. 87:7678.

# FIBRONECTINS AND EMBRYONIC CELL MIGRATION

Charles ffrench-Constant[1], Richard O. Hynes[2],
Janet Heasman[1], and Christopher C. Wylie[1]

[1]Wellcome/CRC Institute of Developmental Biology
and Cancer, Tennis Court Rd, Cambridge.
CB2 1QR, U.K.

[2]Center for Cancer Research and Howard Hughes Medical
Institute, Department of Biology, Massachusetts
Institute of Technology, Cambridge, Massachusetts
MA 02139, USA

## Introduction

Cell migration represents an essential feature in the development of complex multicellular organisms. The extent of this migration is striking - for example the neural crest cells that form the melanocytes of the skin and much of the peripheral and autonomic nervous system arise from the dorsal surface of the neural tube and migrate long distances to their final destinations (Le Douarin, 1982). The mechanisms responsible for the control of this and and other cell migrations must be very precise to ensure normal development, but remain poorly understood. In order to elucidate the role of the extracellular matrix in these mechanisms we have focussed on the adhesive glycoprotein fibronectin (FN). FN is known to stimulate cell migration and is widespread throughout the embryo at the time when many cell migrations are occurring (Hynes, 1990), making it an excellent candidate for an important role in the mechanisms of control.

Recent work on the molecular biology of FN and its cell surface receptors has provided a fairly detailed understanding of the structure of FN and the molecular mechanisms by which FN interacts with the cell. This work has shown that two classes of cell-surface receptor are present. The first class is a family of heterodimeric molecules termed integrins two of which, $\alpha5\beta1$ and $\alpha4\beta1$, have been shown to bind specifically to FN at different sites (Humphries, 1990). A second class of molecules is provided by cell-surface heparan sulphate proteoglycans, which can bind FN at an as yet undefined site within the so-called heparin-binding region (Saunders and Bernfield, 1988). Cell culture experiments have established that all three of these sites are involved in cell adhesion and spreading (Ruoslahti and Pierschbacher, 1986, Humphries et al. 1986, McCarthy et al. 1990, Guan and Hynes, 1990) and have also shown that cells can distinguish between different sites suggesting that each site may play a different role in the control of cell behaviour (Woods et al. 1986, Dufour et al. 1988).

Work on the structure of FN has revealed potential molecular heterogeneity of FN as a result of alternative splicing of the primary gene transcript ( Hynes, 1990). This splicing can include or exclude all or part of three separate exons, each of which is derived from one of the so-called type three repeats that constitute the central part of the molecule. In rats, humans and chickens, two of these exons (designated EIIIB and EIIIA) are either completely included or excluded, while the third exon (designated the variable or V region) can be partially excluded (Fig. 1). The V region of FN has been shown to contain cell binding sites (Humphries et al. 1986) (shown as hatched areas in Fig. 1) and splicing changes may therefore produce forms of FN with differing adhesive properties that may play different roles in the promotion and/or guidance of cell migration.

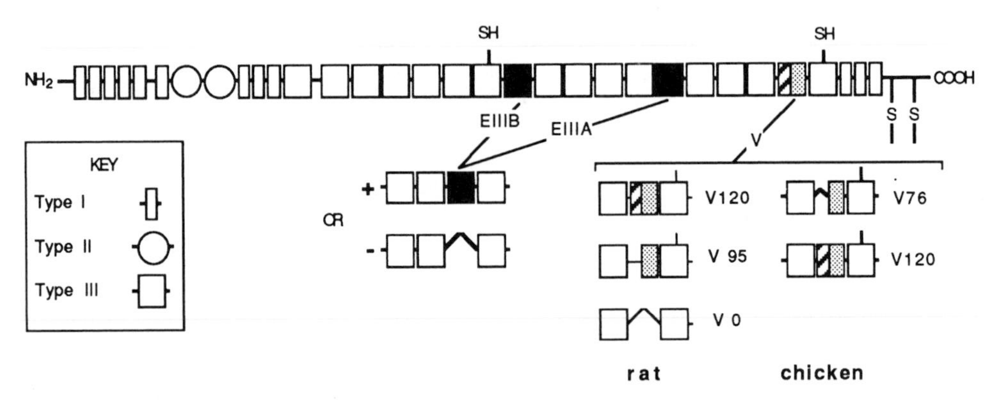


Fig. 1 A diagrammatic representation of FN showing the three regions of alternative splicing and illustrating the potential heterogeneity that can be generated by this mechanism. Note that the V region is spliced differently in rat and chicken.

Despite this detailed understanding of the molecular mechanisms, however, our knowledge of how FN regulates cell behaviour <u>in vivo</u> remains less well understood. This reflects both the potential complexity of the FN within the extracellular matrix and the inaccessibility to study of specific cell populations during the four different stages of migration; timing of start, stimulation of migration, guidance and cessation of movement when the target is reached. In order to overcome these problems we have taken two approaches. First we have chosen to study an embryonic cell population, primordial germ cells, which can be isolated and examined before, during and after migration. Second, we have used in situ hybridization to define the extent of molecular heterogeneity of FN within the developing embryo both before and after a number of different cell migrations.

## Fibronectin and primordial germ cell migration

Primordial germ cells, which will form the gametes of the adult animal, can first be recognized in the extra-embryonic mesoderm of 7 - 7.25 days post coitum (dpc) mouse embryos (Ginsburg et al. 1990). From here they move into their correct location within the genital ridges over the next 4 days. This movement occurs in two distinct phases. The first phase, in which they move from the endoderm into the hindgut between 7.5 and 8.5 dpc, may result from the passive carriage of cells along with the invaginating hindgut endoderm as the cells shown no ultrastructural features of a migratory phenotype (Clark and Eddy, 1975). In contrast, the second phase, in which they move through the dorsal mesentery and into the genital ridge between 9.5 and 11.5 dpc, results from active migration as the PGCs now show ultrastructural features of migration and, in addition, they will migrate when placed in cell culture (Clark and Eddy, 1975, Blandau et al. 1963, Stott and Wylie, 1986).

PGCs can be dissociated from the embryo before, during and after the phase of active migration and can be identified both in these cultures and in intact embryos by their cell-surface alkaline phosphatase activity (Donovan et al. 1986). In addition PGC migration in vitro can be analyzed by preparing explant cultures of appropriate regions of the embryo (ffrench-Constant et al. 1991). Thus if 9.5 dpc hindgut fragments are cultured on a feeder cell layer of STO cell embryonic fibroblasts PGCs will emigrate from the explant and move onto the STO cell feeder layer (Fig. 2). By using carboxyfluorescein diacetate succinimidyl ester (CFSE) to label the explant with fluorescein it can shown that this represents a genuine emigration from the explant rather than spreading of the explant itself associated with passive carriage of PGCs. In contrast, explants cultured on substrates in the absence of the STO cell feeder layer show no such emigration, with PGCs remaining within the spreading explant. Explants prepared from 8.5 dpc embryos show no emigration of PGCs consistent with the ultrastructural observations suggesting that these cells are not actively migratory but are carried passively to the developing hindgut (Clark and Eddy, 1975)

FN stimulates the migration of PGCs from these hindgut fragments; the addition of 10 μg/ml plasma FN to the culture medium significantly increases the number of PGCs leaving the fragments and, as a result, the mean distance travelled by PGCs from each explant (Fig. 3). Taken together with the localization of FN within the dorsal mesentery through which PGCs migrate (Fujimoto et al. 1985) this observation suggests that FN is responsible for stimulating PGC migration in vivo. This stimulation is also seen with many other embryonic cell types (Hynes, 1990), suggesting that this represents an important general property of FN during development.

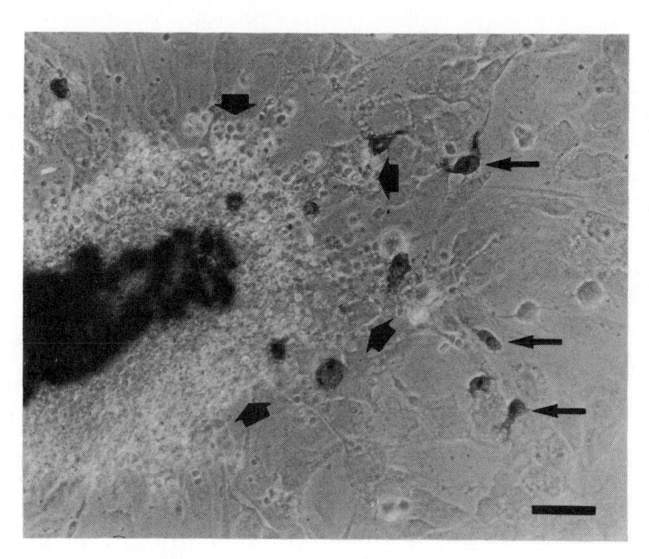

Fig. 2 9.5 dpc primordial germ cells will migrate independently over STO cell monolayers. 9.5 dpc hindgut fragments were cultured on STO cell monolayers; PGCs (thin arrows) can be seen beyond the edge of the hindgut fragment (thick arrows). Scale bar = 50 μm.

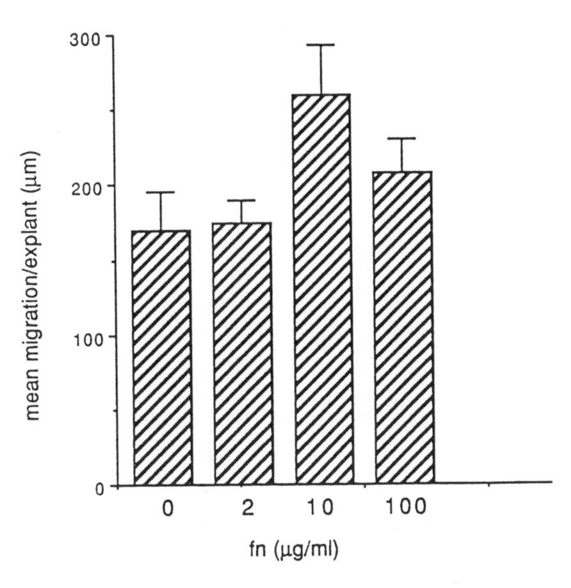


Fig. 3 Primordial germ cell migration is enhanced by fibronectin. The total extent of PGC migration from 9.5 dpc hindgut fragments such as that shown in Fig. 2 was quantified to assess the effects of increasing concentrations of soluble fibronectin in the culture medium.

PGCs do not appear to contribute FN to their migratory substrates. In situ hybridization studies using a double-labelling technique to localise FN mRNA and PGCs in the same section show that PGCs contain little, if any, FN mRNA (ffrench-Constant et al. 1991). FN on which PGCs migrate is apparently synthesized and secreted by the adjacent mesenchymal cells of the mesentery as these cells contain significant levels of FN mRNA (ffrench-Constant et al. 1991) and immuno-electron microscopy studies show FN to be present in the endoplasmic reticulum of these cells (Fujimoto et al. 1985). Taken together with the finding that FN stimulates PGC migration, these observations suggest that the guidance of PGC migration may, at least in part, be determined by the orientation and distribution of the FN produced by the cells of the mesentery. Such a mechanism has been proposed for the guidance of neural crest cell migration, as these cells also do not appear to synthesize FN (Newgreen and Thiery 1980, ffrench-Constant and Hynes, 1988). Interestingly, extracellular matrix fibrils orientated in the direction of migration have been seen in both neural crest cell (Newgreen. 1989) and PGC migratory pathways (Heasman, unpublished observations). Such fibrils are likely to contain FN and may therefore contribute to the guidance of migrating cells.

## Cell-fibronectin interactions and the timing of migration

An analysis of PGC adhesion to FN suggests a role for FN not only in stimulating and guiding migration but also in the timing of movement (ffrench-Constant et al. 1991). When regions of 8.5, 9.5 and 12.5 dpc embryos containing PGCs were dissociated in a non-enzymatic buffer (to preserve cell-surface receptors) very different levels of adhesion were observed (Fig. 4). These ages were chosen as they correspond to developmental stages before, during and after PGC migration; this observation shows therefore that the start of migration is associated with a fall in adhesion to FN, with a further fall observed after migration is complete.

These changes in adhesion could determine the timing of migration from the hindgut.The high level of PGC adhesion to FN would prevent active migration and facilitate the passive carriage of the PGCs into the hindgut by the invaginating endoderm. Once this phase of PGC movement is complete, however, the fall in adhesion seen in PGCs may allow the cells to leave the hindgut by active migration. Finally, the very low levels of adhesion seen in post-migratory PGCs may be insufficient to support migration because the cells are unable to generate sufficient traction.

This hypothesis emphasises that, as with all migrating cell types, the behaviour of PGCs on FN will reflect a complex balance between adhesion and migration. FN may stimulate cell migration but high levels of adhesion to FN, resulting either from high concentrations of FN or from expression of appropriate receptors, may prevent cell movement. In keeping with this, we found that high concentrations of FN (100 μg/ml) did not stimulate cell migration in 9.5 dpc PGCs (Fig. 3). Moreover, it has recently been shown that binding of cell-surface integrin receptors by substratum-bound antibodies of moderate avidity promoted migration more effectively than antibodies with high avidity (Duband et al. 1991).

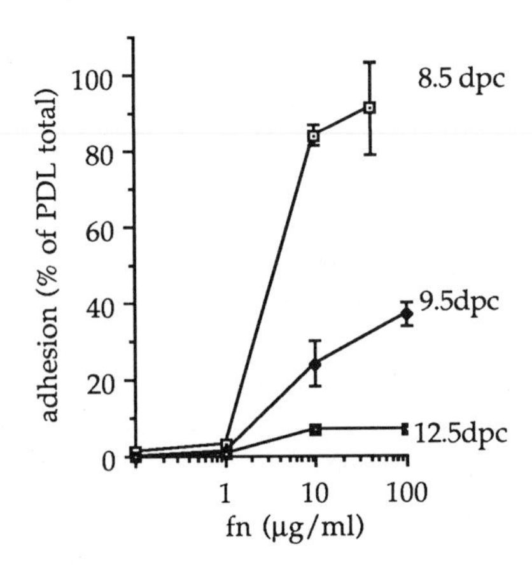


Fig. 4. Primordial germ cell adhesion to fibronectin. Adhesion was measured at four different concentrations of fibronectin. Results are expressed as a percentage of the total number of PGCs initially plated out in that assay. Error bars represent standard errors of the mean, with the errors at 12.5 dpc too small to show clearly at this scale.

A prediction of the hypothesis that the timing of migration is regulated by changes in the level of adhesion to FN is that the start of active migration may be associated with a decrease in the affinity of PGC FN receptors. The integrin family of receptors, which contains at least two well characterized FN receptors (Humphries, 1990), are excellent candidates for a role in the timing of PGC migration and experiments to examine the different integrins on pre-migratory and migratory PGCs will be required to test this prediction.

Alternative splicing of fibronectin during cell migration

Another prediction of this hypothesis is that a migratory substrate which provides a low level of adhesion may be more appropriate for cell movement during development. Our work on the alternative splicing of the FN primary gene transcript during development and repair suggests that molecular heterogeneity of an individual extracellular matrix component such as FN may be one mechanism by which changes in adhesiveness can be achieved. In an in situ hybridization analysis of the patterns of FN mRNA splicing during development in the chicken we found that the FN mRNA present in migrating cells or their immediate neighbours always contained the two spliced exons EIIIA and EIIIB, suggesting that EIII$^+$ EIIIB$^+$ FN is the predominant FN in the migratory substrate (ffrench-Constant and Hynes, 1988). Later in development, however, when migration was largely completed other combinations of EIIIA and EIIIB were seen in different tissues in a cell-type specific manner. In the late embryonic liver, for example, we found three different patterns of splicing. Hepatocytes contained EIIIA$^-$ EIIIB$^-$ FN mRNA which is to be expected as these cells are the main source of soluble plasma FN which is known to lack both these regions . In contrast, the walls of large veins contained EIIIA$^+$ EIIIB$^-$ FN mRNA while in the smaller arterioles the FN mRNA was EIIIA$^+$ EIIIB$^+$. Although not present in liver, the fourth possible combination of EIIIA$^-$ EIIIB$^+$ FN mRNA was found in chondrocytes in the sternum at the same age (ffrench-Constant and Hynes, 1989).

These results suggest that EIIIA$^+$ EIIIB$^+$ FN may be most appropriate in cell migration. In order to test this idea we performed a study of wound healing in the adult animal. The process of healing involves cell migration in a previously differentiated tissue, as keratinocytes migrate to cover the wound( Clark and Henson, 1988). Undisturbed rat skin showed a low level of expression of EIIIA$^-$ EIIIB$^-$ FN mRNA. Wounding resulted in significant increase in the level of FN mRNA. More importantly, however, we observed a reappearance of the embryonic pattern of EIIIA$^+$ EIIIB$^+$ FN mRNA in association with migrating keratinocytes (ffrench-Constant et al. 1989). This observation supports our conclusion from the developmental study that EIIIA$^+$ EIIIB$^+$ FN is most appropriate for migration and suggests that a re-expression of the embryonic FN is an important part of the healing process.

These observations on FN splicing raise two questions. First, why is EIIIA$^+$ EIIIB$^+$ FN most appropriate for cell migration? Second, do EIIIA and

EIIIB have separate functions and, if so, what are these functions? The answer to the first question will come from experiments with purified EIIIA+ EIIIB+ FN prepared in recombinant expression systems (Guan et al. 1990), as this spliced variant cannot be independently purified from tissues or from plasma. At this stage, however, it is an attractive hypothesis that EIIIA+ EIIIB+ FN is most appropriate for migration because it is less adhesive than the other spliced variants. These other variants, which appear later in development, may be more appropriate for maintenance of the differentiated state seen after migration is complete.

The second question, the function(s) of EIIIA and EIIIB also cannot be answered directly at this stage. The in situ hybridization studies do, however, allow testable hypotheses to be generated. EIIIA may have a role in the incorporation of FN into the matrix, as the only two cell types we have found that do not include EIIIA into FN mRNA, hepatocytes and chondrocytes, do not assemble extensive FN-rich pericellular matrices. Against this, however, it has been shown directly using immunoelectron microscopy that EIIIA+ and EIIIA- FNs can be incorporated in the extracellular matrix (Pesciotta Peters et al. 1990). EIIIB, in contrast, may have a role in cell migration and/or proliferation. The evidence for this comes from a study of the chorioallantoic membrane in the chick embryo. This tissue is unusual in that growth stops rather abruptly once the membrane has grown around the inside of the eggshell. This cessation of growth is associated with the disappearance of EIIIB, but not EIIIA, from FN mRNA (ffrench-Constant and Hynes, 1988). These findings suggest that EIIIB may play a more important role than EIIIA in any changes in the properties in FN during and after the migration and/or proliferation associated with growth. Whatever the nature of this role, it is interesting to note that EIIIB is better conserved between species than adjacent regions of FN (Norton and Hynes, 1987). adding further weight to the idea that this is a particularly important functional region of the molecule.

Part of the third spliced region of FN, the V region, has been shown to contain a cell binding site (Humphries et al. 1986) and splicing in this region can therefore be expected to generate functionally different FNs that might regulate cell migration. We have been unable to detect any changes in V region splicing during cell migration in chicken embryos (ffrench-Constant and Hynes, 1988). This may reflect the low sensitivity of the in situ hybridization technique to small changes in the level of V splicing that nonetheless might have important functional properties. Studies in another species, the rat, using ribonuclease protection assays do show differences in V region splicing during development, with an increased degree of inclusion of the cell binding site during development as compared to young adult animals (Pagani et al. 1991). This does suggest that inclusion or exclusion of the cell binding site in this region may have an important role in generating different FNs during development.

## Conclusions

These studies on FN allow two main conclusions. First, FN plays more than one role in cell migration. As well as stimulating migration the interaction of FN with the cell appears to be involved in the timing of migration by alteration of the level of cell adhesion. Second, when considering the function of FN it is important to realise that FN is a family of molecules which differ as a result of alternative splicing and that each of these FNs may have important functional differences. Turning to the specific question of cell migration, these conclusions emphasize that the cell-matrix interaction regulating movement will be modified not only by changes in receptors for FN and other extracellular matrix molecules on the cell surface, but also by changes during migration within the matrix itself. The flexibility provided by such an interaction, in which both cells and matrix can rapidly change their properties, may be essential to provide the necessary precise control of cell movement and fate during development.

References


Blandau, R.J., White, B.J. and Rumery, R.E., (1963) Observations on the movements of the living primordial germ cells in the mouse, Fertil.Steril.,14:482-489.

Clark,J.M. and Eddy,E.M.,(1975) Fine structural observations on the origin and associations of primordial germ cells of the mouse, Devl.Biol.,47:136-155.

Clark,R.A.F. and Henson, P.M., (1988)"The Molecular and Cellular Biology of Wound Repair", Plenum press, New York.

Donovan,P.J., Stott, D., Cairns, L.A., Heasman, J. and Wylie, C.C., (1986) Migratory and postmigratory mouse primordial germ cells behaved differently in culture, Cell,44:831-838.

Duband, J-L, Dufour, S., Yamada, S.S., Yamada, K.M. and Thiery, J.P., (1991) Neural crest cell locomotion induced by antibodies to β1 integrins. A tool for studying the roles of substratum molecular avidity and density in migration, J. Cell Sci., 98:517-532.

Dufour, S., Duband, J-L, Humphries, M.J., Obara, M., Yamada, K.M. and Thiery, J.P., (1988) Attachment, spreading and locomotion of avian neural crest cells are mediated by multiple adhesion sites on fibronectin molecules,EMBO J.,7:2661-2671.

ffrench-Constant, C. and Hynes, R.O.,(1988) Patterns of fibronectin gene expression and splicing during cell migration in chicken embryos, Development, 104:369-382.

ffrench-Constant, C. and Hynes, R.O., (1989) Alternative splicing of fibronectin is temporally and spatially regulated in the chicken embryo,Development, 106:375-388.

ffrench-Constant, C., Van De Water, L., Dvorak, H.F. and Hynes, R.O.,(1989) Reappearance of an embryonic pattern of fibronectin splicing during wound healing in the adult rat, J.Cell Biol.,109:903-914.

ffrench-Constant, C.,Hollingsworth, A., Heasman, J. and Wylie, C.C., (1991) Response to fibronectin of mouse primordial germ cells before, during and after migration, Development,113, 1365-74.

Fujimoto, T., Yoshinaga, K. and Kono, I.,(1985) Distribution of fibronectin on the migratory pathway of primordial germ cells in mice,Anat.Rec.,211:271-278.

Ginsburg, M., Snow, M.H.L. and McLaren, A., (1990) Primordial germ cells in the mouse embryo during gastrulation, Development,110-528. Guan, J. and Hynes, R., (1990) Lymphoid cells recognize an alternatively spliced segment of fibronectin via the integrin receptor α4β1, Cell, 60:53-61.

Guan, J.L., Trevithick, J.E. and Hynes, R.O., (1990) Retroviral expression of different alternatively-spliced forms of rat fibronectin, J.CellBiol.,110:833-847.

Humphries, M.J., Akiyama, S.K., Komoriya, A., Olden, K. and Yamada,K.M., (1986)Identification of an alternatively spliced site in human plasma fibronectin that mediates cell type-specific adhesion, J.Cell Biol.,103:2637-2647.
Humphries, M.J., (1990)The Molecular basis and specificity of integrin-ligand interactions, J.CellSci.,97:585-592.
Hynes, R.O., (1990) "Fibronectins", Springer-Verlag, New York.
Le Douarin, N., (1982) "The Neural Crest", Cambridge University Press,Cambridge, UK.
McCarthy, J.B., Skubitz, A.P.N., Zha, Q., Yi, X., Mickelson, D.J., Klein,D.J.and Furcht, L.T.,(1990) RGD-independent cell adhesion to the carboxy- terminal heparin-binding fragment of fibronectin involves heparin-dependent and-independent activities, J.Cell Biol.,110:777-787.
Newgreen, D. and Thiery, J-P., (1980) Fibronectin in early avian embryos: synthesis and distribution along the migration pathways of neural crest cells, Cell Tissue Res., 211:269-291.
Newgreen, D.F.,(1989) Physical influences on neural crest cell migration in avian embryos: Contact guidance and spatial restriction,Devl.Biol., 131:136-148.
Norton, P.A. and Hynes, R.O., (1987) Alternative splicing of chicken fibronectin in embryos and in normal and transformed cells, Mol.Cell.Biol.,7:4297-4307.
Pagani, F., Zagato, L., Vergani, C., Casar,m G., Sidoli, A. and Baralle, F.E.,(1991)Tissue-specific splicing pattern of fibronectin messenger RNAprecursor during development and aging in rat,J.Cell Biol.,113:1223-1229.
Pesciotta Peters, D.M., Portz, L.M. Fullenwider,J. and Mosher, D.F., (1990) Co-assembly of plasma and cellular fibronectins into fibrils in human fibroblast cultures, J. Cell Biol., 111:249-256.
Ruoslahti, E. and Pierschbacher, M.D.,(1986) Arg-Gly-Asp: A versatile cell recognition signal, Cell, 44:517-518.
Saunders, S. and Bernfield, M.,(1988) Cell surface proteoglycan binds mouse mammary epithelial cells to fibronectin and behaves as a receptor for interstitial matrix, J.Cell Biol., 106:413-430.
Stott, D. and Wylie, C.C.,(1986) Invasive behaviour of mouse primordial germ cells in vitro, J.Cell Sci., 86:133-144.
Woods, A., Couchman, J.R., Johansson, S. and Hook, M.,(1986) Adhesion and cytoskeletal organisation of fibroblasts in response to fibronectin fragments, EMBO. J.,5:665-670.

LOCALIZATION OF ENDOGENOUS LECTINS DURING EARLY CHICK DEVELOPMENT

Sara E. Zalik[1], Eliane Didier[2], Irene M. Ledsham[1], Esmond J. Sanders[3] and Christopher K. Guay[1]

Departments of Zoology [1] and Physiology [3]
University of Alberta Canada and Laboratoire de Biologie Animale[2]
Université de Clermont Blaise-Pascal, France

Our interest in studying the endogenous lectins from the early chick embryo, arose from previous work in which we found that mechanically dissociated cells from primitive streak chick blastoderms, were readily agglutinated by a variety of plant lectins. Among the plant lectins the galactose-specific *Ricinus communis* agglutinin (RCA), exhibited a strong agglutinating effect (Zalik and Cook, 1976, Phillips and Zalik, 1982). Ensuing from this work we decided to investigate whether the embryo contains endogenous lectins that could interact with the plant lectin receptors detected by cell agglutination.

Our early work in the search for the endogenous lectins of the chick embryo consisted in testing blastoderm homogenates for their activity to induce agglutination in trypsinized glutaraldehyde-fixed rabbit erythrocytes (Cook, et al. 1979; Zalik, et al. 1983). In these studies we found that extracts from whole unincubated blastoderms as well as from the area pellucida, and the epiblast and endoblast of the area opaca of primitive streak blastoderms, exhibited an hemagglutinating activity specific for saccharides bearing a β-galactoside configuration. The blastoderm lectin was preferentially inhibited by thiodigalactoside, lactose and asialofetuin, the latter a glycoprotein with terminal galactose groups.

The galactose-specific lectins from primitive blastoderms (Hamburger and Hamilton stages 4 to 6), were isolated by affinity chromatography on lactoside-derivatized Sepharose. Affinity-purified lectin preparations when separated by sodium dodecylsulphate polyacrylamide gel electrophoresis (SDS-PAGE) under reducing conditions followed by silver staining are seen to consist of three main bands: a 70 ± 2 kD band which remains to be characterized, and 2 main bands separating between 14 and 16 kD (Fig 1). By immunoblot analysis the latter two bands have been identified as the 14 and 16kD galactose-binding lectins, since they are recognized by the antibodies to these lectins (Provided by Drs. Y. Oda and K. I. Kasai, Teikyo Univ. Japan) (Fig 2). Also using immunoblot analysis, a fourth band has been identified in the purified blastoderm lectin extracts, this band corresponds to an apolipoprotein (Apo), of the very low density lipoproteins Apo-VLDL II (Fig. 2). This protein is produced by the liver during oogenesis, released into the circulation, and subsequently taken up by the growing oocyte where it is incorporated into yolk (Wallace, 1985). Apo-VLDL II is also known as Apovitellenin I (Dugaiczyk, et al., 1981). In crude lectin extracts the antibody to Apo-VLDL II recognizes the monomer as well as the dimer of this protein (Fig

3), (Nimpf, et al., 1988, Zalik, et al., 1990); the anti-16kD-lectin antiserum reacts mainly with the 16kD band and weakly with the 14kD band. The anti-14kD lectin antiserum recognizes mainly two bands of approximate molecular weights of 25kD and 50 kD which may correspond to storage forms of this lectin.

Lectin localization during gastrulation and early embryogenesis

In the early chick embryo, the lectins as well as Apo-VLDL II are localized intracellularly within the yolk platelets of the ectoderm, mesoderm and endoderm. These lectins are also present in extracellular material present at the lower surface of the endoderm that is emerging and migrating laterally from the streak (Figs 4, 5), (see also Zalik, et al., 1990). Lectin-rich cells are also found in the epiblast within and adjacent to the primitive streak (Zalik, et al., 1990); at the present time we do not know what functional role in the embryo that these lectin-rich cells may have. These cells may migrate through the primitive streak and find their way into the endoderm which is also abundant in lectin. The primordial germ cells also possess a lectin-rich cytoplasm (Figs 4,6 and 7). Using immunoelectronmicroscopy with protein A-gold, the 14kD lectin and Apo-VLDL II were localized in the lucent component of the yolk platelet (Fig 8), while the 16kD lectin was shown to be more abundant in the dense compartment of this organelle (Fig 9). The 16kD lectin and Apo-VLDL II are also present at the cell surface associated with a component with a matrix-like structure (Figs. 10), (see also Sanders, et al., 1990).

Lectin expression during early morphogenesis has been examined using an antiserum to the 16kD lectin purified from the liver of the laying hen (obtained by SEZ) and an antiserum to the 16-17 kD lectin from the embryonic chick pectoral muscle (obtained by ED). In crude extracts from chick blastoderms, the anti-16kD-liver lectin antiserum reacts mainly with the 16kD lectin and exhibits a weak immunoreactivity to the 14kD lectin; the anti- 16kD lectin antiserum from pectoral muscle reacts only with the 16kD lectin. The localization of this lectin was investigated in embryos up to 4 days of incubation. In embryos fixed with paraformaldehyde and processed in polyethylene glycol, the antiserum to the 16kD liver lectin detected extracellular matrix material present at the apical and basal surfaces of the neural tube (Figs 11,12,13 and 14), that surrounding the notochord(Figs 11,12 and 13), as well as material located at the surface of the epimyocardium of the fusing cardiac rudiments (Fig 11). The neural crest cells, immediately above the recently formed neural tube were also recognized by this antiserum (Fig 12). The antiserum also stained intensely the vitelline membrane, material present in the subvitelline space and extracellular material at the apical surface of the epiblast and the neural canal (Fig 14). The matricular material recognized by the anti-16kD liver lectin antiserum is present up to stage 13 and disappears at subsequent stages up to 4 days of incubation. The anti 14kD lectin antiserum displayed a weak reactivity towards the vitelline membrane and the apical surface of the ectoderm (Fig 14).

The endodermal cells of the area opaca of the primitive streak chick embryo are rich in intracellular lectin (Zalik, et al., 1990; Sanders, et al., 1990; Zalik, 1991). Since these cells are the precursors of the yolk sac we examined this tissue for changes in lectin activity during development. Galactose-specific lectin activity is present in early yolk sac extracts (Mbamalu and Zalik 1987), and this activity increases up to 4 days of development and remains high throughout. Purified lectin preparations from chick yolk sacs, when examined under SDS-PAGE, also consist of a 14 and a 16 kD band (Fig 15). Using immunocytochemistry in sections of yolk sac membranes from 5 day old embryos, the 14kD lectin is seen to be present intracellularly in the endodermal cells, where it appears to be associated with the periphery of intracellular yolk (Fig 17). A very similar distribution was observed when sections were reacted with antisera to Apo-VLDL II (Fig 16). The anti-16kD lectin antiserum also stained the endodermal cells of the yolk sac membrane, however, in this case the lectin was present in small intracellular inclusions which

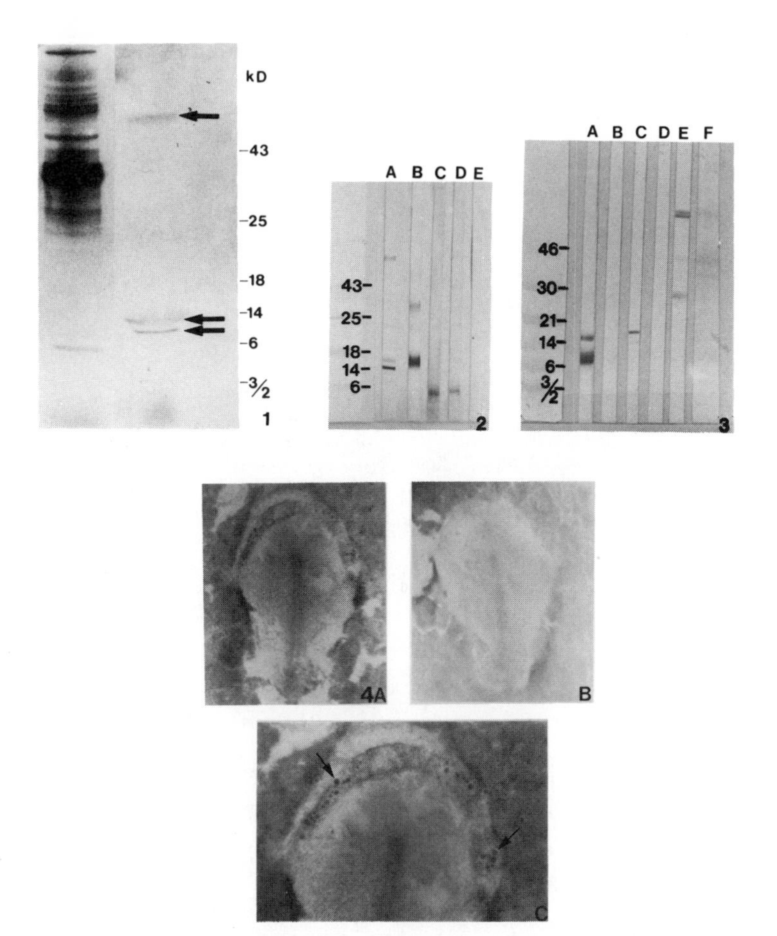


**Fig 1.** SDS-PAGE of extracts of crude blastoderm lectin (left), and of a purified preparation of blastoderm lectin (right). Numbers to the right correspond to the migration of the molecular weight standards. Arrows point to the location of the 14 and 16kd bands as well as the 70 ± 2 kD band. From Zalik, et al. (1990).

**Fig 2.** Immunoblot analysis of a purified preparation of blastoderm lectin. A) reacted with anti-14kD lectin antiserum; B) reacted with anti-16kD liver lectin antiserum; C) reacted with an antiserum to anti-adult Apo-VLDL II; D) reacted with an antiserum to embryonic Apo-VLDL II; E) preimmune serum. Molecular weight markers are to the left. From Zalik, et al. (1990).

**Fig 3.** Immunoblot analysis of a crude lectin extract from chick blastoderm reacted with A) anti- Apo VLDL II antiserum; B) control preimmune serum; C) anti-liver-16kD lectin antiserum; D) control preimmune serum; E) anti-14kD lectin antiserum; F) nitrocellulose transfer stained with Ponceau red. Observe that in A), the monomer and the dimer of Apo-VLD II are present in this preparation; in immunoblots of crude extracts E), the 14kD lectin antiserum recognizes mainly a 25 and a 70 kD band. Molecular weight standards in kD are shown to the left. From Sanders, et al. (1990).

**Fig 4.** Immunoperoxidase staining of whole mounts of primitive streak embryos stained with A) anti-16kD liver lectin antiserum; B) control preimmune serum; C) higher magnification of A. Observe in the area pellucida, staining in the primitive streak as well as staining in the area opaca where immunoreactivity is concentrated in the endoderm (Zalik, et al. 1990). The germinal crescent is also stained, arrows point to some of the primordial germ cells. From Zalik (1991).

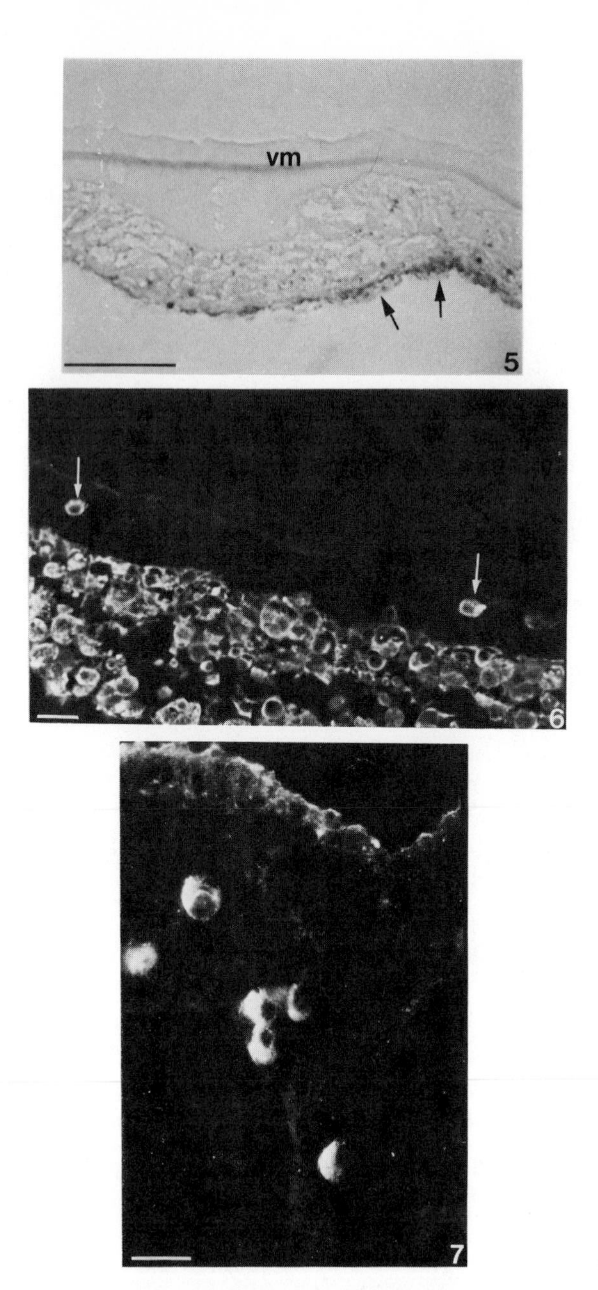


**Fig. 5.** Immunoperoxidase staining of a section of a primitive streak blastoderm reacted with anti-16kD liver lectin antiserum. In this section an area just lateral to the primitive streak is shown. Observe stained material at the lower surface of the endoderm (arrows), (vm) vitelline membrane. Bar is 50 μ.

**Fig. 6.** Immunofluorescent staining of the area of the germinal crescent (gc) of a stage 8 embryo reacted with an antiserum to the 16kD lectin from embryonic muscle. Arrows point to the germ cells. Bar is 50 μ.

**Fig 7.** Immunofluorescent staining of a section of a stage 18 embryo reacted with the antiserum to the 16kD lectin of embryonic pectoral muscle. Observe staining of the primordial germ cells migrating through the dorsal mesentery. Bar is 20 μ.

were also associated with intracellular yolk (Fig 18). These inclusions may correspond in ultrastructure to the pleiomorphic organelles described by Sanders, et al. (1990) in early embryos, which are are rich in 16kD lectin.

The antiserum to the 16-17kD embryonic chick pectoral muscle lectin also recognizes the 16kD lectin in crude extracts of chick blastoderms. Although this antiserum displays negligible immunoreactivity to the matrix-type lectin, it recognizes material at the apical surface of the neural tube, intracellular material in the early germ layers and the somite, as well as in the myotome and developing heart of stages 18-24 embryos. The material recognized by this antiserum is best preserved in paraffin sections of ethanol-fixed tissues, while the material recognized by the anti-16kD-liver lectin is not preserved under these conditions. All of these antisera appear to recognize intracellular material in the extraembryonic endoderm and the germ cells.

## Concluding Remarks

Our studies on the endogenous lectins of the early chick embryo, started with the assumption that in the embryos these lectins could associate with the cell surface lectin receptors that were detectable by agglutination with plant lectins. The results so far indicate that the endogenous galactose lectin(s) in the embryos are present in the apical domains of the epiblast and endoderm, as well as the basal and apical surfaces of the cells of the neural tube. This indicates that at least some of these lectins bind to receptors that are localized regionally at the cell surface. The extracellular lectin, in many instances, appears to be associated with extracellular matrix that surrounds the neural tube and notochord. This matrix has been shown to contain laminin and tenascin; proteins that contain N-acetyllactosamine or poly-N-acetyllactosamine sequences (Thiery, et al. 1982; Crossin, et al. 1986; Duband and Thiery, 1987; Mackie, et al. 1988; Hoffman, et al. 1988; Thorpe, et al. 1988; Drake, et al. 1990). Since several galactose-binding lectins have been shown to bind to poly-N-acetyllactosamine and polylactosaminoglycans (Abbot, et al. 1988; Oda and Kasai, 1984), it is possible that the lectins associate with these extracellular matrix proteins via proteincarbohydrate interactions and could play a role in segregating cells from the early organ primordia and in limiting the mixing of defined cell groups within the embryo. Several galactose-binding lectins can bind to laminin via saccharide protein interactions in a non-integrin manner (Zhou and Cummings, 1990; Woo, et al. 1990; Cooper and Barondes, 1990; Cooper, et al. 1990; Mercurio 1990).

A lectin with a specificity for sulfated fucan or mannan has been found at the external surface of the vitelline membrane (Cook, et al. 1985). In contrast, the 16kD lectin is abundant and appears to have an even distribution throughout this membrane. It is known that the blastoderm will expand and develop when cultured on the internal surface of the vitelline membrane but not on its external surface (New, 1959). The differential localization of these two lectins, may contribute in part to the substrate suitability of this membrane for adhesion and subsequent expansion of the blastoderm. The galactose-binding lectin may be involved in the formation of transitory adhesive bonds necessary for blastoderm expansion (Milos and Zalik, 1982, Zalik,1991).

The primordial germ cells contain abundant lectin in their cytoplasm, and N-acetyllactosamine-rich antigens have been detected in these cells (Didier, et al, 1990; Loveless, et al. 1990). We do not know at this time the significance of these findings, but the presence of intracellular lectin may represent a retention of an early embryonic trait by the germ line. The endodermal cells of the yolk sac produce 14 and 16kD lectins throughout development. Although at the present time we do not know the role that these lectins may have in organ and tissue differentiation, these lectins could be transported into the embryo via the circulation and could serve as growth factors as has been shown in insects (Kawaguchi, et al 1991). In the chick embryo these lectins have

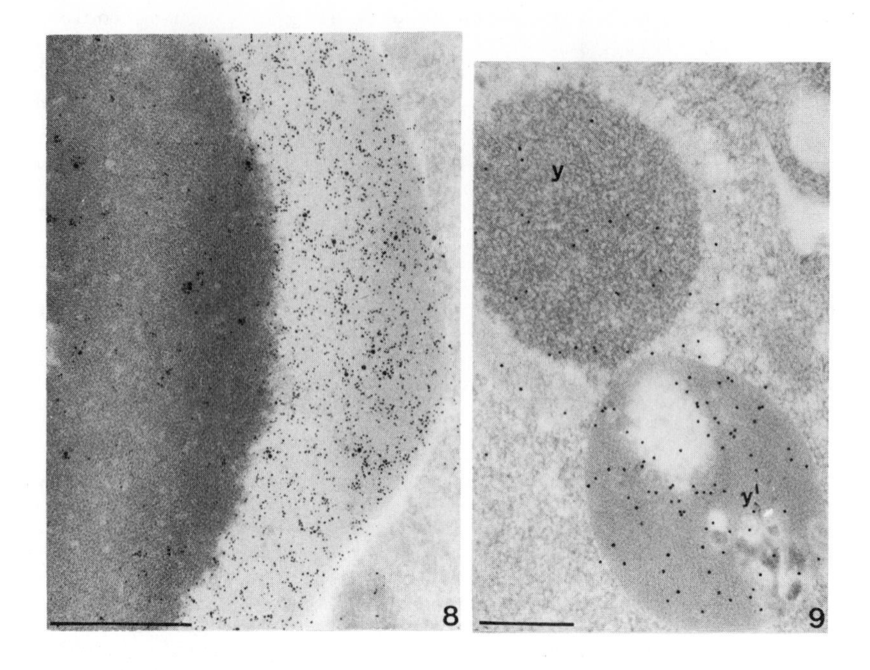


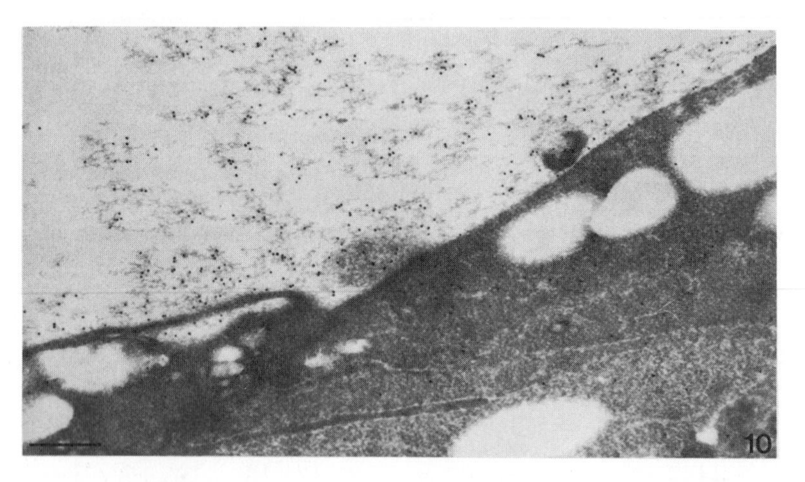


**Fig 8.** Intracellular yolk platelet reacted with anti-Apo-VLDL II antiserum (5 nm gold) and anti-14kD lectin antiserum (15nm gold). Observe that both labels are concentrated mainly in the lucent component with weak label in the dense component . Bar is 0.5 µ. From Sanders, et al. (1990).

**Fig 9.** Intracellular yolk (y), and yolk related organelle (y'), in a section of a primitive streak embryo reacted with anti-16kD liver lectin antiserum and subsequently reacted with protein A-gold. Observe that label predominates in the dense component of the yolk and in the yolk-related organelle. Bar is 0.5 µ. From Sanders, et al (1990).

**Fig 10.** Immunoelectronmicroscopy with Protein A-gold of an area from a section of the epiblast reacted with anti-16kD liver lectin antiserum. The apical region of two epiblast cells is shown. Observe label in the fibrous matrix-like material at the upper surface of two epiblast cells. Bar is 0.5 µ.

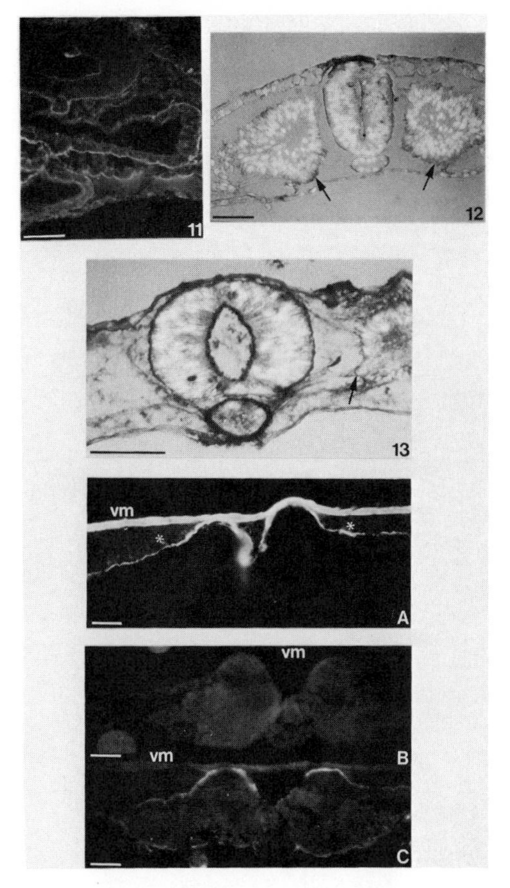


**Fig. 11.** Immunofluorescent staining of a transverse section at the anterior region of a 13 somite chick embryo reacted with anti-16kD-liver lectin antiserum. Staining is present on the internal and external surfaces of the neural tube and the periphery of the notochord. Staining is also present at the basal side of the endodermal cells of the foregut and at the basal surface of the epimyocardial layer. Bar is 50 μ. From (Zalik, 1991).

**Fig. 12.** Immunoperoxidase staining with anti-16kD liver lectin antiserum of a section of a 13 somite chick embryo. Observe staining in the neural crest and at the basal surface of the somite (arrow). Bar is 50 μ. From (Zalik, 1991).

**Fig 13.** Immunoperoxidase staining of the posterior region of a stage 13 embryo using anti-16kD liver lectin antiserum. Staining is present on internal and external surfaces of the neural tube and the external surface of the notochord. Observe staining at the basal surface of the somite (arrow) and in the matrix present in the embryonic space.Bar is 50 μ. From (Zalik, 1991).

**Fig 14.** Immunofluorescent staining of consecutive sections of the posterior region of a stage 13 chick embryo showing the vitelline membrane and the invaginating neural plate stained with A) anti-16kD-liver lectin antiserum, B) preimmune serum of A, and C) anti-14kD lectin antiserum. Observe intense reaction of the 16kD lectin antiserum with the vitelline membrane and the apical surface of the neural plate and the epiblast. Some staining is also present in the perivitelline space (*). A weak immunoreactivity of these structures is also evident in the section stained with the 14kD lectin antiserum (vm) vitelline membrane.
Bar is 50 μ.

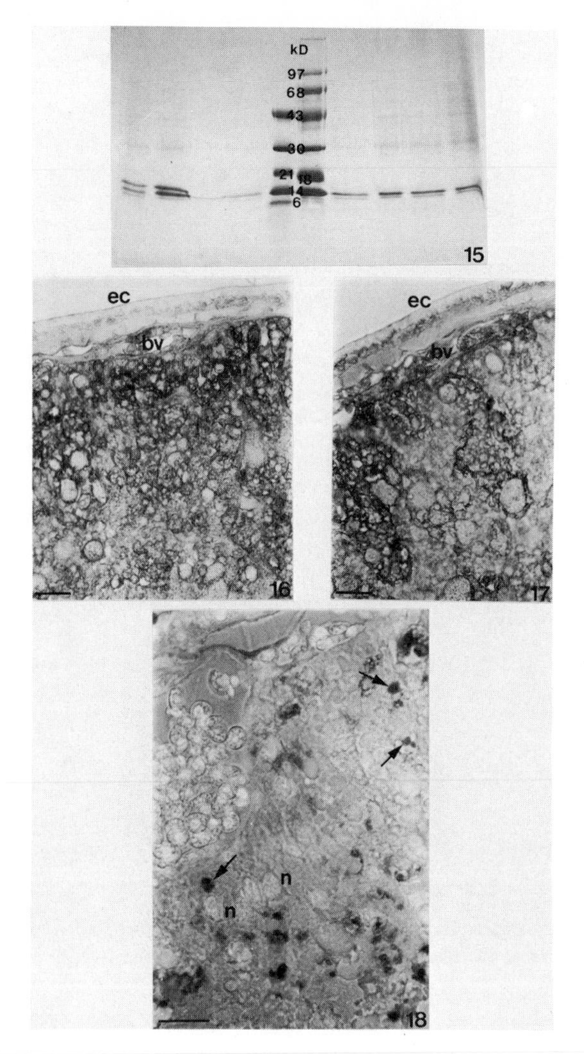


**Fig 15.** Gel run using SDS-PAGE conditions showing successive fractions obtained after purifying the lectin of the 15 day old yolk sac with lactoside-derivatized Sepharose. Fractions with highest activity are shown to the left. Bands of molecular weight standards in kD are present in two wells at the center of the gel.

**Fig 16.** Section of a 5 day yolk sac membrane stained with an antiserum against Apo-VLDL II using immunoperoxidase. Observe intense staining at the periphery of the intracellular yolk platelets present in the yolk sac endoderm; (ec) ectoderm, (bv) blood vessel in the mesoderm of the membrane. Bar is 50 µ.

**Fig 17.** Immunoperoxidase staining of a 5 day yolk sac membrane with anti 14kD lectin antiserum. Observe staining in the intracellular yolk platelets of the yolk sac endoderm. Staining is similar to that obtained with the antiserum to Apo-VLDL II in Fig 16; (ec) ectoderm, (bv) blood vessel in the yolk sac mesoderm. Bar is 50 µ.

**Fig 18.** Immunoperoxidase staining of a 5 day yolk sac membrane reacted with anti-16kD-liver lectin antiserum. Observe staining in discrete organelles associated with the intracellular yolk in the endodermal cell (arrows). Some of the nuclei stained with DAPI are labelled (n). The nucleated blood cells inside a vessel are shown at the upper left hand corner. Bar is 25 µ.

been shown to stimulate limb chondrogenesis (Matsutani and Yamagata, 1982).

Acknowledgement

We thank the Medical Research Council (SEZ and SJS) and the National Sciences and Engineering Research Council of Canada (SEZ) for their support. Our appreciation to Daniele Bayle, Nicole Benay and Sita Prasad for their excellent assistance. We are grateful to Mrs. Arlene Cowan for typing the manuscript.

## References

Abbot, M. W., Hounsell, E. F., and Feizi, T., 1988, Further studies of oligosaccharide recognition by the soluble 13kDa lectin of bovine the heart muscle, Biochem. J., 252:283-287.

Cook, G M W, Zalik S E., Milos N., and Scott,V., 1979, A lectin that binds specifically to β-D-galactoside groups is present at the earliest stages of chick embryo development, J. Cell Sci., 38:293-304.

Cook, G. M. W., Bellairs, R., Rutherford, N. G., Stafford, C. A., Alderson, T., 1985, Isolation, characterization and localization of a lectin within the vitelline membrane of the hen's egg, J. Embryol. Exp. Morph., 90:389-487.

Cooper, D. N. W., Barondes, S. H., 1990, Evidence for export of a muscle lectin from cytosol to extracellular matrix and for a novel secretory mechanism, J. Cell. Biol., 110:1681-1691

Cooper, D. N. W., Maasa, S. M., Barondes, S. H., 1990, L-14 soluble lectin is deposited in muscle extracellular matrix by binding to laminin polylactosamines, J. Cell Biol., 111:13a.

Crossin, K. L., Hoffman, S., Grumet, M., Thiery, J. P., Edelman, G. M., 1986, Site restricted expression of cytotactin during development of the chicken embryo, J. Cell Biol., 102:1917-1930.

Didier, E., Didier, P., Fargeix, N., Guillot, J., Thiery, J. P., 1990, Expression and distribution of carbohydrate sequences in the chick germ cells. A comparative study with lectins and the NC-1/HNK-1 monoclonal antibody, Int. J. Dev. Biol., 34:421-431.

Drake, C. J., Davis, L. A., Walters, L., Little, C. D., 1990, Avian vasculogenesis and the distribution of collagens I, IV, laminin and fibronectin in the heart primordia, J. Exp. Zool., 255:309-322.

Dugaiczyk, A., Inglis, A. S., Strike, P. M., Burley, R. W., Beattie, W. G., and Chan, L., 1981, Comparison of the nucleotide sequence of cloned DNA coding for apolipoprotein (apo VLDL II) from avian blood and the amino acid sequence of an egg yolk protein (apovitellenin I); equivalence of the two sequences, Gene, 14:175-182.

Duband, J. L., Thiery, J. P., 1987, Distribution of laminin and collagens during avian neural crest development, Development, 101:461-478.

Hamburger, V. and H. L. Hamilton 1951 A series of normal stages in the development of the chick embryo. J. Morph. 88: 49-92.

Hoffman, S., Crossin, K. L., Edelman, G. M., 1988, Molecular forms, binding functions and developmental expression of cytotactin and cytotactin-binding proteoglycan, an interactive pair of extracellular matrix molecules, J. Cell Biol., 106:519-532.

Kawaguchi, N., Komano, H. and Natori, S., 1991, Involvement of Sarcophaga lectin in the development of the imaginal discs of Sarcophaga Peregrina in an autocrine manner, Develop. Biol., 144:86-93 (1991).

Loveless, W., Bellairs, R., Thorpe, S. J., Page, M., Feizi, T., 1990, Developmental patterning of the carbohydrate antigen FC10.2 during early embryogenesis in the chick, Development, 108:97-106.

Mackie, E. J., Tucker, R. P., Halfter, W., Chiquet-Ehrismann, Epperlein, H. H., 1988, The distribution of tenascin coincides with pathways of neural crest cell migration, Development, 102:237-250.

Mbamalu, G. M. , Zalik, S. E., 1987, Endogenous β-D-galactoside-binding lectin during expansion of the yolk sac in the developing chick embryo, Roux's Arch. Dev. Biol., 196:176-184.

Matsutani, E., Yamagata, T., 1982, Chick endogenous lectin enhances chondrogenesis of cultured chick limb bud cells, Dev. Biol. 92:544-548.

Mercurio, A. M., 1990, Laminin: multiple forms, multiple receptors. Curr. Op. Cell Biol., 2:845-849.

Milos, N., Zalik, S. E., 1982, Mechanisms of adhesion among cells of the early chick blastoderm. Role of the β-D-galactoside-binding lectin in the adhesion of extraembryonic endoderm cells, Differentiation, 21:175-182.

New, D. A. T., 1959, Adhesive properties and expansion of the chick blastoderm, J. Embryol. Exp. Morphol., 7:146-164.

Nimpf, J., George, R., and Schneider, W. J., 1988, Apolipoprotein specificity of the chicken oocyte receptor for low and very low density lipoproteins: lack of recognition of apolipoprotein VLDL II., J. Lipid Res., 29:657-667.

Oda, Y., Kasai, K. I. 1984, Photochemical cross-linking of β-galactoside-binding lectin to polylactosamino-proteoglycan of chick embryonic skin, Biochem. Biophys. Res. Commun., 123:1215-1220.

Phillips, J. R., Zalik, S. E., 1982, Differential lectin-mediated agglutinabilities of the embryonic and the first extraembryonic cell line in the early chick embryo, Roux's Arch. Dev. Biol., 191:234-240.

Sanders, E. J., Zalik, S. E., Schneider, W. J., Ledsham, I. M., 1990, The endogenous lectins of the chick blastoderm are present in association with an apolipoprotein in distinct organelles and in the extracellular matrix, Roux's Arch. Dev. Biol., 199:295-306.

Thiery, J. P., Duband, J. L., Delouvee, A., 1982, Pathways and mechanisms of avian trunk neural crest cell migration and localization, Dev. Biol., 93:324-343.

Thorpe, S. J., Bellairs, R., Feizi, T., 1988, Developmental patterning of carbohydrate antigens during early embryogenesis of the chick: expression of antigens of the poly-N-acetyllactosamine series, Development, 102:193-210.

Wallace, R . A., 1985, Vitellogenesis and oocyte growth in non-mammalian vertebrates. in. Developmental Biology, A comprehensive synthesis, Vol I, ed. L. W. Browder (Plenum Press, New York) pp 127-178.

Woo, H. J., Shaw, L. M., Messier, J. M., Mercurio, A. M., 1990, The major non-integrin laminin binding protein of macrophages if identical to carbohydrate-binding protein 35 (Mac-2), J. Biol. Chem., 265:7097-7099.

Zalik, S. E., Cook, G. M. W., 1976, Comparison of embryonic and differentiating cell surfaces. Interaction of lectins with plasma membrane components, Biochim. Biophys. Acta., 709:220-226.

Zalik, S. E., Milos, N., and Ledsham, I., 1983, Distribution of two β-galactoside-binding lectins in the gastrulating chick embryo, Cell Differ., 12:121-127.

Zalik, S. E., 1991, On the possible role of endogenous lectins in early animal development, Anat. Embryol., 183:521-536.

Zalik, S. E., Schneider, W. J., Ledsham, I., 1990, The gastrulating chick blastoderm contains 16-kD and 14-kD galactose-binding lectins possibly associated with an apolipoprotein, Cell. Diff. and Devel., 29:217-231.

Zhou, Q., Cummings, R. D., 1990, The S-type lectin from calf heart tissue binds selectively to the carbohydrate chains of laminin, Arch. Biochem. Biophys., 281:27-35.

# INDUCTION AND PATTERN EMERGENCE IN THE MESODERM

Antone G. Jacobson

Department of Zoology
Center for Developmental Biology
The University of Texas
Austin, Texas U.S.A.

## INTRODUCTION

The dorsal mesoderm and the ventral mesoderm are induced prior to gastrulation, and in the frog *Xenopus* the signals responsible for these inductions have influences in the ectoderm beyond the portions that they induce to form mesoderm. These signals bias the prospective ectoderm toward epidermal or neural development and the interaction of the signals tentatively sets the boundary between epidermis and neural plate before gastrulation. Following the early induction of the mesoderm in general, there are additional interactions that specify the various mesodermal organs. The nature and patterning of these inductions, and the early segmentation of the mesoderm are the main topics to be explored here.

## SIGNALS THAT INDUCE THE DORSAL AND THE VENTRAL MESODERM AND THE INFLUENCE OF THESE SIGNALS ON THE ECTODERM

Mesoderm induction and patterning is best understood in amphibian embryos, especially in the frog *Xenopus laevis*. The topic has been reviewed a number of times (Gerhart, *et al.*, 1984; Gimlich and Gerhart, 1984; Dale *et al.*, 1985). Following is a brief summary.

The entrance of the sperm and activation of the egg leads to a rotation of cortical cytoplasm relative to the inner cytoplasm of the zygote. On the side opposite the sperm entry point, the cortical cytoplasm normally rotates upwards, and that side becomes the dorsal side. In the vegetal cytoplasm of the dorsal side of the zygote the ability to later induce dorsal mesoderm is somehow activated by the rotation events. This "vegetal dorsalizing localization" in the zygote cytoplasm is incorporated after a few cleavages into large vegetal cells fated to become endoderm. The embryo by now has formed a blastocoel within. The cells of the walls of the blastocoel are prospective mesoderm and some pharyngeal endoderm. The cells of the roof of the blastocoel are future ectoderm, and the cells of the floor of the blastocoel are future endodermal cells. The dorsal axis is established when the vegetal cells inheriting the vegetal dorsalizing localization send signals that induce contiguous cells toward the animal pole to

*Formation and Differentiation of Early Embryonic Mesoderm*
Edited by R. Bellairs *et al.*, Plenum Press, New York, 1992

become dorsal mesoderm (and pharyngeal endoderm). These contiguous cells are part of a marginal zone around the equator of the egg. The vegetal cells doing dorsal signaling occupy about 60 degrees of the circumference of the egg. Vegetal cells in the remaining 300 degrees of the vegetal region signal contiguous marginal zone cells to form ventral mesoderm. There are thus two signals in the early embryo inducing mesoderm. One signal induces dorsal mesoderm (and pharyngeal endoderm). The dorsal mesodermal products are prechordal plate and notochord along the dorsal midline, and parts of the paraxial somites. The other signal induces ventral mesoderm, and the products of that induction are mainly blood cells, mesothelium and some muscle (Dale, *et al.,* 1985). The numerous other organs that form from mesoderm are not yet determined at these stages. The "three signal model" holds that the dorsal mesoderm then signals the ventral mesoderm to form from some of its cells the intermediate mesodermal structures that would include at least the kidney-gonad complex and the lateral portions of the paraxial mesoderm (Dale and Slack, 1987). I will give evidence below that additional inductions augment this early third signal to specify several of the mesodermal organs.

The induced cells that will form dorsal axial structures occupy the dorsal lip of the blastopore at the beginning of gastrulation. Experiments have shown that at first, dorsal signaling is emitted only by the dorsal vegetal cells. By the 32 cell stage some of the adjacent marginal cells acquire the ability to signal induction of dorsal mesoderm as well (Gimlich, 1986). By stage 9 (when the dorsal lip is just commencing to form), the ability to signal dorsal mesoderm induction has ceased in the vegetal cells and continues in the dorsal marginal cells of the blastopore lip (Gerhart, Doniach and Stewart, 1991).

At the 64-cell stage, about one-fourth of the vegetal cells are signaling to induce dorsal mesoderm and the rest signal to induce ventral mesoderm (Gimlich and Gerhart, 1984). Both dorsal and ventral signaling is through the plane of the contiguous tissue (Fig. 1). An important set of questions is: "How do the boundaries within the induced regions get established?" (Sater and Jacobson, 1988). Boundaries could be set by some threshold level of induction as a signal attenuates with distance, or the dorsal and ventral signals may pervade the entire animal hemisphere, attenuating with distance from the localized sources, and their interactions help set boundaries. This latter view of events in the frog egg would have some similarity to the interactions among localized gene products that specify the anterior-posterior axis or the dorsal-ventral axis of the *Drosophila* egg (reviewed by Nüsslein-Volhard, 1991).

## Evidence that Mesoderm-Inducing Signals Continue through the Ectoderm and Help Define the Epidermis/Neural Plate Boundary

It appears that the dorsal and ventral signals that induce the dorsal and ventral mesoderm may continue through the ectoderm of the animal hemisphere and the interaction of the two signals influences the position of the boundary between epidermis and the neural plate. There is evidence that as early as the 32-cell stage in *Xenopus* the prospective epidermis area is already programmed to later (at early neurula stage 14) produce the epitope *epi-1* (London, *et al.,* 1988), so these regions start to be different very early.

We (Zhang and Jacobson, 1991) have hypothesized that the ventral and dorsal signals from the vegetal cells of the early embryo spread beyond the mesodermal regions that they induce to traverse the entire animal hemisphere, and their interaction biases the ectoderm and positions the epidermal/neural plate boundary. A number of experiments support this hypothesis. As a control, we have removed a plug of cells at the animal pole of a stage 7+ *Xenopus* embryo (a late blastula stage), marked the extracted cells by exposing them to rhodamine (TRITC), then implanted them back into the animal pole. The fluorescent cells are later found in and near the cement gland which occupies a position at the anterior midline of the neural plate, and these embryos develop normally. If our hypothesis is correct then vegetal cells that signal induction of ventral mesoderm should, when transplanted to the animal pole of a stage 7+ embryo, drive the anterior end of the neural plate boundary back toward the blastopore. This is exactly what happens, and the embryos develop with much shortened dorsal axes. The marked implanted vegetal cells become part of the endoderm and are located some distance ventral and posterior to the cement gland. Marking experiments show that a region of ectoderm that should have become located in the anterior neural plate has become epidermis.

We thought that a time should be reached when the boundary would be set and such transplants would have no effect. This proved to be by stage 9, the beginning gastrula stage. Vegetal cells that signal ventral mesoderm induction were marked and implanted into the animal poles of stage 9 embryos. These marked cells were later found in the region of the cement gland and no effect was seen on the length of the dorsal axis. Another control experiment was to implant dorsal lip cells from stage 10+ embryos into the animal pole. These lip cells signal induction of dorsal mesoderm by this stage. Except for the position of the implant, this is the classical Spemann experiment, and the results were in accord with Spemann's. A secondary dorsal axis forms face to face with the normal axis, and the total length of dorsal axis, secondary and primary combined, is longer than normal. Discussion of the anterior-posterior orientation of the dorsal axis is complex and not necessary for the story here.

Another experiment done was to remove at stage 7+ some of the vegetal cells that signal induction of the ventral mesoderm. This should weaken the putative ventral signals traversing the ectoderm and we predicted that the dorsal axis would be longer than normal. That is what happens.

We conclude from these experiments that dorsal and ventral signals from vegetal cells before stage 9 extend into the prospective ectoderm where they interact and tentatively establish the epidermal/neural plate boundary. We have argued in the past that planar signaling

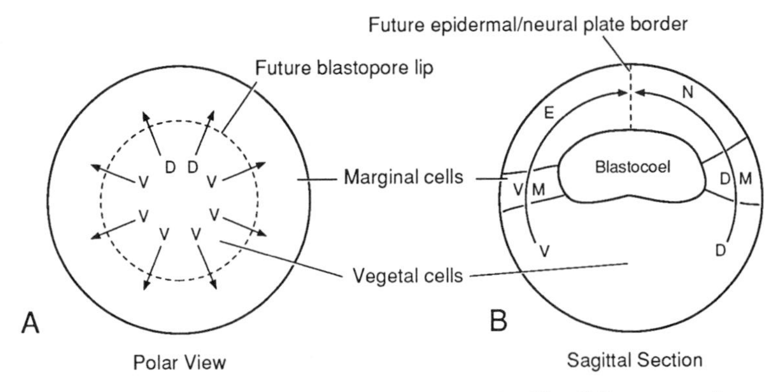


Fig. 1 **A**. This diagram is a vegetal pole view of a 64-cell *Xenopus* embryo. The vegetal cells (future endoderm) lie inside the dotted circle (future blastopore lips). These cells induce dorsal mesoderm (D) and ventral mesoderm (V) from the marginal zone just outside the blastopore line. More cells induce ventral than dorsal. **B**. This drawing of a 64-cell *Xenopus* embryo in sagittal section shows how the vegetal cells signal through the plane of the tissue to induce ventral mesoderm (VM) and dorsal mesoderm (DM), and how these signals may continue through the ectoderm to create a bias toward epidermal (E) or neural (N) development, and how the interaction of these signals may tentatively establish the epidermal/neural plate border.

establishes other boundaries, namely those between chordamesoderm and notoplate, and between notoplate and neural plate, before gastrulation begins (Sater and Jacobson, 1988).

## PATTERNS OF INDUCTION OF LATERAL MESODERMAL ORGANS

The three signal model holds that intermediate mesodermal structures such as the kidneys are induced from ventral mesoderm by a signal from the dorsal mesoderm. I describe here the induction histories of three lateral mesodermal organs, the mesonephros, the heart, and the limb. Each of these organs is indeed induced in part by the dorsal mesoderm, but in each case the underlying endoderm and some other near-by tissues also assist in and may be necessary for the determination of the organ. The role of signals that suppress determination is also noted.

### Induction of the mesonephros

An organ that everyone agrees is formed from the intermediate mesoderm is the mesonephros. The induction history of this organ is one of the most carefully explored of any

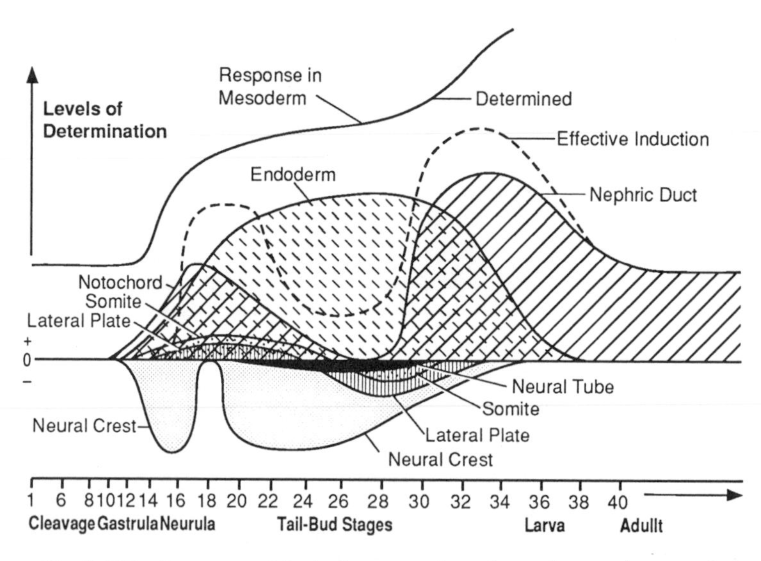


Fig. 2 This diagram models the interpretation of experiments that revealed the ability of various tissues to induce mesonephric tubules (effects above the 0 line), or suppress the determination of mesonephric tubules (effects below the 0 line). The main inductors of the mesonephros are notochord, endoderm, and nephric duct. They have cumulative and additive effects on the responding mesenchyme, while the principal suppressor, the neural crest cells, subtracts from the effects. The neural crest first resides in the neural folds and has effects on the nearby mesonephric mesoderm, but the folds lift away at stages 17-19 and these effects diminish, Then the crest cells migrate into the mesonephrogenic mesenchyme and have lasting suppressive effects until the crest cells differentiate. The "effective induction" (dashed line) is the sum of the inductive effects less the suppressive effects at each time point. The effect of the suppressive interactions is to delay determination of the mesonephros by several stages - from neurula into tail bud stages (from Etheridge, 1968b).

organ (Etheridge,1968a). The mesonephrogenic mesenchyme was explanted either alone in overlying epidermis or in combination with one or another of the surrounding tissues from neurula through tail bud stages (stages 16 to 26) in the newt *Taricha torosa*. The explants were scored for formation of kidney tubules. Several tissues were identified as inductors of mesonephric tubules and a few as suppressors of tubule formation, and the abilities to induce or suppress varied with developmental time. At mid-neurula stage 16, 17% of the cases already are specified for kidney tubule formation. Interactions of the mesonephrogenic mesenchyme with other inductor tissues and suppressor tissues after that time produces a complicated response curve (Fig. 2).

The three signal model suggests that intermediate mesoderm, such as the mesonephros, is induced from ventral mesoderm by a signal from the the dorsal mesoderm. Indeed, the notochord is a strong inductor of the mesonephros throughout neurulation, but this inductive ability is lost by tail bud stages. Somites are weak inductors during neurulation, but become suppressors during tail bud stages (and the same is true of lateral plate mesoderm). The endoderm that underlies the mesonephrogenic mesoderm is a very strong inductor of kidney tubules, and this inductive activity is high throughout neurulation and tail bud stages. The pronephric duct extends itself through the mesonephrogenic mesenchyme during tail bud stages, and it is a powerful inductor of kidney tubules, acting on a mostly determined mesonephrogenic mesenchyme shortly before kidney tubules are formed.

There is a very clear activity by the neural crest cells that suppresses mesonephros determination. This activity is effective both while the crest cells are in the neural folds and especially later after some of the crest cells migrate into and take up residence in the mesonephrogenic mesenchyme at early tail bud stages. Results of experiments that test this suppressive activity of crest cells are illustrated in Figure 3. Suppression of kidney determination by neural crest has been shown in both *Taricha torosa* and *Xenopus laevis* (Etheridge, 1972).

In summary, the amphibian mesonephros is induced in early stages by notochord and endoderm. The endoderm is the most persistent and powerful inductor of kidney tubules. Somites and lateral plate mesoderm initially have a weak inductive effect, but these tissues become weakly suppressive in tail bud stages. The nephric duct encounters the mesonephrogenic mesoderm just before the latter is completely specified, and this duct is a strong inductor of kidney tubules, but the responding tissue is already largely determined. The neural crest, both in the fold and later when crest cells enter the kidney mesoderm, is a powerful suppressor of tubule determination. The crest loses its ability to suppress kidney as the crest cells differentiate. The effect of suppression by the neural crest is to delay considerably the time of appearance of kidney tubules. The induction history of the mesonephros revealed by these experiments does fit in with the three signal model in that the notochord is an early inductor of this portion of the intermediate mesoderm, but there are many additional inductive and suppressive interactions.

**The induction of the heart**

We concluded from our older studies on newt embryos (Jacobson, 1960, 1961, 1966 (review); Fullilove, 1970; Jacobson and Duncan, 1968; Jacobson and Sater, 1988 (review)) that the heart of the California newt *Taricha torosa* is induced during neurulation by the dorsolateral pharyngeal endoderm that underlies the two heart rudiments when they are the anteriormost mesoderm lying just lateral to the neural plate and beneath the prospective ear epidermis. Heart rudiments isolated at the early neurula stage formed beating heart tissue in 10 to 14% of the cases. By removing the entire endoderm at various stages of neurulation, it was found that the endoderm is essential for heart formation during neurulation and continues to have effects later. Recombination experiments (heart rudiments explanted from early neurula embryos with other tissues) identified the anterior dorso-lateral pharyngeal endoderm as the most powerful heart inductor. Explants of heart rudiments in epidermal vesicles together with anterior dorso-lateral endoderm produced beating heart tissue in 100% of the cases. Similar experiments showed anterior ventral endoderm (over which the heart finally resides) to be a strong inductor. The edge of the anterior neural plate and the neural fold (with crest) were suppressors of heart determination.

Four things should be noted about these studies on newt embryos. 1) The position of the two prospective heart rudiments from gastrula to early neurula stages is at the extreme anterior end of the prospective intermediate mesoderm. These mesenchymal rudiments migrate to a midventral position during neurula stages in the salamander embryo. 2) A few heart rudiments at midneurula stages (10 to 14%) are aleady sufficiently determined to form beating heart tissue when isolated alone. 3) The underlying endoderm during neurula stages appears to be the essential inductor tissue to complete determination of all the heart rudiments during neurulation. (Similar results have been obtained for another urodele, the axolotl embryo (Smith and Armstrong, 1990), but these authors found the anterior midventral endoderm to be the strongest inductor.) 4) The anterior neural plate and fold suppresses heart determination.

Muslin and Williams (1991) have used the axolotl neurula to test various growth factors for their heart-inducing ability. They find that TGF-$\beta_1$(transforming growth factor beta) and

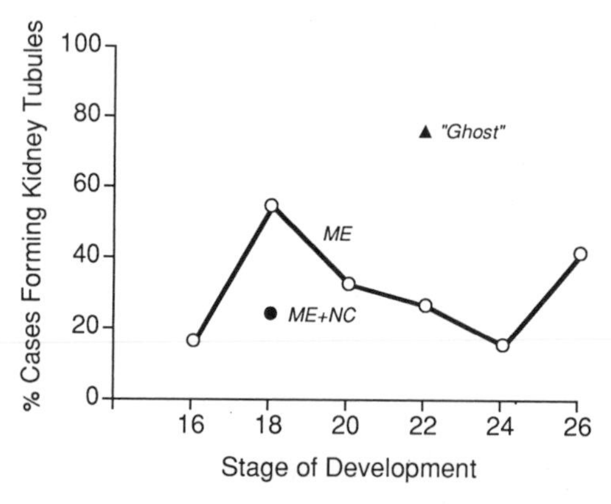


Fig. 3. These data from Etheridge (1968a) show the percentage of cases that form kidney tubules when mesonephrogenic mesenchyme is isolated in vesicles of overlying epidermis (ME) at the stages indicated. The strange decline in the state of determination between stages 18 to 24 was postulated to be due to suppression by the neural crest cells. That this is so was demonstrated by combining stage 18 mesonephrogenic mesenchyme with neural crest in epidermal vesicles (ME + NC) and getting less than half the response seen with ME without neural crest. Also, neural crest was removed from stage 16 embryos. Such deprived embryos fail to form pigment cells and are termed "ghosts". These embryos were reared to stage 22 when the mesonephrogenic mesenchyme was isolated in epidermal vesicles. Kidney tubules formed in 78% of the "ghost" cases rather than in 16% of the cases from embryos that had neural crest.

PDGF (platelet-derived growth factor) can mimic the tissue inductors of heart and that bFGF (fibroblast growth factor) can inhibit heart differentiation. These growth factors are present at appropriate times and places to be heart inductors and suppressors in the amphibian embryo, but whether they actually are remains to be proven.

Amy Sater investigated heart induction in the frog *Xenopus laevis* in my laboratory. We were somewhat surprised to find that heart rudiments from early neurula stage 14 explanted in epidermal vesicles formed beating hearts in 100% of the cases. Similar explants at stage 12.5 (late gastrula) formed hearts in 50% of the cases, and in 70% of the cases if pharyngeal endoderm were also included in the explant (Sater and Jacobson, 1989). Clearly, heart induction occurs much earlier in this frog than in salamanders, perhaps related to its fast development. Further experiments have defined some of the events of this early induction in *Xenopus* and, we suspect, account for the early determination events in the salamanders that have already specified heart formation in 10 to 14% of the cases at early to mid-neurula stages.

Since the late gastrula stage 12.5 is the earliest time that the prospective heart mesoderm can be separated from the other germ layers, somewhat indirect approaches are needed to examine induction before this time. The frogs have organized their pharyngeal endoderm into a superficial layer that is not present in the salamanders. This layer of prospective bottle cells that later becomes the pharyngeal endoderm can be conveniently removed as gastrulation is beginning (stage 10), depriving the embryo of any pharyngeal endoderm, but otherwise not affecting development. This operation did not affect the production of hearts, so pharyngeal endoderm is not essential for heart induction in *Xenopus*. These experiments did not eliminate the possibility of induction by the deep dorsal endoderm (Sater and Jacobson, 1989).

The heart rudiments during gastrula stages lie anterior to and in line with the future intermediate mesoderm which, according to the three signal model (*eg*. Dale and Slack, 1987), is induced from the ventral mesoderm by a signal from the dorsal mesoderm. The heart rudiments arise from deep dorsolateral mesoderm of the early gastrula and the dorsal mesoderm from the dorsal lip. Explant experiments of dorsal, dorso-lateral, and ventral regions of the marginal zone of early gastrulae located the prospective heart mesoderm as between 30° and 45° from the dorsal midline. The dorsal 60° of the early gastrula does not express heart-forming potency. Explant experiments of dorso-lateral mesoderm (including prospective heart mesoderm) with and without future dorsal mesoderm of the dorsal lip revealed that heart formation requires presence of the dorsal lip early in gastrulation since no explants from stage 10 that lacked dorsal lip formed hearts while 60% formed hearts by stage 10.25 and 100% by stage 11. The heart inductive action of the dorsal lip is thus essentially complete by stage 11. When FDA (fluoresceinated dextran amine) labelled dorsal lip of the early blastopore from within this 60° region is excised and implanted into the ventral marginal zone of unlabelled host embryos (the Spemann experiment), hearts are induced from local ventral mesoderm in most of the cases that form secondary axes (Sater and Jacobson, 1990a). The dorsal mesoderm thus initiates specification of the heart in *Xenopus*, and the heart might be considered from its induction pattern to be the most anterior portion of the intermediate mesoderm.

A continued induction of heart by portions of the dorsal endoderm after dorsal lip action is not excluded and is probable. Certainly one aspect of early heart formation in *Xenopus* includes interaction of the mesoderm with underlying endoderm. When the heart is completely specified by the end of neurulation, there is an extended region of mesoderm beyond that which actually later forms the heart that is capable of heart formation . The future heart rudiment of late neurula stage *Xenopus* (stage 20) is located in the anterior ventral region, but the anterior ventro-lateral mesoderm retains the ability to form heart, and gradually loses it by stage 28 when heart morphogenesis has begun. This field characteristic is sustained through this time by interactions with the underlying endoderm, and is lost sooner if the endoderm is lacking (Sater and Jacobson, 1990b).

In summary, the heart of *Xenopus* is induced from the ventral mesoderm largely by the dorsal mesoderm and the role of underlying endoderm as a heart inductor is much less in this form than in the salamanders. Endoderm does have a role in sustaining the heart field. The induction history and position of the heart rudiments suggest that heart mesoderm is the anteriormost portion of the intermediate mesoderm.

### Induction of the Forelimb in Salamanders

The forelimb is a mesodermal structure that arises in the somatic lateral plate mesoderm, but only if additional inductive events act on this ventral mesoderm. My Ph.D. student, Nancy Parker, analysed limb induction in the salamander *Ambystoma tigrinum* for her dissertation research (Parker, 1965), but unfortunately never published the results. I summarize here the main findings of her dissertation.

She explanted the lateral plate and intermediate mesoderm, which contains the prospective limb field, in vesicles of overlying epidermis at neurula stages 15 through 19. She cultured the explants for a month then sectioned them to assay for limb development. Limb buds are easily recognized by the typical bud protrusion and the cartilage nodules within. Twenty percent of the cases formed limbs in explants from stage 15 embryos, and 86% formed limbs in explants from stage 19 embryos. Obviously limb induction is well underway and proceeding rapidly through these neurula stages. Stage 15 embryos were chosen to compare results when other tissues were combined with the lateral and intermediate mesoderm (Fig. 4).

Results indicate that both dorsal mesoderm and underlying endoderm are effective inductors of the limb. The paraxial mesoderm (future somites) and the notochord both increase limb response greatly, but the paraxial mesoderm is the more effective. (Since this work was done, it has become known that the somites contribute myoblasts to the forming limb.) The underlying endoderm is also a very powerful inductor of the limb, and is likely to have been inducing long before the experiments were done. It appears that the three signal model may hold for the limb since the dorsal mesoderm is a strong inductor, but as with the other mesodermal organs we are considering, the endoderm also has a large role in the induction of the organ.

In summary, each of the lateral mesodermal organs considered above is induced from ventral mesoderm in part by signals from dorsal mesoderm, and in part by underlying endoderm and other nearby tissues. The neural crest is a powerful suppressor of determination of the mesonephros and the heart (proper experiments to determine this have not been done for the limb). One must conclude that the three signal model is a partial explanation of the induction of lateral mesodermal organs.

## THE SEGMENTATION OF THE PARAXIAL MESODERM

The usual view has been that the paraxial mesoderm posterior to the ear in the head and that in the trunk is segmented when somites appear in these regions. The paraxial mesoderm of the more rostral parts of the head was considered not to be segmented in amniotes, but much older literature described the segmentation of the mesoderm in the heads of sharks (Balfour, 1878, 1881; de Beer, 1922; Goodrich, 1930).

Meier (1979) discovered that the entire paraxial mesoderm of the chick embryo from the tip of the head to the tail bud is composed of segmentally organized mesenchymal units that he named "somitomeres". These units are squat cylinders consisting of two layers of mesenchymal cells whose intertwined processes make a bulls-eye pattern. The whole unit is either convex or concave, and they are best visualized with stereo scanning electron microscopy after removing the ectoderm and washing away the extracellular matrix. The units have the appearance of being expansion figures when initially formed, and in the chick embryo they form lateral and slightly posterior to Hensen's node. The somitomeres of the trunk progressively condense to become somites, but most of the head somitomeres, the more rostral ones, remain dispersed. Representative embryos of six classes of vertebrates have been examined and all segment their entire paraxial mesoderm into somitomeres during gastrulation. More than thirty papers document the existence and fates of somitomeres (Jacobson and Meier, 1986; Jacobson, 1988 for reviews).

Possibly because it is difficult to visualize somitomeres, or because the mesenchymal cells of somitomeres retain their patterning while they are in motion as mesenchymal cells usually are, some recent authors ignore them. For example, Lumsden and Keynes (1989) dismiss somitomeres as "inconsequential". This is unfortunate because the segmentation pattern represented by somitomeres appears to offer additional insights into the arrangements of cranial nerves and the expression of HOX and Krox genes in the hindbrains of advanced chick and mouse embryos.

Detwiler (1934) showed experimentally that the segmentation of the spinal nerves is imposed by the positions of the somites, and more recently this has become better explained.

Ventral roots from the neural tube will enter only the anterior portion of the sclerotome of each somite (Keynes and Stern, 1984), and migration of neural crest cells that will form the dorsal root ganglion is restricted to the anterior portion of the sclerotome of each somite (Rickman, *et al.*, 1985; Teillet *et al.*, 1987). All somites condense from somitomeres, and somitomeres have

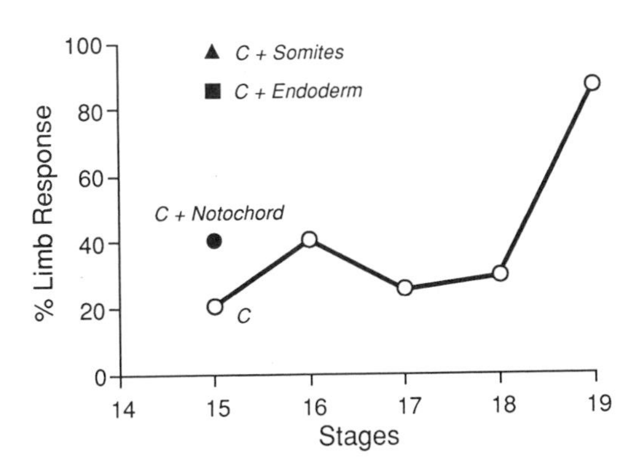


Fig. 4. These data from Parker (1965) show the percentage of cases that form limbs when lateral plate and intermediate mesoderm containing the fore-limb field was explanted in overlying epidermis alone (C) or together with other tissues. Somites and endoderm are powerful limb inductors, and notochord also is a moderately strong inductor.

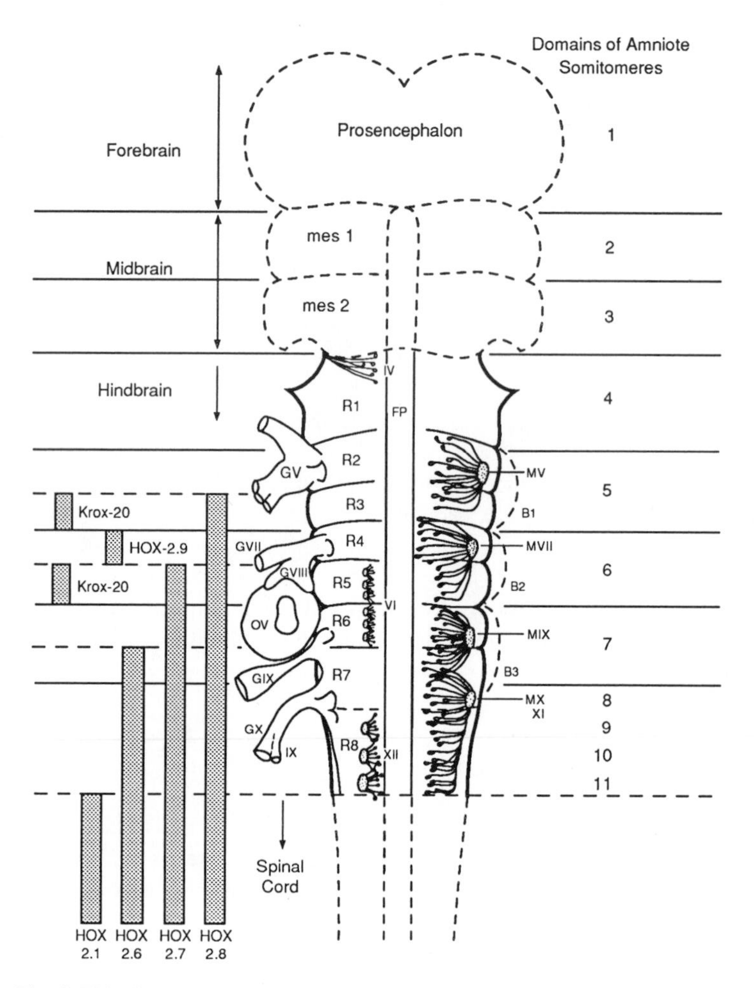


Fig. 5. This diagramatic representation of the amniote brain combines cytological data relating cranial nerve roots and ganglia to brain neuromeres in the stage 21 chick embryo (central part of figure from Lumsden and Keynes, 1989), the positional levels of expression of homeobox-containing genes (HOX-2 series) and a zinc-finger gene (Krox-20) in the 9.5 day mouse embryo (from Wilkinson *et al.*, 1989a,b), and the positions of the head somitomeres of the amniote embryo (1-11, right side)(from Jacobson, 1988). Somitomeres eight to eleven condense to become the four occipital somites. The mesencephalic neuromeres (mes 1.2) and the rhombomeres are numbered. The cranial sensory ganglia (GV-X), branchial and somatic motor nuclei (IV-XII), and the combined roots of the sensory and branchial motor nerves (MV-MXI) are shown. The prosencephalon and mesencephalon are symbolized rather than accurately drawn to show the relationships of these parts to the first three somitomeres. The branchial arches (B1-3) coincide with the positions of somitomeres five, six, and seven. Note that the roots and ganglia of the cranial nerves appear to be positioned in the rostral halves of the somitomeres. FP, floor plate.

anterior and posterior portions defined by lines of cell processes that bisect each somitomere perpendicular to the axis. The somitomeres of the head that do not condense have these divisions as well. It should be possible that the cranial nerves are restricted to form only in anterior portions of the somitomeres.

In all species the neuromeres of the brain arise after the somitomeres have underlain the neural plate. Initially there is a correspondence between the positions of somitomeres and the positions of the first neuromeres causing one to suspect that the initial segmentation of the somitomere is impressed upon the nervous system to position the early neuromeres. After a time, neuromeres subdivide so that the hindbrains of advanced embryos have two neuromeres next to each somitomere. When they do subdivide, the line that subdivides them lies adjacent to the line that subdivides the corresponding somitomere.

In Figure 5, I have redrawn the data from Lumsden and Keynes (1989) that shows the relationships between cranial nerve roots and ganglia to neuromeres of the hindbrain, and have indicated the positions of expression of homeobox containing HOX genes and a zinc-finger gene (Krox-20) after Wilkenson, *et al.*, 1989a,b. To their data I have added, on the right, the positions of the somitomeres. It does appear that cranial nerve motor roots and cranial ganglia emerge and form in the anterior portions of the somitomeres.

## REFERENCES


Balfour, F.M. (1878) *A Monograph on the Development of Elasmobranch Fishes.* Theil 1, Königsberg.

Balfour, F.M. (1881) *A Treatise on Comparative Embryology.* Vol. 2., Macmillan, London.

de Beer, G.R. (1922) The segmentation of the head in *Squalus acanthias. Quar. J. Micr. Science* 66:457-474.

Dale, L. & Slack, J.M.W. (1987) Regional specification within the mesoderm of early embryos of *Xenopus laevis. Development* 100:279-295.

Dale, L., Smith, J.C. & Slack, J.M.W. (1985) Mesoderm induction in *Xenopus laevis:* a quantitative study using a cell lineage label and tissue-specific antibodies. *J. Embryol. Exp. Morph.* 89:289-312.

Detwiler, S.R. (1934) An experimental study of spinal nerve segmentation in *Amblystoma* with reference to the plurisegmental contribution to the brachial plexus. *J. Exp. Zool.* 67:395-441.

Etheridge, A.L. (1968a) Determination of the mesonephric kidney. *J. Exp. Zool.* 169:357-370.

Etheridge, A.L. (1968b) Determination of the mesonephric kidney in the newt, *Taricha torosa.* Ph.D. Dissertation, The University of Texas at Austin.

Etheridge, A.L. (1972) Suppression of kidney formation by neural crest cells. *Roux' Arch. EntwMech. Org.* 169:268-270.

Fullilove, S.L. (1970) Heart induction: Distribution of active factors in newt endoderm. *J. Exp. Zool.* 175:323-326.

Gerhart, J.C., Doniach, T. & Stewart, R. (1991) Organizing the *Xenopus* Organizer. In: *Bodega Marine Laboratory Marine Sciences Series* Vol. 3, Keller, R., Clark, W., Jr. & Griffin, F. (eds), Plenum Press, New York (in press).

Gerhart, J.C., Vincent, J-P., Scharf, S.R., Black, S.D., Gimlich, R.L., & Danilchek, M. (1984) Localization and induction in early development of *Xenopus. Phil. Trans. R. Soc. Lond. B* 307:319-330.

Gimlich, R.L. (1986) Acquisition of developmental autonomy in the equitorial region of the *Xenopus* embryo. *Dev. Biol.* 115:340-352.

Gimlich, R.L. & Gerhart, J.C. (1984) Early cellular interactions promote embryonic axis formation in *Xenopus laevis*. *Dev. Biol.* 104:117-130.

Goodrich, E.S. (1930) *Studies on the Structure and Development of Vertebrates*. Macmillan, London. (Republished by Dover, New York, 1958.)

Jacobson, A.G. (1960) The influences of ectoderm and endoderm on heart differentiation in the newt. *Dev. Biol.* 2:138-154.

Jacobson, A.G. (1961) Heart determination in the newt. *J. Exp. Zool.* 146:139-152.

Jacobson, A.G. (1966) Inductive processes in embryonic development. *Science* 152:25-34.

Jacobson, A.G. (1988) Somitomeres: mesodermal segments of vertebrate embryos. *Development* 104 (Supplement):209-220.

Jacobson, A.G. & Duncan, J.T. (1968) Heart induction in salamanders. *J. Exp. Zool.* 167:79-103.

Jacobson, A.G. & Meier, S. (1986) Somitomeres: The primordial body segments. In: *Somites in Developing Embryos,* R. Bellairs, D.A. Ede, and J. Lash (Eds.), Plenum Publ. Corp. New York, NATO ASI Series. Series A, Life Sciences, v. 118. pp. 1-16.

Jacobson, A.G.&Sater, A.K. (1988) Features of embryonic induction. *Development* 104:341-359.

Keynes, R.J & Stern, C.D. (1984) Segmentation in the vertebrate nervous system. *Nature* 310:786-789.

London, C., Akers, R. & Phillips, C.R. (1988) Expression of Epi 1, an epidermal specific marker, in *Xenopus laevis* embryos is specified prior to gastrulation. *Dev. Biol.* 129:380-389.

Lumsden, A. & Keynes, R. (1989) Segmental patterns of neuronal development in the chick hindbrain. *Nature* 337:424-428.

Meier, S. (1979) Development of the chick mesoblast. Formation of the embryonic axis and establishment of the metameric pattern. *Dev. Biol.* 73:25-45.

Muslin, A.J. & Williams, L.T. (1991) Well-defined growth factors promote cardiac development in axolotl mesodermal explants. *Development* 112:1095-1101.

Nüsslein-Volhard, C. (1991) Determination of the embryonic axes of *Drosophila. Development Supplement* 1:1-10.

Parker, N.J.R. (1965) *Determination of the Urodele Forelimb*, Ph.D. dissertation, University of Texas at Austin.

Rickman, M., Fawcett, J.W., & Keynes, R.J. (1985) The migration of neural crest cells and the growth of motor axons through the rostral half of the chick somite. *J. Embryol. Expt. Morph.* 90:437-455.

Sater, A.K. & Jacobson, A.G. (1988) Features of embryonic induction. *Development* 104:341-359.

Sater, A.K. & Jacobson, A.G. (1989) The specification of heart mesoderm occurs during gastrulation in *Xenopus laevis*. *Development* 105:821-830.

Sater, A.K. & Jacobson, A.G. (1990a) The role of the dorsal lip in the induction of heart mesoderm in *Xenopus laevis*. *Development* 108:461-470.

Sater, A.K. & Jacobson, A.G. (1990b) The restriction of the heart morphogenetic field in *Xenopus laevis*. *Dev. Biol.* 140:328-336.

Smith, S.C. & Armstrong, J.B. (1990) Heart induction in wild-type and cardiac mutant axolotls (*Ambystoma mexicanum*). *J. Exp. Zool.* 254:48-54.

Teillet, M-A., Kalcheim, C., & Le Douarin, N. (1987) Formation of the dorsal root ganglia in the avian embryo: Segmental origin and migratory behavior of neural crest progenitor cells. *Dev. Biol.* 120:329-347.

Wilkenson, D.G., Bhatt, S., Chavrier, P., Bravo, R., & Charney, P. (1989a) Segment-specific expression of a zinc-finger gene in the developing nervous system of the mouse. *Nature* 337:461-464.

Wilkenson, D.G., Bhatt, S., Cool, M., Bonicelli, E., & Krumlauf, R. (1989b) Segmental expression of HOX-2 homeobox-containing genes in the developing mouse hindbrain. *Nature* 341:405-409.

Zhang, J. & Jacobson, A.G. (1991) Evidence that the border of the neural plate is positioned by the interaction between signals that induce ventral and dorsal mesoderm. (submitted).

# SEGMENTATION AND COMPARTMENTS IN THE VERTEBRATE EMBRYO

Keith M. Bagnall

Department of Anatomy and Cell Biology
University of Alberta
Edmonton
Alberta
Canada
T6G 2H7

## INTRODUCTION

Knowledge of segmentation and compartments in the vertebrate embryo is important because it is fundamental to understanding the basic vertebrate body plan (Lawrence, 1990). Understanding the development of the rostrocaudal (vertebral) axis is also of primary importance as it is the structure around which the rest of the embryo is built (de Robertis et al., 1990). Consequently, over the past few years, we have concerned ourselves with studying the development of segmentation and compartments along the rostrocaudal axis using the chick embryo as our animal model.

## THE THEORY OF RESEGMENTATION IN VERTEBRAL FORMATION

### Supporting Evidence

The most obvious segmentation pattern in the early vertebrate embryo centres on the sequential development of pairs of somites along the rostrocaudal axis in the mesoderm layer adjacent to the neural tube and notochord. These somitic cells soon disperse with some cells (sclerotome) migrating medially to surround the neural tube and notochord where they eventually form the vertebral column. The obvious segmentation pattern of the somites is replaced at this stage by the equally obvious pattern of the individual vertebrae. However, the transition from somite pairs to vertebrae does not appear to be straightforward and Remak (1866) proposed a theory of 'resegmentation' of vertebral development in which it was said that a single vertebra was formed by a combination of cells from the caudal half of one pair of somites with cells from the rostral half of the next pair of somites. A single vertebra was NOT thought to form from a single pair of somites as might have been expected. This conclusion was drawn from the study of serial sections of staged embryos and similar conclusions from a variety of vertebrates, including humans, has been reached by other workers using similar methodology (Williams, 1910; Piiper, 1928; Dawes, 1930; Sensenig, 1943, 1949; Shaner, 1985). The work is particularly convincing in studies of autotomous vertebrae (Winchester and Bellairs, 1977) where the potential site for division of the vertebrae can be identified at the time of somitic cell dispersal and subsequently followed during development. However, all of these studies are subjective in nature as the cells from the original somites cannot be truly identified in the later stage embryos. This has led to individual interpretation of the results and although most workers have concurred with the original conclusions of Remak, there have been some significant alternatives which have proposed that the vertebral column develops from an unsegmented, perichordal tube formed from the migrating sclerotome (Baur, 1969; Verbout, 1976; 1985; Dalgleish, 1985). Moreover, even within the work that supports the theory of resegmentation, there are differences in descriptions of the extent of contribution of the two pairs of somites to each vertebra, particularly to the neural arches and the intervertebral disc (for review see O'Rahilly and Meyer, 1979). Such claims are not easily refuted because the subjective method of studying serial sections of staged embryos allows for individual interpretation of the results.

Beresford (1983), in a study of the origin of brachial muscles, transplanted equivalent quail somites into host chick embryos, and made a brief observation that quail cells were found in consecutive vertebral bodies as well as in the intervening disc. In 1987, Stern and Keynes also made a similar brief observation while studying the interactions between the cells of the rostral and caudal halves of the chick somite. In 1988, we studied the question in more detail and effectively provided the first experimental evidence in support of the theory of resegmentation (Bagnall *et al.*, 1988). We replaced individual chick somites with equivalent quail somites and observed the subsequent distribution of the quail cells in older chick embryos. Quail cells were found on only one side of the vertebral column and were found in the caudal half of one vertebral body, the intervertebral disc, the rostral half of the adjacent vertebral body, and in the neural arches of both affected vertebrae. Lance-Jones (1988) later confirmed our results using similar methodology. However, the method of somite transplantation has been criticised. For example, the surgery involved might disrupt the microenvironment and alter cell migration patterns similar to that found for neural crest cells (Newgreen and Erickson, 1986). Similarly, the difference in cell adhesiveness found between chick and quail cells (Sanders, 1986) might promote abnormal cell migration. Furthermore, it is also possible that correct orientation of the transplanted somite is critical to normal development and this is difficult to achieve when only one somite is transplanted. Therefore, it has been necessary to continue collecting experimental evidence related to the formation of vertebrae using different methodologies.

Cells in the caudal half of the sclerotome have been found

to stain positively with peanut agglutinin (PNA) whereas cells in the rostral sclerotome half remain unstained (Stern *et al.*, 1986). We showed that this differentiation between the two sclerotome halves remained at least until the stage at which vertebrae could be clearly identified and, by staining serial sections of staged embryos with PNA, we provided further evidence in support of the theory of resegmentation (Bagnall and Sanders, 1989). The unstained cells of the rostral sclerotome half appeared to form the caudal half of one vertebra while the stained cells of the caudal sclerotome half appeared to form the intervertebral disc and the rostral half of the next consecutive vertebra. The neural arches were a mixture of stained and unstained areas (Bagnall and Sanders, 1989). This work can be criticised, however, as the original somitic cells were not labelled in a permanent manner and could only be assumed to be the same cells in the later stages by the similarity of staining reaction. The consistency of the banded pattern throughout the later stages of development, first seen at the sclerotome stage, was very convincing although it was assumed that some cells did lose their ability to react with PNA in the later stages (Bagnall and Sanders, 1989).

Similar results in support of the theory of resegmentation in vertebral formation were obtained from studies using the carbocyanine dye, DiI (Bagnall, 1991 - In press). Following injection, this fluorescent dye is readily absorbed into the plasma membrane of cells and can be seen as a bright fluorescence. With development, the dye remains in the cell membrane and is only transferred to daughter cells. This allows the development of specific groups of cells to be analysed (Honig and Hume, 1989). However, with proliferation, the intensity of the dye becomes less as it is diluted but can still be seen clearly after several days of development. After injection of the dye into the somitocoele (enclosed cavity at the centre of the epithelial somite) the surrounding somite cells became fluorescent. Examination of a series of staged embryos (up to 10 days old) that had all been injected with DiI at 2 days of incubation, showed that fluorescent cells could be followed throughout development of the rostrocaudal axis. Fluorescent cells were eventually found in adjacent neural arches, the intervertebral disc, the connective tissue between the neural arches, and the connective tissue surrounding the developing vertebrae. Surprisingly, there was no fluorescence in any precartilaginous structures suggesting that the fluorescence was being masked in some way or that the cells forming the precartilaginous vertebral elements were proliferating at such a high rate that the fluorescence was being diluted beyond the ability to be seen.

The experiments outlined above, all provide experimental evidence in support of the theory of resegmentation although none of them are entirely conclusive as the methodologies used can all be criticised. Before any conclusion can be reached, however, the contradictory evidence must also be considered.

Opposing Evidence

Baur (1969), Verbout (1976, 1985), and Dalgleish (1985) all studied serial sections of staged, vertebrate embryos and basically described the vertebral column as developing from an unsegmented perichordal tube formed from the migrating sclerotome. The methodology can be criticised, however, in a manner similar to that of other workers whose results have supported the theory of resegmentation in vertebral formation (Williams, 1910; Sensenig, 1949; Shaner, 1985). The methodology requires subjective interpretation of the results as the cells in the original somite were not labelled and could not be reliably identified in the later stages of development.

Congenital anomalies can often be explained by the failure of one piece of tissue to fuse appropriately with another during development (Sadler, 1990). If the theory of resegmentation in vertebral formation is correct, then the caudal half of one sclerotome must fuse with the rostral half of the next sclerotome to form the vertebral body. Consequently, vertebral anomalies might be expected where this fusion has not occurred and several possibilities exist for the development and arrangment of such an anomaly. For example, the neural arch might be attached to a thin section of vertebral body with another thin section of vertebral body, devoid of neural arch, lying directly caudal. No such anomaly appears to exist in the literature although it is entirely possible that such an anomaly is not conducive to life. The vertebral anomalies discussed in the literature (Tsou *et al.*, 1980) describe the absence of various whole quadrants of the vertebral body which appear to reflect the evolutionary development of vertebrae and the complete lack of development of specific contributing elements (Kent, 1978). The fact that these anomalies affect the full thickness of the vertebral body and do not appear to be restricted to either the rostral or caudal half might be viewed as providing evidence refuting the theory of resegmentation. This has not been studied in detail and warrants further consideration.

One anomaly that is clearly described in the orthopaedic literature is that of 'hemivertebrae'. In these cases, there is often a complete absence of the lateral half of one vertebra at a specific level. Moe *et al.* (1978), in their classification of hemivertebrae, considered the hemivertebra to be the most common of all congenital spinal anomalies, either singly or in association with other hemivertebrae or other anomalies. With the malleability of bone, the hemivertebra tends to become wedge-shaped during development but clearly it appears to have received cellular contribution from only one side of the body. Some orthopaedic literature (Tsou *et al.*, 1980) ascribes the formation of this anomaly to a 'misalignment' of the pairs of somites such that at some point the pairing goes out of alignment and one somite remains isolated. The subsequent development of this single somite, in the absence of a complementary somite on the opposite side, is thought to produce the hemivertebra (Tsou *et al.*, 1980). If the theory of resegmentation is correct and normal development of the single somite in the absence of its complement on the other side is possible, then the anomaly produced by the single somite should affect two consecutive vertebrae. In a series of experiments in which we removed single somites from chick embryos and allowed them to develop further (Bagnall *et al.*, 1986), the vertebral anomaly that developed was a hemivertebra bracketed both rostrally and caudally by apparently normal, whole vertebrae. This strongly suggested that a hemivertebra developed from the absence of a single somite and was evidence against the theory of resegmentation. Furthermore, if several, consecutive somites were removed on one side then an equivalent number of hemivertebrae developed on the opposite side presumably from the unmatched somites (Bagnall *et al.*, 1986). However, the length of time between somite removal and vertebral assessment, coupled with the plasticity and malleability of the developing vertebrae, cast doubt on some of the vertebral assessments, and small traces of additional vertebral elements could be found on closer examination in some cases. The identity of these additional elements could not be made with any degree of certainty. Nevertheless, the hemivertebra anomaly is difficult to explain if the theory of resegmentation is correct as two pairs of somites would be involved in producing the anomaly. This is an area that warrants further investigation as it is perhaps the most substantial experimental evidence to date against the theory of resegmentation in vertebral formation.

The question of the validity of the theory of resegmentation in vertebral formation has not been answered with any degree of certainty although there appears to be more experimental evidence to support the theory than to refute it. Certainly, the question can withstand being examined further particularly if methods of cell labelling are used which can be applied to the cells in the original somites and their subsequent development followed. In this regard, the use of retroviral mediated gene transfer techniques (Sanes, 1989) would appear to be a method of promise.

## COMPARTMENTS ALONG THE ROSTROCAUDAL AXIS

During development, an organism becomes divided into groups of cells called compartments (Crick and Lawrence, 1975) that function semiautonomously and give rise to specific structures (Caveney, 1985). The concept of developmental compartments being the basic building blocks during development is well accepted in insects where compartments have been identified on the basis of cell lineage (Lawrence, 1973; Garcia-Bellido *et al.*, 1973, 1976), gene expression criteria (Garcia-Bellido *et al.*, 1973, 1976), and communication between cells shown by the passage of fluorescent dyes through gap junctions (Warner and Lawrence, 1973, 1982: Blennerhassett and Caveney, 1984). These three different methodologies may identify the same cells as belonging to the same developmental compartment (Caveney, 1985). The presence of compartments along the segmented rostrocaudal axis in vertebrates is thought to be fundamental to the basic vertebrate body plan (Lawrence, 1990) but, as yet, there is little experimental evidence supporting their presence. However, there is reason to believe that they do exist in vertebrates (Lo and Gilula, 1979a, 1979b; Guthrie, 1984; Kalimi and Lo, 1988) and evidence for compartments in the chick embryo in particular is accumulating, especially in relation to the development of rhombomeres in the neuroectoderm based on studies of cell lineage (Fraser *et al.*, 1990) and gene expression (Lumsden and Keynes, 1989; Patel *et al.*, 1989).

## Cell Lineage Studies

Further studies involving the transfer of single, donor quail somites into host chick embryos (Bagnall *et al.*, 1989) have shed light on the question of cell lineage of the somite cells and formation of the vertebral column. It was found (Bagnall *et al.*, 1989) that the sclerotome cells from a single somite remained as a clearly delimited group during their migration to surround the notochord and neural tube and that they did not mix to any great extent with cells from adjacent somites. Furthermore, the sclerotome cells from a single somite appeared to form all the tissues within a segment of the vertebral column except for the neural tissue (neurites and ganglia) and notochord. These results were confirmed by our experiments involving the labelling of complete somites by the fluorescent, carbocyanine dye DiI (Bagnall, 1991 - In press) where the fluorescent cells were found to remain as a clearly defined group during migration and further development and contributed to adjacent neural arches, the intervening disc, the connective tissue around the vertebrae, and all the connective tissue surrounding the vertebral column including blood vessels. The PNA studies described previously (Bagnall and Sanders, 1989) were able to refine the descriptions of somite cell contributions to the actual vertebrae as the rostral and caudal sclerotome halves were able to be identified throughout the period of development. This enabled the specific contributions of the rostral and caudal sclerotome halves to be identified to some extent. The unstained cells in the rostral sclerotome half appeared to form most of the vertebral body and part of the neural arch, while the stained cells from the caudal sclerotome half formed part of the vertebral body, part of the neural arch and the intervertebral disc. Precise determination of individual somitic cell lineage awaits the development of adequate labelling techniques.

The data concerning cell lineage suggest that the sclerotome cells might form a developmental compartment based on a definition by Lawrence (1990) which stated that the descendents of a group of founder cells form the tissues within a compartment. The sclerotome cells and their descendants remain within clearly defined borders and do not appear to mix with cells from adjacent somites. The sclerotome cells also appear to form all of the tissues within a segment apart from the neural tissue and the notochord which was there prior to formation of the somites. This assessment does not exclude the possibility that the sclerotome can be subdivided into smaller compartments with the two separate rostral and caudal sclerotome halves being individual compartments. While the sclerotome cells remain as a close group during further development, the cells in the rostral and caudal halves do not appear to mix with each other (Stern and Keynes, 1986) and our experiments using PNA (Bagnall and Sanders, 1989) appear to confirm these results. In addition, there have been several studies in which the cells in the rostral and caudal sclerotome halves have been shown to differ based on different criteria: neural crest cell migration (Rickmann *et al.*, 1985; Newgreen and Erickson, 1986); neurite outgrowth (Keynes and Stern, 1984); differential distribution of cytotactin or tenascin (Tan *et al.*, 1987; Mackie *et al.*, 1988); differential PNA staining (Stern *et al.*, 1986); and, more recently, it has been shown that the two groups of cells in the separate sclerotome halves are not dye-coupled to each other (Bagnall *et al.*, 1991 - submitted). Perhaps it is also significant that even during the initial formation of the somitomeres in the segmental plate, paraxial cells can be identified along a line perpendicular to the rostrocaudal axis, bisecting the somitomere into rostral and caudal halves (Meier, 1979). It is also interesting to note that it is only after the formation of at least the somites that the distribution of cells appears to be limited, for Stern *et al.* (1988) found that the progeny of single cells in the segmental plate that had been injected with fluorescent dye could be found in different somites. Although Stern *et al.* (1988) concluded from their results that somites did not constitute developmental compartments, it is significant that they injected cells in only the segmental plate and not cells in the actual somite or sclerotome. It is possible that the founder cells for the compartments are not established until at least after the somites have been formed and possibly not until the sclerotome has been identified, perhaps similar to the delayed establishment of founder cells in rhombomeres described by Fraser *et al.* (1990).

If the rostral and caudal sclerotome halves can be identified as separate compartments, then the fundamental segment along the rostrocaudal axis in the vertebrate might be the caudal sclerotome half of one somite coupled with the rostral sclerotome half of the next, in accordance with the theory of resegmentation. The segmental arrangement of the vertebrate body might then be analogous to the parasegmental arrangement described for *Drosophila* (Martinez-Arias and Lawrence, 1985). Morata and Kerridge (1981) found that the function of the homoeotic gene *Ultrabithorax* in *Drosophila* was not delimited by a border of an obvious segment but by a border within a segment. Martinez-Arias and Lawrence (1985) called these out-of-register metameres 'parasegments', where a parasegment consisted of the posterior compartment of one segment and an anterior compartment of the next. Subsequently, Martinez-Arias and Lawrence (1985) have continued to advance the idea that it is the parasegment and not the more obvious segment which might be the fundamental metamere both in the ectoderm and mesoderm. Originally, the concept of parasegments was based on fields of gene expression for the most obvious anatomical landmarks defined the borders of the segments. However, there are less obvious anatomical structures which can be used to define the parasegments. For example, in *Drosophila*, grooves appear in the extended germband of the embryo and these coincide with parasegmental boundaries (Martinez-Arias and Lawrence, 1985; Ingham *et al.*, 1985). Furthermore, in the mesoderm, deeper grooves are formed and these also demarcate the parasegments (Martinez-Arias and Lawrence, 1985). In the vertebrate body, it is possible that the somite represents the most obvious segmental pattern in the early embryo but that the fundamental metamere is represented by the caudal sclerotome half of one segment combined with the rostral half of the next segment. It is this combination that appears to produce the vertebral segmentation pattern and the intrasclerotomal fissure (of von Ebner) (Stern and Keynes, 1986) is a clear anatomical border for such an arrangement. Determination of the true segmental boundaries would be useful in identifying the possible fields of gene expression associated with segmentation along the rostrocaudal axis.

Future studies of cell lineage of somitic cells might focus on the cells found in the somitocoele. Their origin is obscure (Langman and Nelson, 1968; Stern, 1979; Stern and Bellairs, 1984,) and they have different characteristics when compared with the cells of the surrounding epithelial ball. For example, when exposed to PNA, the cells in the somitocoele stain positively before the cells in the surrounding epithelium (Bagnall and Sanders, 1989) and fibronectin production by these cells is far in advance of the surrounding cells even during the early stages of sclerotome migration (Bagnall - study in progress). It is assumed that the cells in the somitocoele mix freely during dispersal of the somite with the cells from the wall of the epithelial somite that have now become mesenchymal. However, if they retain a central position in the sclerotome mass, they would be positioned ideally to form the intervertebral discs as they would surround the intrasclerotomal fissure. Furthermore, the study to determine the boundaries of the communication compartments (Bagnall *et al.*, 1991 - see later) identified a group of cells lining the intrasclerotomal fissure (of von Ebner) which was separate from the independent rostral and caudal sclerotome halves. It is possible that these cells originate from the somitocoele, form an entirely separate group, and eventually form the intervertebral disc.

## Gene Expression Studies

Unfortunately, studies to determine the extent of gene expression along the rostrocaudal axis in vertebrates have been unable to identify compartment boundaries and so support for the second criterion for a developmental compartment (Lawrence, 1990) is lacking. Studies using *in situ* hybridisation techniques have shown that the individual segments of the vertebrate appear to recognise the patterns of gene expression available to them rather than any one specific gene product, making it difficult to identify specific segmental boundaries (Gaunt, 1987, 1988; Gaunt *et al.*, 1988; de Robertis *et al.*, 1990). However, Kessel *et al.* (1990) produced indirect evidence to suggest that the expression of a single homoeobox gene affects the development of one vertebral segment. They developed gain-of-function mutants by introducing genomic sequences of the *Hox1.1* gene into mice and, among other mutations, these transgenic mice manifested an additional vertebra, the proatlas, at the craniocervical transition. Coupled with evidence in support of the theory of resegmentation (Bagnall *et al.*, 1988) it would seem that the boundaries for the field of gene expression for this gene are the intrasclerotomal borders associated with the two sclerotomal halves. A clearer idea of the genetic influence and its relationship to the development of compartments during development of the rostrocaudal axis will emerge as more genetic studies are completed.

## Communication Compartments

The concept of communication compartments is especially

interesting as the restriction of dye spread at a compartment border could reflect a site where the cell-cell exchange of possible morphogens (small signal molecules) might be restricted and directly suggests a method whereby the developmental boundaries can be maintained, allowing groups of cells within the individual compartments to develop a common identity. This has been found previously in insects where, for example, cells that lie on either side of a segment border that forms a developmental boundary show restricted transfer of Lucifer Yellow (Warner and Lawrence, 1973, 1982; Blennerhassett and Caveney, 1984), whereas cells that lie within the same segment freely exchange such small molecules.

Recently, we have been able to determine the boundaries of communication compartments during development of the rostrocaudal axis (Bagnall *et al.*, 1991 - submitted). Lucifer Yellow fluorescent dye, which is able to pass freely through gap junctions between adjacent cells, was microinjected into a variety of cells along the rostrocaudal axis in chick embryos. Cells in the segmental plate were dye coupled as were cells forming the epithelial somites. However, dye coupling was not observed between somites nor was it observed between outer epithelial cells and the cells in the somitocoele (Fig.1). On dispersal of the somite, dermatome cells were found to be dye coupled. However, sclerotome cells were found to be divided into rostral and caudal compartments separated by the intrasclerotomic fissure (of von Ebner). Furthermore, the cells bordering the intrasclerotomal fissure also exhibited dye coupling that was restricted primarily to cells along the fissure. These results show that there is a sequential restriction of communication between the cells of the original

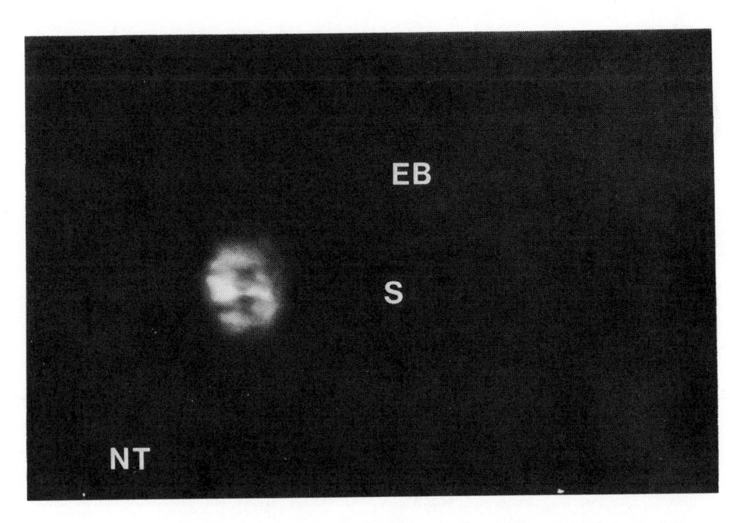


Fig 1. A fluorescence micrograph illustrating the spread of Lucifer Yellow dye following injection into a single cell in the epithelial ball. Note that there is no dye spread to the cells of the somitocoele. NT = neural tube S = somitocoele EB = epithelial ball

somite as the mesoderm cells mature. The borders of the communication compartments were also similar to the borders of compartments identified by our earlier cell lineage studies (Bagnall *et al.*, 1988, 1989, 1991; Bagnall and Sanders, 1988).

There is morphological evidence for the presence of gap junctions between the cells of the segmental plate which would account for the initial dye transfer pattern described for the segmental plate in our study (Bagnall *et al.*, 1991 - submitted). Revel *et al.* (1973) found the surfaces of the mesenchymal cells in the segmental plate to be decorated with small gap junctions on either the cell bodies or on processes which link the cells. These findings support the results of Sheridan (1968) who reported electrical coupling (presumably through gap junctions) in the early chick embryo, in particular between cells of the segmental plate. Unfortunately, confirmation of the presence and distribution of gap junctions during subsequent development of the mesoderm is not available. Trelstad *et al.* (1967) found intimate cell-cell contact between cells in the somite and sclerotome stages but these contacts were not completely characterised and Revel *et al.* (1973) found small collections of membrane particles reminiscent of gap junctions, but which could not be readily identified as such, between the epithelial cells of the epithelial somite.

The method by which dye spread is restricted at the compartment borders is not clear. However, it would seem reasonable to suggest that the development of a basement membrane surrounding the epithelial somite probably accounts for the apparent isolation of each somite. The basement membrane would involve a reduction in cell-cell contact between cells within the epithelial ball and surrounding structures even though the basement membrane is patchy and incomplete and has processes from the mesodermal cells extending through it, contacting other structures (Solursh *et al.*, 1979). Similarly, our results suggest that gap junctions may be absent or reduced in numbers between the cells of the epithelial ball and the somitocoele where contact between these two groups of cells can occur only at the apical surface of the epithelial cells. Revel *et al.* (1973) and Bellairs *et al.* (1975) found gap junctions in a similar epithelium of the chick epiblast, but only on the basolateral surface of the cells. The absence of dye coupling at other compartment boundaries is more difficult to explain.

The sclerotome cells consist of loose mesenchyme, similar to the cells in the segmental plate and the somitocoele, which appears not to restrict dye coupling. The intersclerotomic fissure between the somites contains a blood vessel but neither the fissure nor the vessel is sufficiently extensive to provide a barrier to cell-cell contact between adjacent sclerotomal masses (Shaner, 1985) (Fig.2) and yet no dye transfer occurs across this border. Similarly, the intrasclerotomic fissure (of von Ebner) is a narrow gap between the rostral and caudal sclerotome halves where the cells adjacent to the fissure are elongated with their long axis at right angles to the neural tube (Shaner, 1985). The extent of cell-cell contact across both of these fissures appears to be quite extensive (Fig.2) and it is not immediately apparent why dye coupling should not occur to an extent similar to that found between cells within the

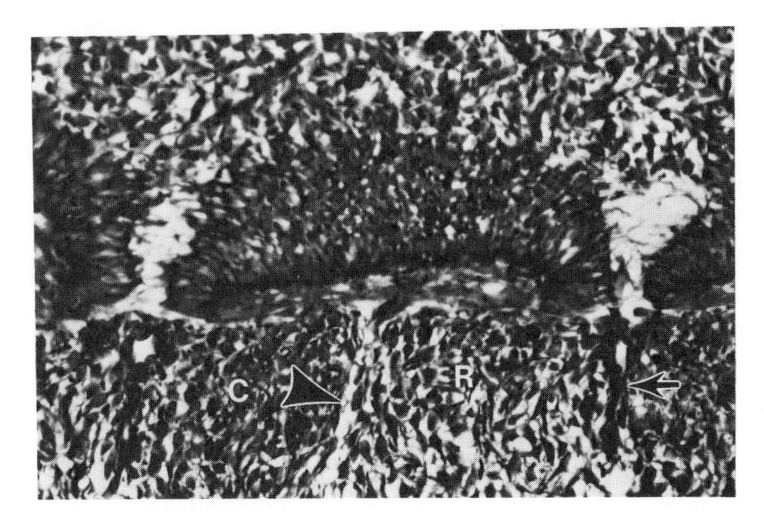


Fig 2. A photomicrograph of the sclerotome region showing the intrasclerotomic fissure (of von Ebner) (large arrowhead) and the intersomitic fissure containing a blood vessel (small arrow). Note the extent of apparent contact between the cells across these fissures. There also appears to be extensive contact between cells in the rostral and caudal sclerotome halves and the cells lining the intrasclerotomic fissure. R = rostral sclerotome C = caudal sclerotome

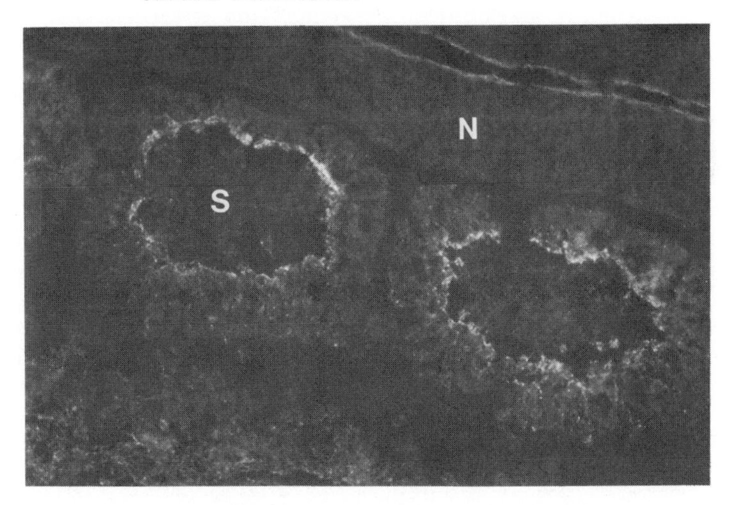


Fig. 3. A photomicrograph showing the distribution of connexin 43 in the developing somite of the chick embryo. The fluorescence reveals that the gap junctions are concentrated near the apical ends of the epithelial cells. Note that there appears to be little staining among the cells of the somitocoele. S = somitocoele N = neural tube

rostral and caudal sclerotome halves. The lack of dye transfer between the cells lining the intrasclerotomic fissure and cells in either of the two sclerotome halves is also difficult to explain. Extensive cell-cell contact is seen between the cells lining the intrasclerotomic fissure and those in either of the sclerotome halves (Fig.2). In other systems where similar communication borders have been identified, cell contact across borders has been maintained but other mechanisms have been found to be employed in preventing the passage of dye. For example, in the insect *Oncopeltus fasciatus* communication across compartment borders has been shown to be under the control of a discrete population of border cells (Blennerhassett and Caveney, 1984). Similarly, in the wing imaginal disc of *Drosophila*, the dorsal and ventral compartments are partitioned by a zone of non-dividing cells (O'Brochta and Bryant, 1985). It is not known whether the cells lining the intrasclerotomic fissure represent such a population in the chick sclerotome. However, initial results from a study in which we are examining the spatiotemporal distribution of gap junctions along the rostrocaudal axis of the chick embryo using a variety of antibodies to various connexins have not revealed any obvious reduction in the distribution of gap junctions at any of the compartment boundaries described in the sclerotome (Fig.3).

## ACKNOWLEDGEMENTS

The author would like to thank the Natural Sciences and Engineering Research Council of Canada for providing a grant to complete this work. The author would also like to thank Dr. D. Paul, Harvard University, and Dr. B. Nicholson, State University of New York at Buffalo, for kindly donating the connexin antibodies.

## REFERENCES

Bagnall K. The migration and distribution of somite cells after labelling with the carbocyanine dye DiI: the relationship of this distribution to segmentation in the vertebrate body. Anat. Embryol. (In press).

Bagnall K., Sanders E., Higgins S., Cheung E. and Leam H. (1986) The effects on vertebral development of removing a single somite from a 2-day old chick embryo. in: Bellairs, R, Ede D, Lash J (eds) Somites In Developing Embryos. Plenum Press, New York, pp 1-16.

Bagnall K., Higgins S., and Sanders E. (1988) The contribution made by a single somite to the vertebral column: experimental evidence in support of resegmentation using the chick-quail chimaera model. Development 103: 69-85.

Bagnall K., Higgins, S., and Sanders, E. (1989) The contribution made by a single somite to tissues within a body segment and assessment of their integration with similar cells from adjacent segments. Development 107: 931-943.

Bagnall K. and Sanders E. (1989) The binding pattern of peanut lectin associated with sclerotome migration and formation of the vertebral axis in the chick embryo. Anat. Embryol. 180: 505-513.

Baur R. (1969) Zum Problem der Neugliederung der Wirbelsaule. Acta Anat. 72:321-356.

Bellairs, R., Breathnach, A., and Gross, M. (1975). Freeze-fracture replication of junctional complexes in unincubated and incubated chick embryos. Cell Tiss. Res., 162, 235-252.
Beresford B. (1983) Brachial muscles in the chick embryo: the fate of individual somites. J. Embryol. Exp. Morphol. 77: 99-116.
Blennerhassett, M. and Caveney, S. (1984). Separation of developmental compartments by a cell type with reduced junctional permeability. Nature, 309, 361-364.
Caveney, S. (1985). The role of gap junctions in development. Ann. Rev. Physiol., 47, 319-335.
Crick, F. and Lawrence, P. (1975). Compartments and polyclones in insect development. Science, 189, 340-347.
Dalgleish A. (1985) A study of the development of thoracic vertebrae in the mouse assisted by autoradiography. Acta Anat. 122: 91-98.
Dawes B. (1930) The development of the vertebral column in mammals as illustrated in its development in Mus musculus. Phil. Trans. Roy. Soc. B 218: 115-170.
de Robertis E, Oliver G, and Wright C (1990) Homeobox genes and the vertebrate body plan. Sci. Amer. 263: 46-52.
Fraser S, Keynes R, and Lumsden A (1990) Segmentation in the chick embryo hindbrain is defined by cell lineage restrictions. Nature 344: 431-435.
Garcia-Bellido, A., Ripoll, P., and Morata, G. (1973). Developmental compartmentalisation of the wing disc of Drosophila. Nature, 245. 251-253.
Garcia-Bellido, A. Ripoll, P., and Morata, G. (1976). Developmental compartmentalization in the dorsal mesothoracic disc of *Drosophila*. Dev. Biol., 48, 132-147.
Gaunt S (1987) Homoeobox gene Hox 1.5 expression in mouse embryos: earliest detection by in situ hybridization is during gastrulation. Development 101: 51-60.
Gaunt S (1988) Mouse homeobox gene transcripts occupy different overlapping domains in embryonic germ layers and organs: a comparison of Hox 3.1 and Hox 1.5. Development 103: 135-144.
Gaunt S, Sharpe P, and Duboule D (1988) Spatially restricted domains of homeo-gene transcripts in mouse embryos: relation to a segmented body plan. Development 104 (suppl): 169-179.
Guthrie, S. (1984). Patterns of junctional communication in the early amphibian embryo. Nature, 311, 149-151.
Honig M, and Hume R (1989) DiI and DiO: versatile fluorescent dyes for neuronal labelling and pathway tracing. Trends NeuroSci. 12: 333-341.
Ingham P., Martinez-Arias A., Lawrence P., and Howard K. (1985) Expression of engrailed in the parasegment of Drosophila. Nature 317: 634-636.
Kalimi, G. and Lo, C. (1988) Communication compartments in the gastrulating mouse embryo. J. Cell Biol., 107, 241-255.
Kent G. (1978) Comparative Anatomy of the Vertebrates. 4th Ed. C.V. Mosby Co., St. Louis.
Kessel M., Balling R., and Gruss P. (1990) Variations of cervical vertebrae after expression of a *Hox1.1* transgene in mice. Cell 61:301-308.
Keynes R., and Stern C. (1984) Segmentation in the vertebrate nervous system. Nature 310: 786-789.
Lance-Jones C. (1988) The somitic level of origin of embryonic chick hindlimb muscles. Dev. Biol. 126: 394-407.

Langman J. and Nelson G. (1968) A radioautographic study of the development of the somite in the chick embryo. J. Embryol. exp. Morph. 19: 217-226.

Lawrence, P. (1973). Maintenance of boundaries between developing organs in insects. Nature, 242, 31-32.

Lawrence P. (1990) Compartments in vertebrates. Nature 344: 382-383.

Lo, C. and Gilula, N. (1979a). Gap junctional communication in the preimplantation mouse embryo. Cell, 18, 399-409.

Lo, C. and Gilula, N. (1979b). Gap junctional communication in the post-implantation mouse embryo. Cell, 18, 411-422.

Lumsden, A. and Keynes, R. (1989). Segmental patterns of neuronal development in the chick hindbrain. Nature, 337, 424-429.

Mackie E., Tucker R., Halfter W., Chiquet-Ehrismann R., and Epperlein H. (1988) The distribution of tenascin coincides with pathways of neural crest cell migration. Development 102: 237-250.

Martinez-Arias A., and Lawrence P. (1985) Parasegments and compartments in the Drosophila embryo. Nature 313: 639-642.

Meier S. (1979) Development of the chick embryo mesoblast. Dev. Biol. 73: 25-45.

Moe J., Winter R., Bradford D., and Lonstein J. (1978) Scoliosis and Other Spinal Deformities. W.B. Saunders Co., Toronto.

Morata G., and Kerridge S. (1981) Sequential functions of the bithorax complex of Drosophila. Nature 290: 778-781.

Newgreen D., and Erickson C. (1986) The migration of neural crest cells. Int. Rev. Cytol. 103: 89-145.

O'Brochta, D. and Bryant, P. (1985). A zone of non-proliferating cells at a lineage restriction boundary in *Drosophila*. Nature, 313, 138-141.

O'Rahilly R. and Meyer D. (1979) The timing and sequence of events in the development of the human vertebral column during the embryonic period proper. Anat. Emb. 157: 167-176.

Patel, N., Kornberg, T., and Goodman, C. (1989). Expression of Engrailed during segmentation in grasshopper and crayfish. Development, 107, 201-212.

Piiper, J. (1928) On the evolution of the vertebral column in birds, illustrated by its development in *Larus* and *Strutio*. Phil. Trans. Roy. Soc. B 216: 285-351.

Revel, J-P., Yip, P., and Chang, L. (1973). Cell junctions in the early chick embryo - a freeze etch study. Dev. Biol., 35, 302-317.

Rickmann M., Fawcett J., and Keynes R. (1985) The migration of neural crest cells and the growth of motor axons through the rostral half of the chick somite. J. Embryol. Exp. Morph. 90: 437-455.

Sadler T. (1990) Langman's Medical Embryology. 6th Ed. Williams and Wilkins, London.

Sanders E. (1986) A comparison of the adhesiveness of somitic cells from chick and quail embryos. in: Bellairs, R, Ede D, Lash J (eds) Somites In Developing Embryos. Plenum Press, New York, pp 191-200.

Sanes J. (1989) Analysing cell lineage with a recombinant retrovirus. Trends NeuroSci. 12: 21-28.

Sensenig E. (1943) The origin of the vertebral column in the deer-mouse, Peromyscus maniculatus rufinus. Anat. Rec. 86: 123-141.

Sensenig E. (1949) The early development of the human vertebral column. Contrib. Embryol. 33: 21-41.

Shaner, D. (1985). in: Development of the Human Vertebral Column, M.Sc. Thesis, University of Alberta.

Sheridan, J. (1968). Electrophysiological evidence for low-resistance intercellular junctions in the early chick embryo. J. Cell Biol., 37, 650-659.

Solursh M., Fisher M., Meier S., and Singley C. (1979) The role of extracellular matrix in the formation of the sclerotome. J. Embryol. exp. Morphol. 54: 75-98.

Stern C. (1979) A re-examination of mitotic activity in the early chick embryo. Anat. Embryol. 156: 319-329.

Stern C. and Bellairs R. (1984) Mitotic activity during somite segmentation in the early chick embryo. Anat. Embryol. 169: 97-102.

Stern C., and Keynes R. (1986) Cell lineage and the formation and maintenance of half somites. in: Bellairs, R, Ede D, Lash J (eds) Somites In Developing Embryos. Plenum Press, New York, pp 147-160.

Stern C., Sisodiya S., and Keynes R. (1986) Interactioons between neurites and somite cells: inhibition and stimulation of nerve growth in the chick embryo. J. Embryol. exp. Morphol. 91: 209-226.

Stern, C. and Keynes, R. (1987). Interactions between somite cells: the formation and maintenace of segment boundaries in the chick embryo. Development, 99, 261-272.

Stern C., Fraser S., Keynes R., Primmett D. (1988) A cell lineage analysis of segmentation in the chick embryo. Development 104 (suppl) 231-244.

Tan S., Crossin K., Hoffman S., and Edelman G. (1987) Asymmetric expression in somites of cytotactin and its proteoglycan ligand is correlated with neural crest distribution. Proc. Nat. Acad. Sci. 84: 7977-7981.

Trelstad R., Hay E., Revel J-P. (1967) Cell contact during early morphogenesis in the chick embryo. Dev. Biol. 16: 78-106.

Tsou P., Yau A., and Hodgson A. (1980) Embryogenesis and prenatal development of congenital anomalies and their classification. Clin. Orth. Rel. Res. 152: 211-228.

Verbout A. (1976) A critical review of the 'Neugliederung' concept in relation to the development of the vertebral column. Acta Biotheor. 25: 219-258.

Verbout A. (1985) The development of the vertebral column. Adv. Anat. Embryol. Cell Biol. 90: 1-22.

Warner, A. and Lawrence, P. (1973). Electrical coupling across developmental boundaries in insect epidermis. Nature, 245, 47-48.

Warner, A. and Lawrence, P. (1982). Permeability of gap junctions at the segmental border in insect epidermis. Cell, 28, 243-252.

Williams L. (1910) The somites of the chick. Amer. J. Anat. 11: 55-100.

Winchester L., and Bellairs A. (1977) Aspects of vertebral development in lizards and snakes. J. Zool. 181: 495-525.

GLYCOSYLATION MODIFIERS AND

THE FIRST CELLULAR MIGRATIONS/INDUCTIONS IN EARLY CHICK EMBRYO*

Nikolas Zagris and Maria Panagopoulou

Division of Genetics and Cell and Developmental Biology
Department of Biology
University of Patras, Patras, Greece

## INTRODUCTION

Glycoproteins are integral components of membranes and of the extracellular matrix (ECM) and are involved in several aspects of cellular behaviour such as cellular migration, recognition, adhesion and differentiation (Manasek, 1975; Sanders, 1983; Edelman, 1984, 1986, 1988; Thiery, 1984; Liotta et al., 1986; West, 1986; Ruoslahti and Pierschbacher, 1987). The oligosaccharide chains of glycoproteins can be classified into oligosaccharides attached to the polypeptide by an O-glycosidic linkage from N-acetylgalactosamine to serine or threonine and oligosaccharides linked-N glycosidically from N-acetylglucosamine to the amide nitrogen of asparagine. There are two types of N-linked glycans referred to as high-mannose and complex type. Biosynthesis of N-linked glycans of the complex type proceeds via a high-mannose intermediate. Oligosaccharides of the complex type are diverse differing mainly in the number of terminal branches and in the pattern of addition of sialic acid and fucose residues. Cellular behaviour may be modulated by variations in glycosylation of glycoproteins and/or glycolipids. The availability of specific inhibitors acting at different stages in the glycosylation process has permitted the synthesis of protein without oligosaccharide moieties or protein with altered oligosaccharide structures (Fuhrmann et al., 1985; Elbein, 1987).

Inhibitors of N-linked oligosaccharide processing allow the manipulation of N-linked glycan structures in a very precise manner and make it possible to assess the importance of the glycoconjugates in the early developmental processes. Tunicamycin (TN) inhibits the first reaction in the lipid-linked saccharide pathway, the transfer of GlcNAc-1-P from UDP-GlcNAc to dolichyl-P to form dolichyl-PP-GlcNAc (N-acetylglucosaminyl pyrophosphodolichol) thereby preventing the synthesis of any of the oligosaccharides normally transfered to asparagine residues. Glucosidase and mannosidase inhibitors prevent the normal processing of the $(Glc)_3(Man)_9(GlcNAc)_2$ intermediate. For instance, 1-deoxynojirimycin (dNM) inhibits glucosidase I and II causes a decrease in complex type oligosaccharides and an increase in high mannose oligosaccharides while deoxymannojirimycin (dMM) is potent inhibitor of a mannosidase IA/B blocking the conversion of high mannose to complex type oligosaccharides. Endoglycosidase

*This work is dedicated to Prof. Leo Lemez (Charles University, Praha, Czechoslovakia) on the occasion of his retirement.

*Formation and Differentiation of Early Embryonic Mesoderm*
Edited by R. Bellairs *et al.*, Plenum Press, New York, 1992

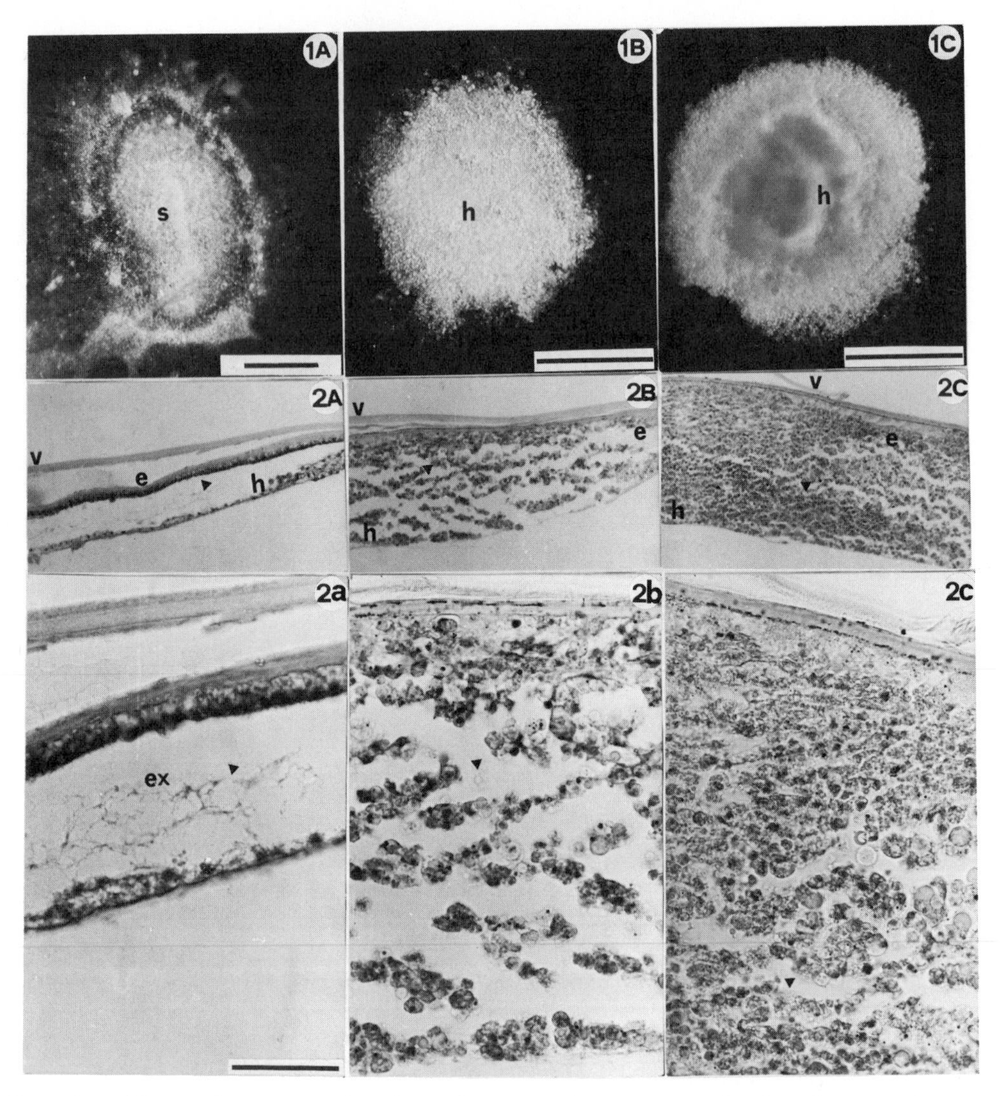


Fig. 1. Effect of TN on induction of embryonic axis in chick embryo. Embryos at stage X (A,B) and stage XIII (C) culturedin plain egg albumen (A) or in albumen containing $5.9 \times 10^{-6}$M TN (B,C) for 2h, transferred to plain egg albumen and photographed 23h after beginning of culture. h, hypoblast; s, primitive streak. Bar,1 mm.

Fig. 2. Transverse sections (5 μm) through the embryo presented in Fig. 1B (2B,b) and the embryo presented in Fig. 1C (2C,c). Non-treated embryo at stage XIII is presented for comparison (2A,a). Sections stained with Alcian blue stain. Arrowheads show location of sections under magnification (capital letters) presented under higher magnification (lower case). e, epiblast; ex, extracellular matrix; h, hypoblast; v, vitelline membrane. Bar, 50 μm.

H (endo H) is an enzyme that catalyzes the hydrolysis of the chitobiose core of high mannose and certain hybrid oligosaccharides.

The first morphogenetic events in the early chick embryo depend on processes involving cellular adhesion, interaction and migration. The primary hypoblast (lower layer) which is formed as a result of cell poly-ingression from the epiblast (upper layer) interacts with the epiblast and induces the first extensive cellular migrations and formation of the primitive streak (PS) in the epiblast. The PS is the site through which cells migrate to form mesoderm and definitive endoderm. The mesodermal cells interact with the overlying ectodermal cells which are induced to form the neural plate (Vakaet,1970; Eyal-Giladi,1984; Bellairs,1986). We are interested in the study of the first cellular migrations and inductions in the chick embryo. Our approach is to use enzyme and inhibitors of glycosylation to study the implication of the N-linked oligosaccharide moiety and the aberrant glycan structures of glycoproteins in the early developmental processes.

## MATERIALS AND METHODS

### Embryo Culture

Embryos at stages X and XIII (Eyal-Giladi and Kochav, 1976) were explanted and cultured as described previously (Zagris and Matthopoulos,1985). Embryos were placed on plain Ringer solution (control) or on Ringer solution containing one of the inhibitors TN ($2.3x10^{-6}$ to $11.9x10^{-6}$M), dNM ($5x10^{-4}$ to $2x10^{-3}$M), dMM ($2x10^{-3}$M), endo H (12.5 to 25mIU/ml) were incubated for various times from 2h to 17h at 37°C, and culture was continued on plain egg albumen. Over 60 embryos in each group were used in the course of this investigation. The rate of survival under these culture conditions is more than 80%.

### Differential Staining of Extracellular Material

Embryos were fixed, dehydrated through graded ethanol sulutions,embedded in paraffin and sectioned. The ECM was stained with Alcian blue 8GX-pH 2.5 (AB) which reveals the presence of glycosaminoglycans and acid glycoprotein showing a turquoise color (Humason, 1972; Zagris et al., 1989; Zagris and Panagopoulou, 1991).

## RESULTS AND DISCUSSION

The formation of the PS is the first major morphogenetic event which signals gastrulation and involves cellular adhesion, interaction and migration. The cellular movements for the formation of the PS are inhibited when chick embryo at stages X and XIII are exposed to TN. These embryos produced a prominent atypical lower layer (Eyal-Giladi and Kochav, 1976) (Fig. 1B, C), but the PS was not produced and, as a consequence, the embryonic axis was not induced. Thus, the target molecules which respond to the stimuli that induce the first cellular migrations and formation of the PS appear to be glycoproteins containing N-linked oligosaccharides. Control embryos cultured in the absence of TN form the embryonic axis in parallel cultures (Fig. 1A). Sections through treated embryos stained with AB show that the ECM is scarce and the epiblast and the hypoblast have lost their organization (Fig. 2Bb, Cc). Section through a normal embryo at stage XIII shows the ECM as an intricate network of fibrilar material in the space between the epiblast and the hypoblast and is presented for comparison (Fig. 2Aa).

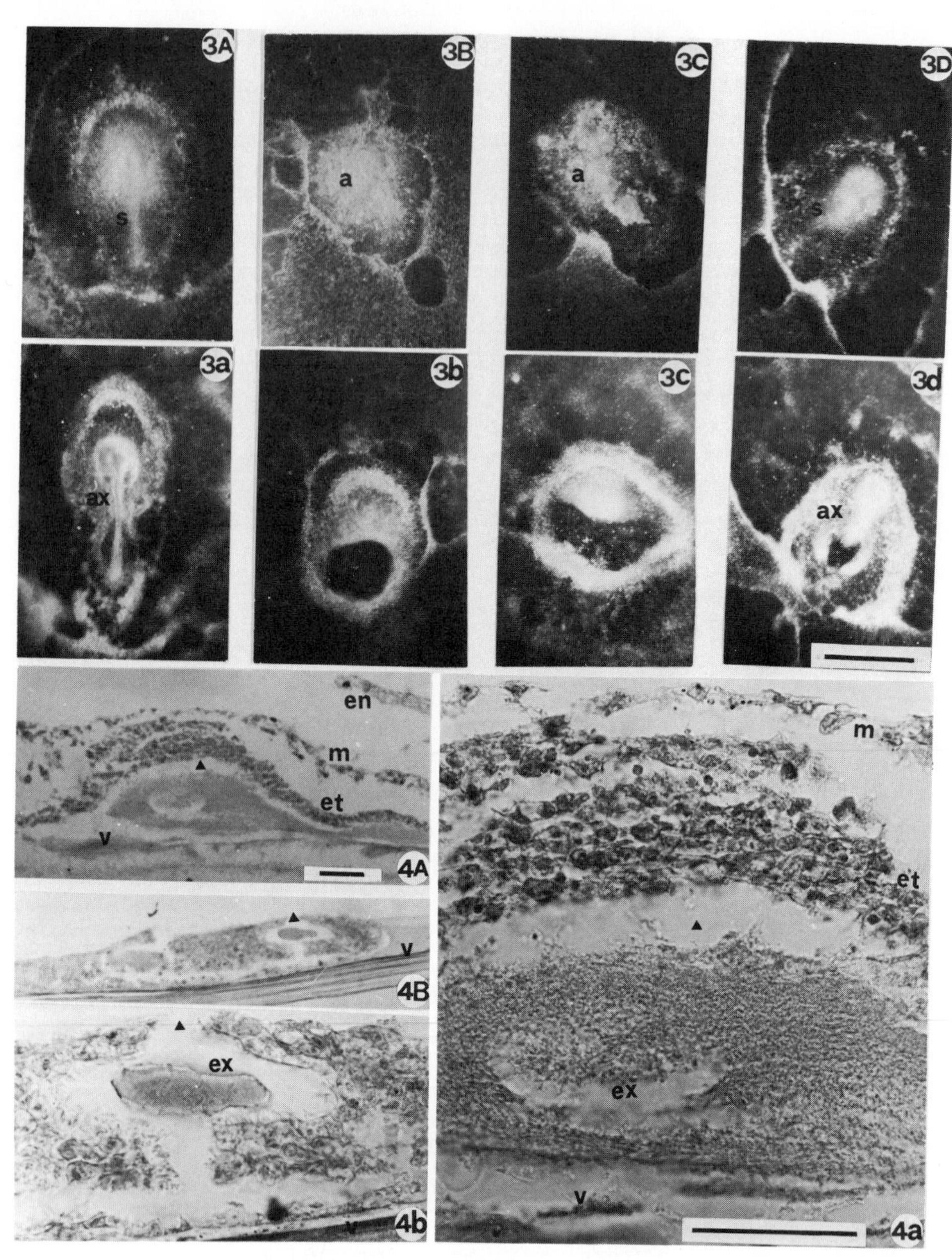


Fig. 3. Effect of glycomodifiers on induction of embryonic axis in chick embryo. Embryos at stage XIII were cultured in plain albumen (A,a) in dNM ($2x10^{-3}$M -B,b), dMM ($2x10^{-3}$M -C,c), endo H (25mIU/ml -D,d) for 6h, transferred to plain albumen and photographed 22h (capital letters) and 48h (lower case) after beginning of culture. Notation as in Fig. 1. Bar, 1 mm.

Fig. 4. Transverse sections through the embryos presented in 3c (4A,a) and 3d (4B,b). Conditions and notation as in Figs.1 and 2. a, cellular agglomeration; ax, embryonic axis; en, endoderm; et, ectoderm; m, mesoderm. Bar, 50 µm.

The prospect of being able to modify the structure of N-linked glycans tempted us to use dNM, dMM, and endo H. These modifiers of glycosylation permit the investigation of the effects of removing or altering sugar residues on the ability of the embryo to start morphogenetic movements. Deoxynojirimycin and dMM did not prevent the first cellular migrations and produced a PS which resembles a cellular agglomeration while endo H produced an atypical PS (Fig. 3B, C, D). Neural induction was not inhibited and, in the case of endo H, the embryo progressed to form an atypical embryonic axis (Fig. 3b, c, d). Control embryo has formed the embryonic axis in a parallel culture (Fig. 3Aa). Sections stained with AB show striking differences in the amount and in the organization of ECM in dMM-treated (Fig. 4Aa) and endo H-treated embryos (Fig. 4Bb) as compared to control embryo (Fig. 2Aa). Similar results to these of dMM are observed when embryos are treated with dNM. Embryos treated with the glycan modifiers show impressively abundant, disorganized EC material compared to control. This result implies that normal glycosylation confers specificity to protein secretion. Polypeptide analysis by one-and two-dimensional polyacrylamide gel electrophoresis and fluorography (manuscript in preparation) shows that tunicamycin and the modifiers of protein glycosylation produce changes in mobility of several polypeptides and induce formation of multiple charged species (isoforms) of polypeptides different from these present in control embryos.

The interference of the inhibitors/modifiers with the migration and induction mechanisms is permanent even when embryos are treated at stage X. Then, it would seem that the bulk of glycoproteins must be accumulated before the first cellular migrations and inductions and be used later. Our results with TN and the glycomodifiers show that elimination of N-linked glycosylation or manipulation of glycan trimming may have different effects on early embryonic processes. It seems that the N-linked units of glycoproteins are required while the complex glycan chains are not required for the migration of cells which form the PS and for neural induction but are necessary for morphogenesis of the embryonic axis.

Acknowledgements: This work was supported by grants from the General Secretariat of Research and Technology of Greece and grant ST2J-0324-C(TT) from the European Communities to N.Z.

REFERENCES

Bellairs, R.(1986) The primitive streak,Anat. Embryol. 174:1-14.

Edelman, G.M. (1984) Cell-adhesion molecules:a molecularbasis for animal form, Sci. Am. 250:80-91.

Edelman, G.M. (1986) Cell adhesion molecules in the regulation of animal form and tissue pattern, Ann. Rev. Cell Biol. 2:81-116.

Edelman, G.M. (1988) Morphoregulatory molecules, Biochem. 27:3533-3543.

Elbein, A.D. (1987) Inhibitors of the biosynthesis and processing of N-linked oligosaccharide chains, Ann. Rev. Biochem. 56:497-534.

Eyal-Giladi, H. (1984) The gradual establishment of cell commitments during the early stages of chick development, Cell Differ. 14: 245-255.

Eyal-Giladi, H., Kochav, S. (1976) From cleavage to primitive streak formation: A complementary normal table and a new look at the first stages of the development of the chick. I. General morphology, Dev. Biol. 49:321-337.

Fuhrmann, V., Bause, E., Ploegh, H. (1985) Inhibitors of oligosaccharide processing, Bioch. Biophys. Acta 825:95-110.

Humason, G.L. (1972) "Animal tissue techniques", Freeman and Co., San Francisco.

Liotta, L.A., Rao, C.N., Wewer, U.M. (1986) Biochemical interactions of tumor cells with the basement membrane, Ann. Rev. Biochem. 55:1037-1057.

Manasek, F.J. (1975) The extracellular matrix: A dynamic component of the developing embryo. In: "Current topics in developmental biology", A.A.Moscona, A.Monroy, eds, Vol. 10. Academic Press, New York, pp 35-102.

Ruoslahti, E., Pierschbacher, M.D. (1987) New perspectives in cell adhesion: RGD and integrins, Science 238:491-497.

Sanders, E.J. (1983) Recent progress towards understanding the roles of the basement membrane in development, Can. J. Biochem. Cell Biol. 61:949-956.

Thiery, J.P. (1984) Mechanisms of cell migration in the vertebrate embryo, Cell Differ. 15:1-15.

Vakaet, L. (1970) Cinematographic investigations of gastrulation in the chick blastoderm, Arch. Biol. (Liege) 81:387-426.

West, C.M. (1986) Current ideas on the significance of protein glycosylation, Mol. Cell Biochem. 72:3-20.

Zagris, N., Matthopoulos, D. (1985) Patterns of protein synthesis in chick blastula: a comparison of the component areas of the epiblast and the primary hypoblast, Dev. Genet. 5:209-217.

Zagris, N., Panagopoulou, M. (1991) Monensin inhibits the first cellular movements in early chick embryo, Roux's Arch. Dev. Biol. 199:335-340.

Zagris, N., Panagopoulou, M., Anastasopoulos, V. (1989) Extracellular matrix organized in embryonic cavities during induction of the embryonic axis in chick embryo, Cell Biol. Int. Rep. 13:833-843.

# SEGMENTALLY REGULATED PATTERNS OF CELL DEATH IN THE HINDBRAIN AND TRUNK OF THE DEVELOPING CHICK EMBRYO

Pete Jeffs and Mark Osmond

Department of Anatomy
Downing Street
Cambridge UK, CB2 3DY

## Introduction

In recent years, there has been much interest in the process of both migration and differentiation of the avian neural crest (NC) (Serbedzija *et al.*, 1989; Couly *et al.*, 1988; Fraser *et al.*, 1991; Stern *et al.*, 1991). Despite this, the nature of the processes governing the differentiation of these pluripotent cells remains enigmatic. In particular, the literature on this subject has become polarised into two camps: a number of authors have favoured a central role for environmental cues in the differentiation of neural crest cells (Fraser *et al.*, 1991; Stern *et al.*, 1991), while others have suggested that certain progenitor cells may become restricted in developmental potential during early migration (Le Douarin, 1986; Le Douarin and Smith, 1988).

Since evidence exists for cell death during the later development of certain neural crest-derived structures (Carr and Simpson, 1978a,b), we have asked whether there is any indication of dying cells during the stages of very early neural crest migration and differentiation. Although cell death is of increasing interest to workers from a number of fields, studies of cell death in the avian embryo have so far largely focussed on examination of the development of the limb-buds (Saunders *et al.*, 1962; Ede and Agerbak, 1968; Ede *et al.*, 1974; Ede and Flint, 1972), the tail-bud (Schoenwolf, 1981; Ooi *et al.*, 1986; Mills and Bellairs, 1989; Osmond, 1989); to later development of the dorsal root ganglion (DRG) (Carr and Simpson, 1978a,b) and to motor neuron outgrowth (Oppenheim, 1985,1989; Oppenheim *et al.*, 1990). Indirect reference has been made to segmental patterns of cell death in the developing chick mesoderm (Saunders *et al.*, 1962; Hinchliffe and Thorogood, 1974; Tosney, 1985, 1988; Sulik *et al.*, 1988), but no systematic study has been devoted to the phenomenon of cell death in the chick embryo at the developmental stages (and tissues) associated with the very early development of the neural crest. Should cells of the neural crest die during early ontogeny, certain predictions may be made with respect to environmental and selectional models of neural crest differentiation.

In this paper, we discuss our recent findings (Jeffs and Osmond, 1992; Jeffs *et al.*, 1992) concerning the distribution of dying cells, as detected with the vital dye Nile Blue sulphate, in the developing chick hindbrain between stages 9-11 (Hamburger and Hamilton, 1951) and in the trunk between stages 15 and 23. We have labelled the cells of the neural crest with HNK-1 antibody (Tucker *et al.*, 1984; Rickmann *et al.*, 1985; Bronner-Fraser, 1986; Loring and Erickson, 1987) in hindbrain preparations and with DiI for neural crest cells in the trunk (Serbedzija *et al.*, 1989), to determine whether a spatial and temporal correlation exists between their presence and the appearance of dying cells. Our results suggest that cell death is extensive during the early development of the neural crest in both the hindbrain and the trunk. Dying cells first appear in the trunk after neural crest

*Formation and Differentiation of Early Embryonic Mesoderm*
Edited by R. Bellairs *et al.*, Plenum Press, New York, 1992

migration from the neural folds. In the hindbrain however, cell death occurs predominantly on the dorsal midline before, or just as neural crest migration commences and is most extensive in discrete regions which do not give rise to a component of the neural crest. Thus, whilst the pattern of dying cells is secondary to neural crest migration in the trunk, cell death may possess an important role in the early determination of neural crest migration patterns in the developing avian head.

## Patterns of cell death in the developing chick hindbrain

Nile Blue sulphate (NBS) vital stain has been used by several authors for studying regions of dying cells (Saunders *et al.*, 1962; Sulik *et al.*, 1988). We have prepared a series of embryos at the developmental stages during which the neural crest migrates, using NBS staining to assess changing patterns of developmental cell death.

Dead cells first appear prior to stage 9 of development (5-7 somites) as a thin band lying on the dorsal midline between the neural folds of the mesencephalon (Figure 1a). This cluster of cells extends from the cranial half of the mesencephalon to the region of the mesencephalon-rhombencephalon border. All morphological designations for hindbrain development are derived from Vaaga (1969).

By stage 10 (10-12 somites), two domains of cell death are seen: a rostral mesencephalic domain and a caudal domain in the region destined to form the first proto-rhombomere RhA1 (Vaaga, 1969) and rhombomere rh3 (Figure 1b). These regions were contiguous in some specimens examined. NBS staining is not apparent caudal to RhA1/rh3 until the somitic region is reached,where stained cells appear in the neural epithelium adjacent to fully-formed somites.

In embryos of stage 11 (14-16 somites), mesencephalic NBS staining is restricted to only the most caudal part of the mesencephalon. However, as in stage 10 embryos, many NBS-stained cells are detected in the rhombencephalon spanning RhA1 and rh3, a region we have termed the rostral necrotic zone, CDr (Figure 1c), which ceases abruptly at the boundary with rhombomere rh4. Rh4 itself is devoid of NBS staining, but caudal to this, NBS-stained cells are detected in the dorsal midline tissue between the cranial part of the developing otic vesicles, corresponding to rh5 (Figure 1c). This we have termed the caudal necrotic zone (CDc). After this stage, extensive staining with NBS is no longer apparent in the dorsal region of the rhombencephalon.

**Figure 1** (Facing page). Nile Blue sulphate staining at stages during early avian development. (a) Pattern of NBS staining in dorsal region of the chick mesencephalon at stage 9 (5-6 somites). (b) Hindbrain region of stage 10 (10-11 somite) embryo showing NBS staining in dorsal neural midline in the region of rhombomeres rh2-rh3. (c) Hindbrain region of stage 11 (15-16 somite) embryo, showing two domains of NBS staining, a rostral domain (CDr) corresponding to the region between rh1 and rh3 and a caudal domain (CDc) corresponding to rhombomere rh5. (d) Rostral pattern of NBS staining in the trunk of chick embryo at stage 17, between somites 12-16. NBS-stained cells appear only in the rostral half of each somite. The segmental borders are shown with black arrowheads, and rostral (r) and caudal (c) halves are clearly marked. (e) Pyramidal pattern of NBS staining in a stage 22-23 embryo in the trunk region between the leg bud and the wing bud.The intersegmental boundaries are marked with black arrowheads. (f) Transition between rostral (r) and pyramidal (p) patterns of cell death: the pyramidal pattern always follows the rostral pattern in a rostrocaudal progression.

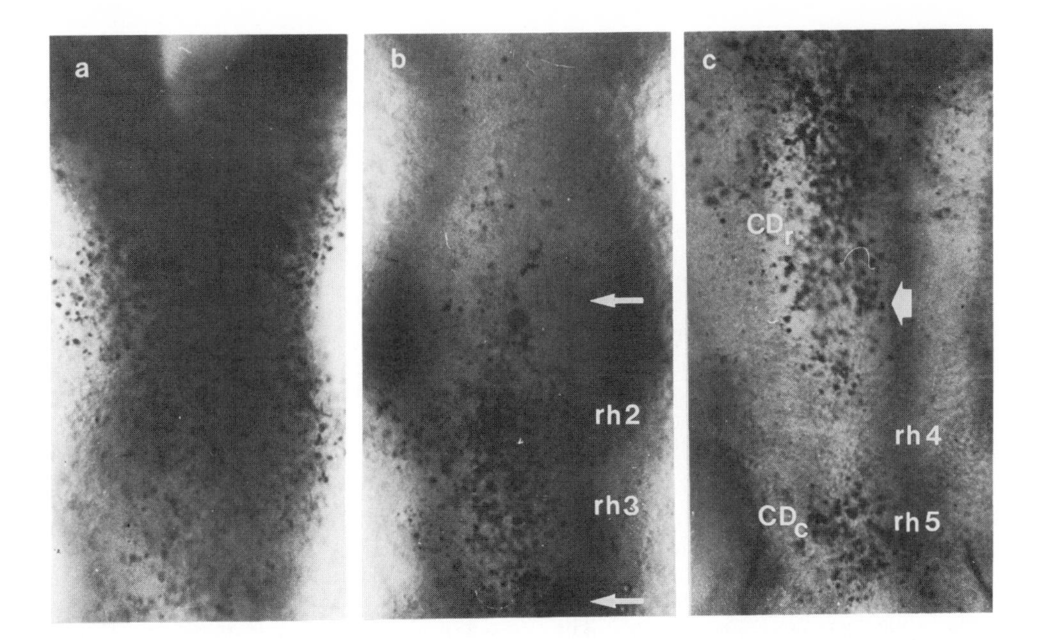
a
b
rh2
rh3
c
$CD_r$
rh4
$CD_c$
rh5

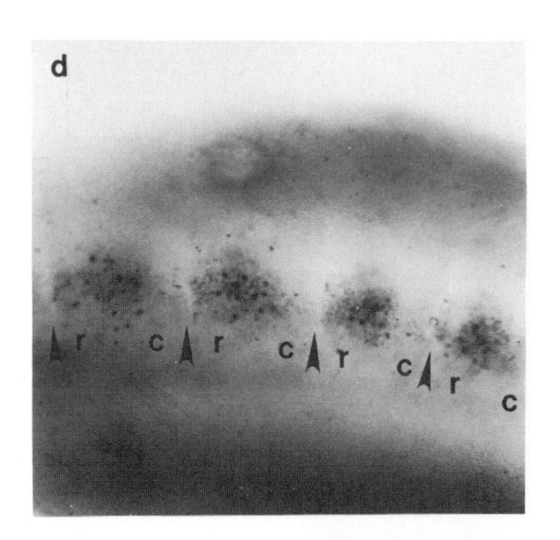
d
r
c
r
c
r
c
r
c

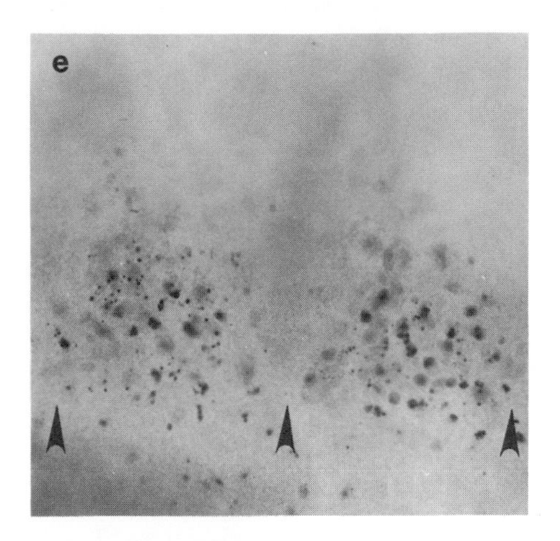
e

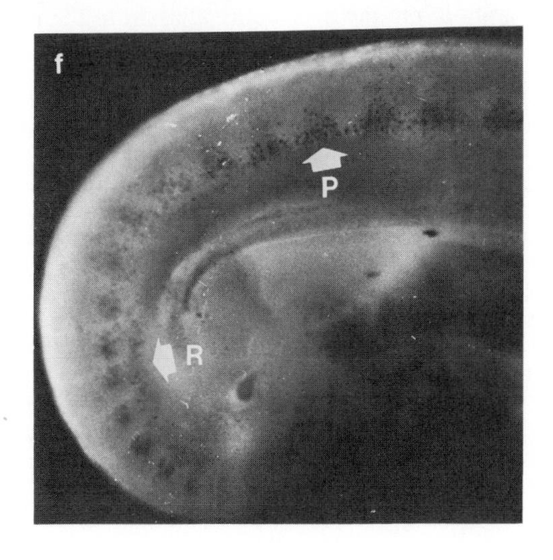
f
P
R

## Patterns of cell death in the developing chick trunk

Having determined the patterns of cell death in the developing hindbrain, we extended the analysis of NBS staining patterns to the trunk region of the embryo. Segmentally restricted regions of NBS staining are first observed in the trunk of the chick embryo during stages 15-16, (24-28 somites). By these stages, the lateral mesoderm adjacent to somites 1-7 contains many staining cells. The somites themselves, in this region, are free of staining cells, in contrast to those caudal to somite 8 in which NBS staining cells are seen restricted to the rostral half of each somite. This forms a periodic pattern referred to as the *rostral pattern*, extending as far as somite 14 (Figure 1d).

At stage 17 (29-32 somites), there are still NBS-stained cells in the lateral mesoderm at the level of somites 1-6, while the rostral pattern can be seen between somites 12 and 16. Somites 6 to 12 now possess a second pattern which is both lateral and superficial to the rostral pattern and forms a region or zone that is pyramidal in shape. This has thus been termed the *pyramidal pattern* (Figure 1e). Somites caudal to somite 16 appear to be free of staining cells.

By stage 18 (35 somites), the pyramidal pattern (Figure 1e) occupies the region between somites 14-22, and the rostral pattern between somites 23-27. These two staining patterns both progress caudally, so that by stage 19, the pyramidal pattern extends to somite 20 and the rostral pattern extends to somite 31. By stage 22-23, the pyramidal pattern can be seen in the region of somites 14-27. The necrotic regions continue to move in a caudal direction and eventually into the tail somites, the rostral pattern always preceeding the pyramidal pattern (Figure 1f). The developmental profile of the NBS staining patterns, in both hindbrain and trunk, is summarised in Figure 2.

## Correlation of cell death with position of neural crest cells in the developing hindbrain

To determine whether a spatial and temporal correlation exists between the patterns of NBS staining cells and the rhombencephalic neural crest, embryos were stained with HNK-1 between stages 9-11. This antibody (a gift from C.D.Stern, Oxford), a well characterised marker for the neural crest (Tucker *et al.*, 1984; Rickmann *et al.*, 1985; Bronner-Fraser, 1986; Loring and Erickson, 1987), was used in whole-mount preparations to facilitate comparison with NBS-stained material.

By stage 9 (5-7 somites), when there are NBS-stained cells present in the mesencephalon, very strong expression of the HNK-1 epitope is seen in the mesoderm surrounding both prosencephalon and mesencephalon (data not shown). Electron microscope studies (see Anderson and Meier, 1981; Meier and Packard, 1984) have also reported the onset of neural crest migration into the mesoderm adjacent to the mesencephalon and the cranial part of the rhombencephalon at stages 8-9.

At stage 10 (10-12 somites), there is NBS staining only adjacent to primary rhombomeres RhA1 and rh3 and, by stage 11, the region of cell death restricted to the region of rh5 appears. At this time, HNK-1 binding appears both rostral and caudal to the otic vesicle (Figure 3a). This pattern is striking because of the absence of labelling in the mesoderm adjacent to rhombomeres rh3 and rh5. Electron microscopy (Anderson and Meier, 1981) shows that whilst the cephalic neural crest remains as a continuous sheet over the midline, there are gaps in the pattern of neural crest migration in the axial segments corresponding to rh3 and rh5. The midline neural crest cells at the level of rh3 and rh5 thus remain condensed and evidently non-migratory at this stage.

Following stage 11 (14 somites onwards), the neural crest migrates ventrally beneath the ectoderm (Noden, 1984) and, rostral to the otic vesicle, cells migrate from rh1-2 and rh4 toward the branchial arches (Jeffs *et al.*, 1992). HNK-1 labelling is seen on either side of the otic vesicle, in close association with the cranial ganglia and the paths of the facial (VII) and glossopharyngeal (IX) nerves, but is still not seen next to rh3 and rh5.

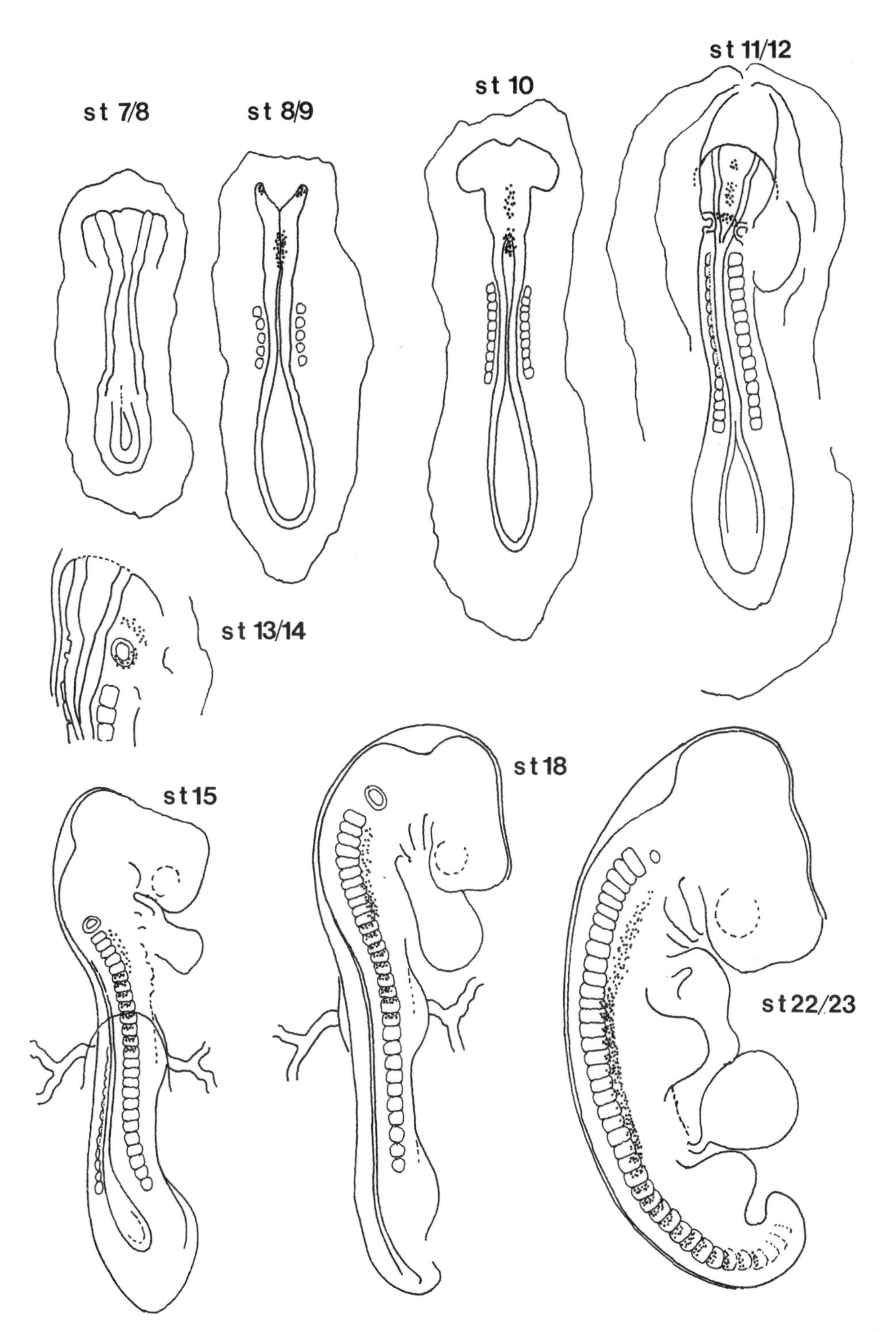


**Figure 2.** An atlas of axial cell death, showing the developmental progression of NBS staining patterns in the cranial and trunk regions of the developing chick embryo. Cranial cell death takes place during stages 9-11 of development (between 5 and 15 somites) and occurs chiefly over the neural midline. Trunk cell death, in contrast, takes place from stage 15 (25 somites) onwards and occurs in the differentiating mesoderm, along the length of the trunk. Regions of cell death are represented by stippled areas.

## Correlation of cell death with position of neural crest cells in the developing trunk

In order to investigate the correlation between neural crest migration pathways and the patterns of cell death in the trunk, neural crest cells were labelled with DiI at stages prior to the onset of segmental cell death in this region. The embryos were observed in whole-mount and most were also stained with NBS in order to compare the positions of labelled neural crest cells with the dying cells. DiI was injected into the lumen of the neural tube of embryos between stages 16-19 (Serbedzija *et al.*, 1989) and injected embryos were incubated for between 24 and 48 hours. Two cases were examined: (1) neural crest migration through the rostral half of the somite and (2) the dorsal migration path, which follows a route over the sclerotome, proximate to the dermomyotome.

In the first case, DiI-labelled neural crest cells were consistently seen in the rostral part of somites which were also positive for NBS staining (data not shown). The onset of neural crest migration through the rostral part of the somite always preceded the onset of cell death. This is also apparent from the fact that the rostral pattern of NBS staining is always seen approximately 10-15 somites rostral to the somite most recently formed. The neural crest is known to migrate into the trunk mesoderm at about the level of the third most recently formed somite (Newgreen *et al.*, 1990).

In the case of the neural crest cell population migrating along the dorsal path, staining with NBS revealed a correlation between the pyramidal pattern of dying cells and individual DiI-labelled neural crest cells (Figure 3b, c). In order to avoid the possibility of NBS-induced autofluorescence, specimens were photographed under fluorescence prior to staining with NBS. In these preparations, single DiI-labelled cells could be visualised, and these collectively formed a pyramidal pattern which subsequently mapped onto the pyramidal pattern visualised with subsequent NBS staining of the material (Figure 3b, c). Amongst the fluorescently labelled cells in this region were cells which possessed the same rounded, blebbed morphology characteristic of NBS-stained cells, in contrast to other surrounding cells which appeared to have a healthy migratory appearance.

## Discussion

Two conclusions can be reached from the results of the experiments described above (see Figure 2 for summary). The first set of experiments, involving NBS staining at various developmental stages, demonstrates that the presence of cell death in cranial and trunk regions of the chick embryo is associated, both temporally and spatially, with areas of early neural crest development. The second set of experiments, combining NBS staining with HNK-1 labelling of the neural crest, in the head, and DiI labelling, in the trunk, enable us to make a direct correlation between neural crest migration pathways and patterns of cell death.

## What is the function of neural crest - associated cell death?

Although cell death in cranial structures has been described (for example, see Martín-Partido *et al.*, 1986), no previously published studies have described discrete patterns of cell death in the rhombencephalon/mesencephalon during the stages of neural crest migration. Further, although it is known that the cranial neural crest migrates as a discontinous sheet of cells over the cranial mesoderm (Anderson and Meier, 1981; Meier and Packard, 1984; Hall, 1987; Hunt *et al.*, 1991) no plausible explanation has so far been proposed for this discontinuity. A tacit inference is that this discontinuity may result from non-permissive interactions with the cranial mesoderm (Anderson and Meier, 1981). EM studies have shown that the neural crest cells adjacent to rh3 and rh5 do not migrate out, but remain over the dorsal midline, presumably to be eliminated in these regions (Anderson and Meier, 1981). In the current study, histology of sectioned specimens shows necrotic figures in these same regions (see Jeffs *et al.*, 1992), thus supporting the whole-mount NBS staining patterns observed. Cell death may therefore represent a suitable mechanism for achieving this discontinuity.

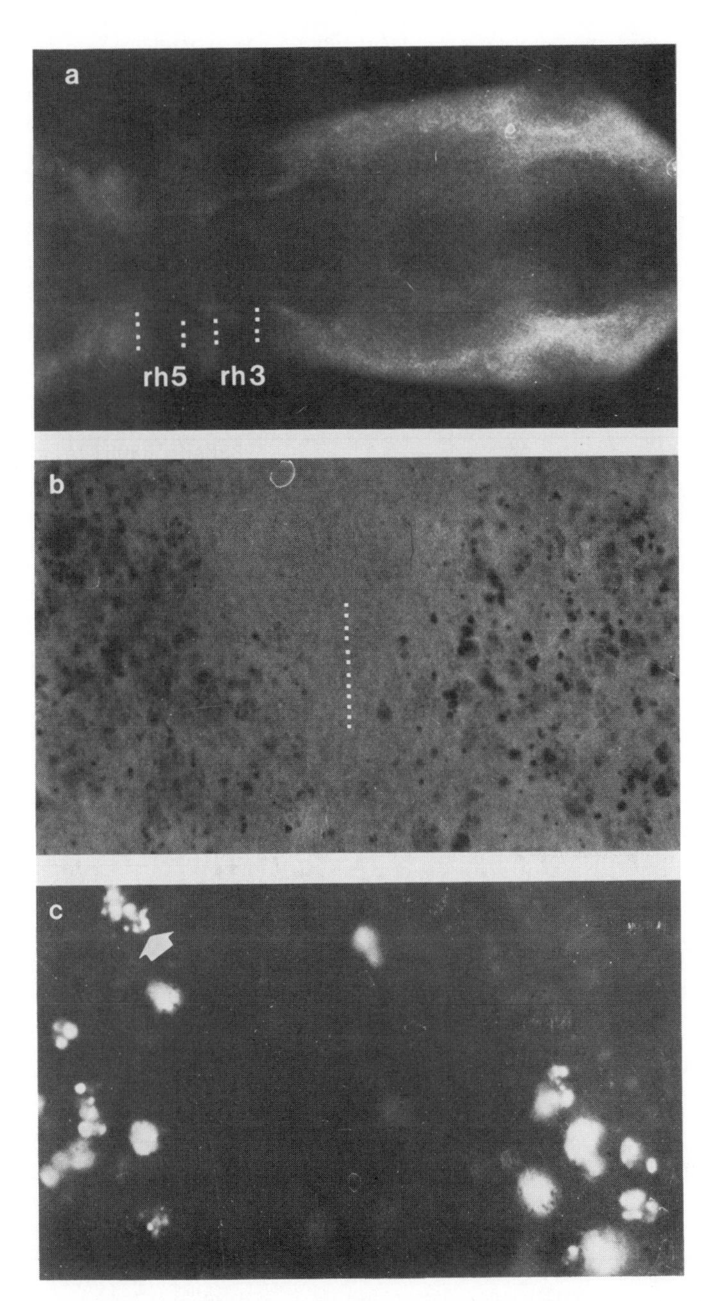


**Figure 3.** (a) Pattern of neural crest migration in the cranial region of the chick embryo detected by HNK-1 antibody labelling. This whole-mount preparation of a stage 11 chick embryo shows the absence of HNK-1 reactivity in the regions of rhombomeres rh3 and rh5. (b) Trunk region of a stage 23 chick embryo (seen under bright-field illumination) showing the cleft between two adjacent pyramidal patterns following staining with NBS. The intersomitic boundary is marked with a dotted line. (c) Same field, viewed under fluorescence, showing a series of highly blebbed cells, labelled with DiI, in the pyramidal region.

These results provide the first evidence that cell death may contribute to the absence of neural crest cells adjacent to rhombomeres rh3 and rh5 and may thus contribute to patterning the initial neural crest migration routes in the developing vertebrate head.

Whilst the caudal region of cranial cell death (CDc) correlates precisely with the absence of neural crest cells in the adjacent mesoderm, the rostral domain of cranial cell death (CDr) is wider than rh3. This, in addition to the appearance of dying cells in the mesencephalon during stage 9, may represent either a response to an overproduction of cranial neural crest cells, or alternatively, may be associated with an early selection of cranial neural crest precursors. Are any cells of the cranial neural crest committed early in development? Tissue culture analysis of quail cranial neural crest cells (Barroffio *et al.*, 1991) suggests that approximately one fifth of cultured clones give rise to monopotent lines. If a similar restriction of fate occurs *in vivo*, it would be interesting to determine whether any relationship exists between such monopotent precursors and the pattern of cell death.

In the trunk, patterns of cell death, as visualised by NBS staining, may be divided into two main classes: a rostral pattern and a pyramidal pattern (a third, minor class has been discussed in Jeffs and Osmond, 1992). Unlike the situation in the rhombencephalon, each pattern follows a rostrocaudal progression, each stage of somite development possessing a characteristic pattern of associated cell death (Figure 1d-f). As in the case of cranial cell death, however, each pattern corresponds both spatially and temporally with the known migration patterns of the neural crest (see Serbedzija *et al.*, 1989; Bronner-Fraser, 1986). As already stated, segmentally regulated cell death commences between 10-15 somites rostral to the most recently formed somite, a level at which neural crest migration is already well established. Furthermore, histological analysis (see Jeffs and Osmond, 1992) confirms that the positions of necrotic cells are appropriate to a neural crest origin.

These results may be interpreted in one of three ways. In the simplest case, the NBS staining cells could be of somitic origin, involved in patterning the route of neural crest migration. This is unlikely because (a) cell death occurs after the onset of neural crest migration and (b) the pyramidal pattern of cell death, for example, can be visualised with DiI labelling alone, after injection into the lumen of the neural tube. Thus, the pyramidal pattern cells must be of neurectodermal origin.

An alternative view suggests that there is an overproduction of neural crest cells in the trunk: cell death thus non-selectively eliminates members of this population. This is a more difficult argument to dismiss, but circumstantial evidence from two sources may be relevant. First, if cell death is a random event, one would expect to effectively visualise all paths of neural crest migration with NBS. This is not the case: rather, one visualises a series of specific necrotic zones. Neural crest cells thus die in a number of restricted environments. Second, in the hindbrain, the patterns of cell death are also far from random: for example, whilst cell death occurs adjacent to rh5, no cell death whatsoever occurs adjacent to rh4, a region clearly associated with neural crest production.

A further hypothesis suggests that a proportion of neural crest cells might undergo cell death at specific developmental stages. This is of relevence to models in which the trunk neural crest is thought of as being composed of different classes of cell from an early stage in development. Thus, cells which are initially equivalent in terms of their destined lifespan, but of different class, could migrate through specific non-permissive zones of the embryo. Contact with these necrotic regions might affect a proportion of the migrating neural crest (ie. cells of one class) and such cells would succumb to this adverse influence and die. In the hindbrain, these zones are concentrated by stage 11, in rh3 and rh5, and may eliminate a considerable proportion of the neural crest. The necrotic zones are, however, less effective in the trunk: thus only a proportion of cells can undergo cell death in the rostral half of the somite during DRG formation.

Such an explanation may be of relevence to the cell line segregation model of gangliogenesis proposed by Le Douarin (1986) which suggests that the trunk neural crest is composed of two classes of progenitor, sensory and autonomic, during its early migration from the neural tube. During gangliogenesis, the sensory precursor cells are able to survive only in the DRG, close to the CNS, where they may benefit from the presence of a growth factor. In the autonomic ganglia the sensory progenitors disappear, leaving only autonomic precursor cells. Cell death may represent a suitable mechanism for sifting between these precursor types.

## Differences between cell death in the hindbrain and the trunk in vertebrate evolution

As stated above, cell death in the trunk is associated with neural crest cells which have already migrated out from the neural tube. In contrast, cell death takes place during cranial development before the neural crest has started to migrate from the midline in regions both where the crest cells will migrate out and, in particular, where they will not. How is this seeming disparity explained? To answer this, it is helpful to consider certain aspects of the evolution of chordate development. Gans and Northcutt (1983) point out that many of the features which distinguish higher vertebrates from their protochordate ancestors concern the behaviour of the muscular hypoderm, the ectodermal placodes and the neural crest. Morphologically, the most important modifications to the evolving vertebrate head involved development of bilaterally paired sense organs necessary for predation, muscular pharynx/gill arch structures for respiration, and the facial modifications capable of sustaining a muscular jaw. It is interesting that the neural crest of higher vertebrates participates in the development of all three of these structures (see Figure 4). In terms of patterning the vertebrate head, a fundamental requirement, highlighted by the neural crest transplantation experiments of Noden (1984, 1988), is that the neural crest from a particular axial level reaches the correct pharyngeal arch. In the embryos of many vertebrate species (including the avian embryo), the neural crest migrates in a discontinuous stream in the head region, pre-otically from rh1-2 and rh4, and post-otically from rh6. This streaming is believed to help direct neural crest into the correct pharyngeal arch.

This specificity is particularly important if the neural crest does indeed possess a role as a carrier of patterning information, such as the hox-code which Hunt *et al.* (1991) have suggested. Specific Hox genes are thought to be expressed early in the development of the neuroepithelium and the cranial neural crest (see Holland *et al.*, 1988; Wilkinson and Krumlauf, 1990). This expression is maintained by the cranial neural crest throughout migration, resulting in the transfer of a hox-code to the cranial ganglia and branchial mesenchyme, reflecting the cranial neural crest's rhombomeric origin (Hunt *et al.*, 1991). Early rhombomere-specific cell death signals may thus contribute to early pattern formation in the head.

These data concerning the discrete domains of cell death, CDr and CDc adjacent to rh3 and rh5, might suggest that the initial patterning of the streams of pre- and post-otic neural crest is specified prior to the cells leaving their particular rhombomeric site of origin. Moreover, the role of cell death in this case would be to help shape the streams of neural crest cells that will migrate over the cranial mesoderm. Presumably, this system arose in evolution at a time when the neural crest was acquiring its role in facial patterning. This model predicts that such a system would not be of importance in species where there was little differentiation between the pharyngeal arches. Only the necessary phylogenetic experiment can answer this.

This situation, however, is in complete contrast to that occuring after stage 15, in the trunk. In this region, cells die along the paths of the neural crest's migration routes. These cells are thought not to possess a role in patterning their surrounding tissues. On the contrary, a number of authors have suggested that these cells are in fact patterned by their environment. Whether or not this is the case, the role of cell death cannot be interpreted to be involved in shaping mass migration patterns, since the neural crest is continuous throughout the extent of the trunk neural tube. Thus, at this level, cell death in the trunk appears to have a different phylogenetic significance to that in the hindbrain. Moreover, it may well be phylogenetically older than the discrete patterns seen in the hindbrain, because the spinal ganglia are older, in evolutionary terms, than the complex structures of the head. We would suggest that if pluripotency is an ancestral feature of the chordate neural crest, the means of its regulation may also be intrinsic (and ancestral) to this tissue.

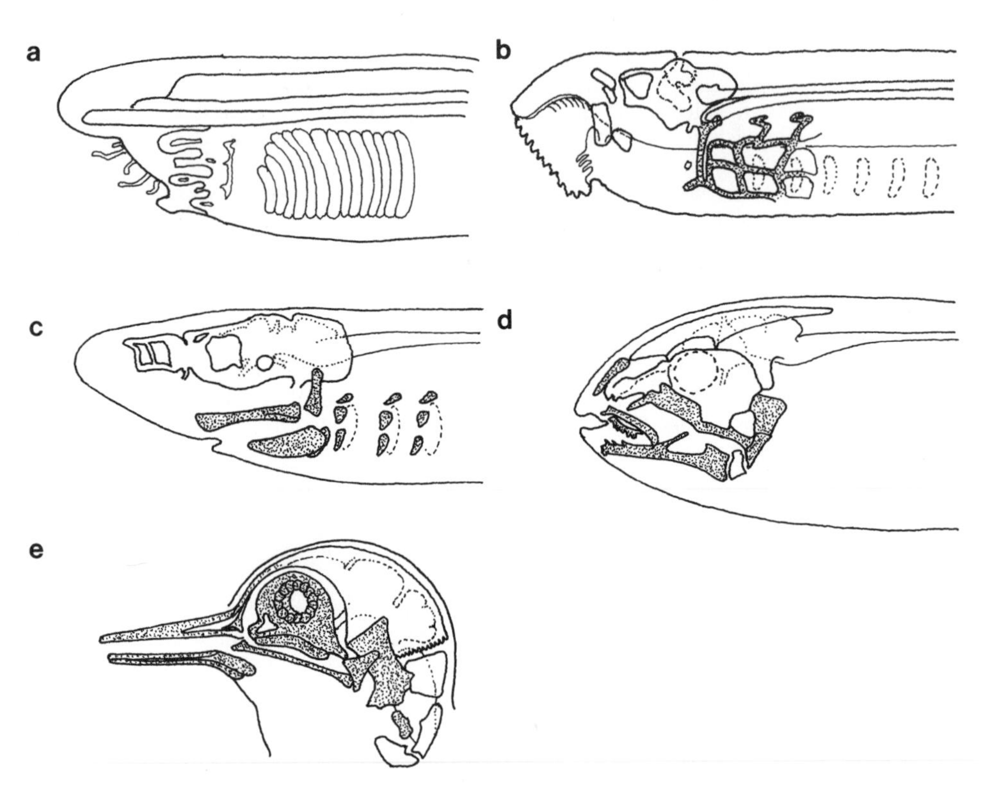


**Figure 4**. Summary of cartilagenous/bone structures which are derived from the neural crest in (a) Amphioxus, (b) Petromyzon, (c) Scylla, (d) Salmo and (e) Gallus. The figure summarises the progressively greater contributions which the neural crest makes to the development of cartilaginous and bony structures of the vertebrate head during evolution. In the higher vertebrates, increasingly complex structures are derived from the cranial neural crest and this is most clearly manifest in the complex architecture of the frontal cranial bones. Neural crest-derived structures are shaded. Figure compiled from many sources, but principally Hall (1987), de Beer (1951) and Goodrich (1930).

## Acknowledgements

We would like to acknowledge the help and support of the following people during the course of the work described here: Professor R. Bellairs, Professor N. Le Douarin, Drs. Marianne Bronner-Fraser, Karen Jaques, Roger Keynes, Claudio Stern, Tom Weaver and Rob White. We also thank John Bashford for help with photography, Jill King and Marie Watkins for histology. PSJ was supported by Action Research, and MKO by the AFRC.

## References

Anderson, C. B. and Meier, S. (1981). The influence of the metameric pattern in the mesoderm on migration of cranial neural crest cells in the chick embryo. *Devl. Biol.* **85**: 385 - 402.

Baroffio, A., Dupin, E, and Le Douarin, N. (1991). Common precursors for neural and mesectodermal derivatives in the cephalic neural crest. *Development* **112**: 301 - 305.

Bronner-Fraser, M. E. (1986). Analysis of the early stages of the trunk neural crest migration in avian embryos using monoclonal antibody HNK-1. *Devl. Biol.* **115**, 44 - 55.

Carr, V. M. and Simpson, S. B. (1978a). Proliferative and degenerative events in early development of chick dorsal root ganglia. I: Normal development. *J. Comp. Neur.* **182**: 727 - 740.

Carr, V. M. and Simpson, S. B. (1978b). Proliferative and degenerative events in early development of chick dorsal root ganglia. II: Responses to altered peripheral Fields. *J. Comp. Neur.* **182**: 741 - 756.

Couly, G. and Le Douarin, N. (1988). The fate map of the cephalic neural primordium at the presomitic to the 3-somite stage in the avian embryo. *Development* **103** (Suppl.): 101 - 113.

de Beer, G. (1951). Vertebrate Zoology. Sidgewick and Jackson, London.

Ede, D. A. and Agerbak, G. S. (1968). Cell adhesion and movement in relation to the developing limb pattern in normal and talpid$^3$ mutant chick embryos. *J. Embryol. exp. Morph.* **20**: 81 - 100.

Ede. D. A., Bellairs, R. and Bancroft, M. (1974). A scanning electron microscope study of the early limb-bud in normal and talpid$^3$ mutant chick embryos. *J. Embryol. exp. Morph.* **31**: 761 - 785.

Ede, D. A. and Flint, O. P. (1972). Patterns of cell division, cell death and chondrogenesis in cultured aggregates of normal and talpid$^3$ mutant chick limb mesenchyme cells. *J. Embryol. exp. Morph.* **27**: 245 - 260.

Fraser, S. E. and Bronner-Fraser, M. (1991). Migrating neural crest cells in the trunk of the avian embryo are multipotent. *Development* **112**: 913 - 920.

Gans C. and Northcutt R. G. (1983). Neural crest and the origin of vertebrates: a new head. *Science* **220**: 268 - 274.

Goodrich, E. S. (1930). Studies on the structure and development of vertebrates. MacMillan, London.

Hall, B. K. (1987). Tissue interactions in head development and evolution. In "Developmental and evolutionary aspects of the neural crest" (ed. P. A. Maderson), pp 215-259. John Wiley and sons, New York.

Hamburger, V. and Hamilton, H. L. (1951). A series of normal stages in the development of the chick embryo. *J. Morph.* **88**: 49 - 82.

Hinchliffe, J. R. and Thorogood, P. V. (1974). Genetic inhibition of mesenchymal cell death and the development of form and skeletal pattern in the limbs of talpid$^3$ (ta$^3$) mutant chick embryos. *J. Embryol. exp. Morph.* **31**: 747 - 760.

Holland, P. W. H. and Hogan, B. L. M. (1988). Expression of Homeobox genes during mouse development: a review. *Genes & Dev.* **2**: 773 - 782.

Hunt, P., Whiting, J., Muchamore I., Marshall, H. and Krumlauf, R. (1991). Homeobox genes and models for patterning the hindbrain and branchial arches. *Development* **1**(Suppl.): 187 - 196.

Jeffs, P. S. and Osmond, M. (1992). A segmented pattern of cell death during development of the chick embryo. *Anat. Embryol.* In press.
Jeffs, P. S., Jaques, K. and Osmond, M. (1992). Cell death and the development of the cranial neural crest. *Anat. Embryol.* In press.
Le Douarin, N. (1986). Cell line segregation during Peripheral Nervous System ontogeny. *Science.* **231**: 1515 - 1522.
Le Douarin, N. and Smith, J. (1988). Development of the peripheral nervous system from the neural crest. *Ann. Rev. Cell. Biol.* **4**: 375 - 404.
Loring, J. F. and Erickson, C. A. (1987). Neural crest migratory pathways in the trunk of the chick embryo. *Dev. Biol.* **121**: 220 - 236.
Martín-Partido, G., Alvarez, I. S., Rodríguez-Gallardo, L. and Navasscués, J. (1986). Differential staining of dead and dying embryonic cells with a simple new technique. *J. Microsc.* **142**: 101 - 106.
Meier, S. and Packard, D. S. (1984). Morphogenesis of the cranial segments and distribution of neural crest in the embryos of the snapping turtle, *Chelydra serpentina. Devl. Biol.* **102**: 309 -323.
Mills, C. L. and Bellairs, R. (1989). Mitosis and cell death in the tail of the chick embryo. *Anat. Embryol.* **180**: 301 - 308.
Newgreen, D. F., Powell, M. E. and Moser, B. (1990). Spatiotemporal changes in HNK-1/L2 glycoconjugates on avian embryo somite and neural crest cells. *Dev. Biol.* **139**:100 - 120.
Noden, A. (1984). Craniofacial development: new views on old problems. *Anat. Recd.* **208**: 1 - 13.
Noden, A. (1988). Interactions and fates of avian craniofacial mesenchyme. *Development* **107** (Suppl.): 121 - 140.
Ooi, V. E. C., Sanders, E. J. and Bellairs, R. (1986). The contribution of the primitive streak to the somites in the avian embryo. *J. Embryol. exp. Morph.* **92**: 193 - 206.
Oppenheim, R. W. (1985). Naturally occurring cell death during neuronal development. *TINS* **8**: 487 - 493.
Oppenheim, R. W. (1989). The neurotrophic theory and naturally occuring motorneuron death. *TINS* **12**: 252 - 255.
Oppenheim, R. W., Prevette, D., Tytell, M. and Homma, S. (1990). Naturally occuring and induced neuronal cell death in the chick embryo in vivo requires protein and RNA synthesis: evidence for the role of cell death genes. *Devl. Biol.* **138**: 104 - 113.
Osmond, M. K. (1989). The effects of vitamin A on the development of mesodermal tissues in the chick embryo. Thesis submitted to the University of London.
Rickmann, M., Fawcett, J. W. and Keynes, R. J. (1985). The migration of neural crest cells and the growth of motor axons through the rostral half of the chick somite. *J. Embryol. exp. Morph.* **90**: 437 -455.
Saunders, J. W., Gasseling, M. T. and Saunders, L. C. (1962). Cellular death in morphogenesis of the avian wing. *Devl. Biol.* **5**: 147 - 178.
Schoenwolf, G. C. (1981). Morphogenetic processes involved in the remodelling of the tail region of the chick embryo. *Anat. Embryol.* **162**: 183 - 197.
Serbedzija, G. N., Bronner-Fraser, M. and Fraser, S. (1989). Vital dye analysis of the timing and pathways of avian trunk neural crest migration. *Development* **106**: 806 - 816.
Stern, C. D., Artinger, K. B. and Bronner-Fraser, M. (1991). Tissue interactions affecting the migration amd differentiation of neural crest cells in the chick embryo. *Development* **113**: 207 - 216.
Sulik, K. K., Cook, C. S. and Webster, W. S. (1988). Teratogens and craniofacial malformations: relationships to cell death. *Development* **103** (Suppl.): 213 - 232.
Tosney, K. W. and Landmesser, L. T. (1985). Development of the major pathways for neurite outgrowth in the chick hindlimb. *Devl. Biol.* **109**: 193 - 214.
Tosney, K. W., Schroeter, S. and Pokrzywinski, J. A. (1988). Cell death delineates axon pathways in the hindlimb and does so independently of neurite outgrowth. *Devl. Biol.* **130**: 558 - 572.
Tucker, G. C., Aoyama, H., Lipinski, M., Turz, T. and Thiery, J. -P. (1984). Identical reactivity of monoclonal antibody HNK-1 and NC-1: conservation in vertebrates on cells derived from neural primordium and on some leucocytes. *Cell. Diffn.* **14**: 223 - 230.

Vaaga, S. (1969). The segmentation of the primitive neural tube in chick embryos (*Gallus domesticus*). Springer-Verlag, Heidelberg.
Wilkinson, D. G. and Krumlauf, R. (1990). Molecular approaches to the segmentation of the hindbrain. *TINS* **13**: 335 - 339.

# A MONOCLONAL ANTIBODY THAT REACTS WITH THE VENTRO-CAUDAL QUADRANT OF NEWLY FORMED SOMITES

James W. Lash*, Dukhee Rhee*, Joseph T. Zibrida*,
and Nancy Philp**

Departments of Cell and Developmental Biology*
and Microbiology**
School of Medicine University of Pennsylvania
Philadelphia, Pennsylvania 19104-6058
USA

## INTRODUCTION

In the late nineteenth century and continuing into the twentieth century, embryology textbooks frequently portrayed somites in a saggital plane showing a distinct morphological difference between the rostral and the caudal portions of the somite. These portrayals were most prominent in the chapter dealing with the development of the spinal cord, and the differences were associated with spinal nerves. It was not until the 1980s that these differences were examined in closer detail. One of the first instances of correlating molecular differences with nerve outgrowth was that of Stern, et al. (1986), who demonstrated a regional specification with relation to the binding of PNA (peanut agglutinin), which localized in the caudal region of the developing somites of chick embryos.

Most reports concerning rostral/caudal localization of antigens do not examine the embryological initiation of these regional differences. Do they appear at the onset of somite formation, or later? Is there any evidence of regional specification in the segmental plate of chick embryos? Do these antibodies react specifically with other tissues in the embryo? We report here the appearance of a specific epitope present in the caudal portion of the somites as soon as the somite is formed. This molecule is detected with a monoclonal antibody (F22) that was derived from 18 day chick embryo retinal pigment epithelial (RPE) cells. Its subsequent distribution and correlation with developmental events will be described.

## MATERIALS AND METHODS

Embryos. Incubated White Leghorn chicken embryos were staged (HH stages) according to the staging series of Hamburger and Hamilton (1951)

Antibodies. Female Balb/c J mice between 4 and 6 weeks old were immunized with RPE cells isolated from chick embryos of 18 days incubation (HH 44). At this age the RPE cells are both functionally and morphologically differentiated. Each immunization consisted of about 5 x $10^6$ cells. The cells were mixed with an equal volume of Freund's complete adjuvant and injected intraperitoneally. Subsequent immunizations were given at two week intervals and the cells were mixed with Freund's incomplete adjuvant. The final injection was given intraperitoneally. Four days after the final immunization the spleens were removed and the cells were fused with NS-1 myeloma cells as described by Köhler and Milstein (1976). Fused cells were plated at a density of $10^5$ cells/well in a 96-well plate. Supernatants from wells containing a single positive clone were screened by binding to lightly fixed RPE cells in an ELISA assay and on frozen sections of RPE cells from E6 and E18 chick embryos. Positive clones were subcloned by limiting dilution at least twice.

Purification of MAb F22 and coupling to Sepharose. The F22 antibody was partially purified from concentrated hybridoma supernatants by ammonium sulfate precipitation. The concentration of the antibody was determined, then the antibody was coupled to CNBr-activated Sepharose 4B (Sigma, St. Louis, MO) in 0.5 M sodium phosphate buffer, pH 7.5 (5 mg of antibody/ ml of sepharose).

SDS-PAGE and Western Blots. Proteins eluted from the F22-Sepharose were electrophoresed on microslab gels using the buffer system of Laemmli (1970). Gels were fixed and stained using a silver reagent. For Western blots the proteins separated on SDS-PAGE were electrophoretically transferred to nitrocellulose membrane (BioRad Laboratories, Richmond, CA) or Immobilon-P transfer membrane (Millipore Corporation, Bedford, MA) at 100 V for 1 hr using a Mini-Transblot apparatus (BioRad Laboratories, Richmond, CA). The blots were incubated in a buffer containing 5% BSA, 20 mM Tris, 500 mM NaCl, 0.1% Tween-20, 1 mM $CaCl_2$, 0,1 mM $MnCl_2$ and 2% azide, pH 7.5 for one hour to block nonspecific binding sites. The blots were then incubated with biotinylated lectins for 1 hour. They were washed three times with this buffer and incubated with alkaline phosphatase conjugated avidin at a dilution of 1:3000 (Sigma) for 45 minutes, followed by two more washes before developing with 0.3 mg/ml NBT (p-nitro blue tetrazolium chloride) and 0.15 mg/ml BCIP (5-bromo-4-chloro-3 indolyl phosphate p-toluidine salt) in a buffer of 0.1M $NaHCO_3$, 1 mM $MgCl_2$, pH 9.8.

Immunocytochemistry. Tissues from staged embryos were fixed for 60 min. in 2% paraformaldehyde containing 1 mM $MgCl_2$ and 3% sucrose in PBS. The tissues were washed with 3 changes of PBS with 3% sucrose, then put in PBS with 30% sucrose overnight in the cold. They were embedded in Tissue Tek II O.C.T. Compound (American Scientific Products, Edison, NJ) and frozen in liquid nitrogen. Six to eight micrometer cryostat sections were cut and mounted on gelatin coated slides. The sections were fixed for 1 minute in 2% formaldehyde generated from the polymer paraformaldehyde, 3% sucrose, and 1mM $MgCl_2$ in PBS, and non-specific binding sites on the tissue were blocked for 30 minutes in 1% goat serum in PBS. All washes and antibody dilutions were made in 1% BSA and 0.05% Tween-20 in PBS. The sections were incubated with the primary antisera for 1 hour at 37°C. After thorough washing, rhodamine conjugated goat anti-rabbit F(ab')2 fragment (Jackson Laboratories, Westchester, PA) was applied for 45 minutes at 37°C. Sections were washed, and coverslips were mounted with gelvatol (Air Products and Chemicals, Inc., Allentown, PA). The tissues were examined on a Zeiss Research microscope equipped with epi-florescence. Photographs were taken with Kodak T-Max black and white (ASA 400), or Kodak Ektochrome (ASA 400) film. The black and white film was developed with HC-110 developer (Tri X) or TMax developer (TMax). The color film was developed by Kodak.

RESULTS

The F22 antigen was purified from bovine interphotoreceptor matrix (IPM) on an antibody affinity column. The material eluted from the column with high salt was analyzed by SDS-PAGE on a 3-15% gradient gel. The eluted material migrated with a reduced molecular weight of about 250 kDa. The protein was further characterized by the binding of biotin-labeled lectins on Western blots. The protein bound wheat germ agglutinin (WGA) Figure 1.

The F22 antibody was shown by initial screening on HH 44 embryos to recognize a component of the interphotoreceptor matrix (IPM) in the retina. Futher screening on frozen sections of whole eyecups showed a specific binding of the F22 MAb to the cartilaginous plate in the sclera and to a region between the inner limiting membrane and the vitreous as well as to the IPM (Figure 2). The distribution of F22 was determined in the developing eye by indirect immunofluorescence using a rodamine conjugated secondary antibody. F22 was first localized at the vitreo-retinal junction between embryonic day 4 and 5 (HH stages 22-25). There was also a discrete staining of some cells in the peri-ocular mesen-chyme. Staining in the the interphotoreceptor space occurred at the time the inner segments of the photoreceptor cells appear between embryonic day 9-10. .

Unlike some of the monoclonal antibodies, F22 cross reacts with mammalian tissue. It has been shown by indirect immunofluorescence (data not shown) that this MAb stains the interphotoreceptor matrix in bovine (Figure 2) as well as the dog, cat and monkey (data not shown) The water insoluble interphotoreceptor matrix was isolated from the bovine retina (Figure 2A and 2C) and the F22 antigen was purified from this material.

In other regions of the embryo a most striking distribution was seen in various tissues. Most prominent was the binding of F22 in all regions of the body containing epithelial structures. In these regions the distribution was always subepithelial and extracellular. In the ectoderm the distribution was coincident with the cells contributing to the dermis (i.e. the protodermis tissue) (Figure 3).

A highly localized distribution was seen in the somites. In a HH 19

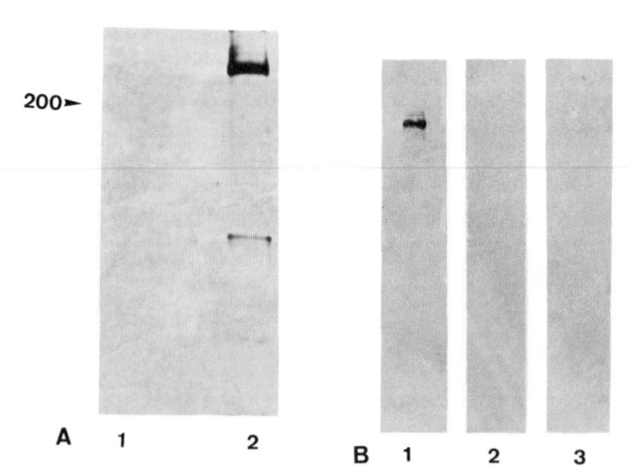


Fig. 1. The F22 antigen was purified from bovine interphotoreceptor matrix on an antibody affinity column. The material isolated was separated on a 3-15% gradient gel and silver stained (A) or western blotted (B) and probed with biotinylated wheat germ agglutinin (B1), WGA and N-acetyl glucosamine (B2) or with peanut agglutinin (B3).

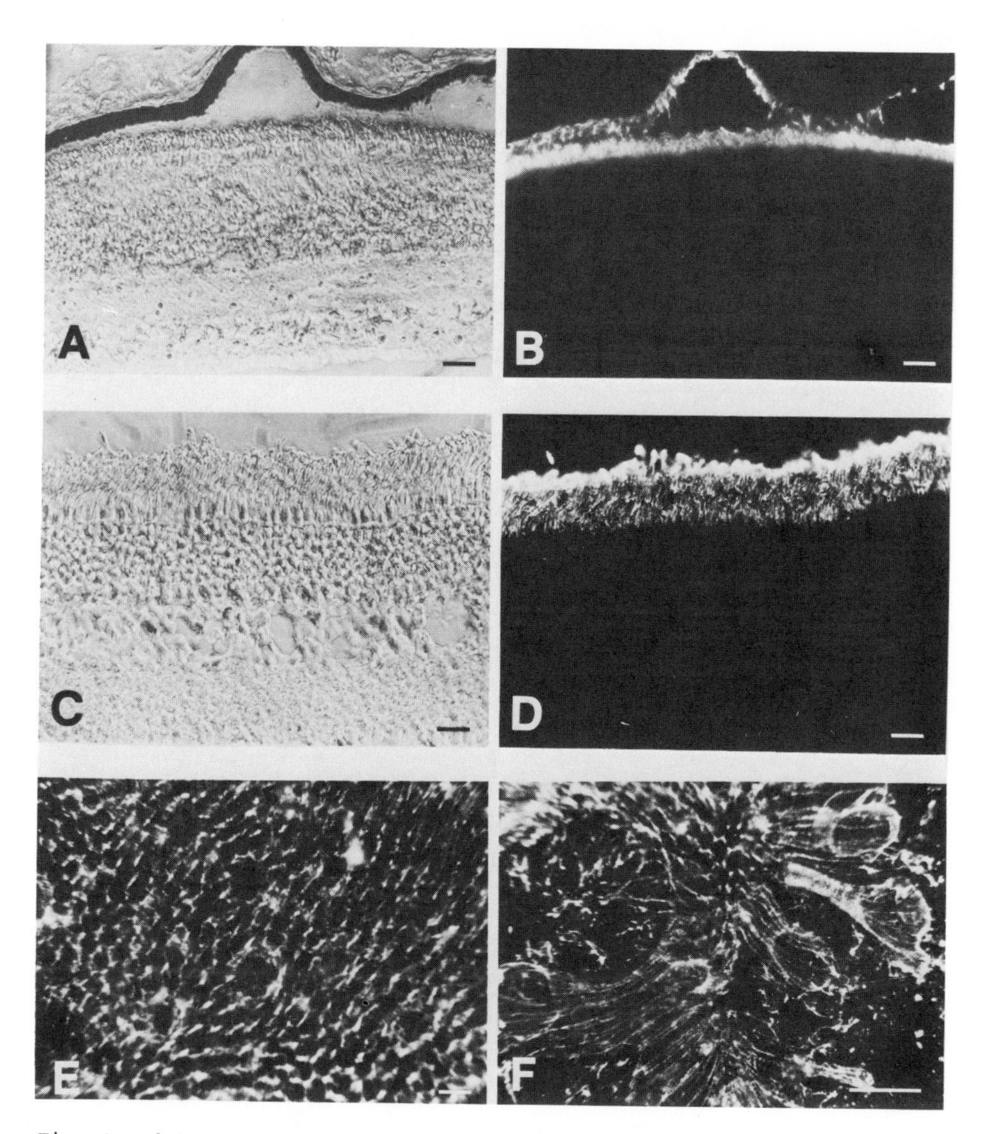


Fig. 2. Light and fluorescence micrographs showing the distribution of F22 binding in frozen sections of embryonic chicken and bovine retina. The F22 antibody shows specific binding to a region the interphotoreceptor matrix (IPM) between the photoreceptor cells and the apical from 18 day chick embryo (A and B) and bovine retina (C and D). Sheets of interphotoreceptor matrix isolated from bovine retina place in 2 mM $CaCl_2$ for 5 min at $4^{O}C$. This method results in the separation of intact sheets of interphotoreceptor matrix from the retina. Matrix isolated using this procedure binds F22 (E and F). Bar=50μm.

embryo, the antibody was found in the caudal quadrant of a newly formed somite (Figure 3). Moving in a cephalad direction in the embryo, the antibody binding spread through the somite dorsally until it exhibited its presence in the entire caudal half of the somite (Figure 4). There was never any evidence of the antibody binding to anything in the rostral half of the somite. Nor has there yet been any evidence that the F22 Mab binds to the segmental plate. Thus within the same embryo, the antigen binding the F22 Mab is seen to appear first in the ventro-caudal portion of a newly formed somite, and then moving to more rostral somites, it spreads throughout the caudal half of the somite, and is not detected in the rostral portion of the somite.

As the myotome migrates underneath the dermatome, it separates tissue reacting with the F22 antibody from non-reacting tissue, the future muscle of the myotome. This reacting tissue (presumably the protodermis) is seen beneath the ectoderm, and beneath this in the non-reacting future segmental muscles. (not shown).

In addition to this regional specificity within the somite, there was also a regional specificity in the axial tissues, the neural tube and the notochord. The notochordal sheath and the chondrogenic cells apposed to it strongly reacted with the F22 antibody. The sheath surrounding the neural tube had a discontinous antibody localization. The alar and basal regions, and the lateral margins of the neural tube displayed strong antibody localization. In the dorsolateral and ventrolateral regions, where neurones were either entering (dorsal sensory nerves) or exiting (ventral motor nerves) there was a total absence of F22 binding. Their is also an absence of binding in the region where neural tube closure is being completed. It is not yet known whether this is an artifact. These finding suggest that there is an incompatibility between regions containing F22-binding molecule(s) and the presence or growth of nerve fibers. Further evidence of this incompatibility can be seen in the lateral margins of the differentiating somites. There, nerves can be seen entering or leaving the rostral portion of the somite (i.e. where there is no F22 antibody binding), but are not seen in the caudal portion of the somite (where F22 antibody binding is markedly obvious) (Figure 4).

The distribution of F22 binding suggests that it is localized to areas that are non-innervated. Additionally, the F22 binding in the avascular scleral region of the eye, the avascular perinotochordal area, and the avascular chondrogenic areas of HH stage 24 limb buds suggests that the molecule binding the F22 Mab is an extracellular matrix molecule. The chondrogenic humerus anlagen is shown in Figure 7.

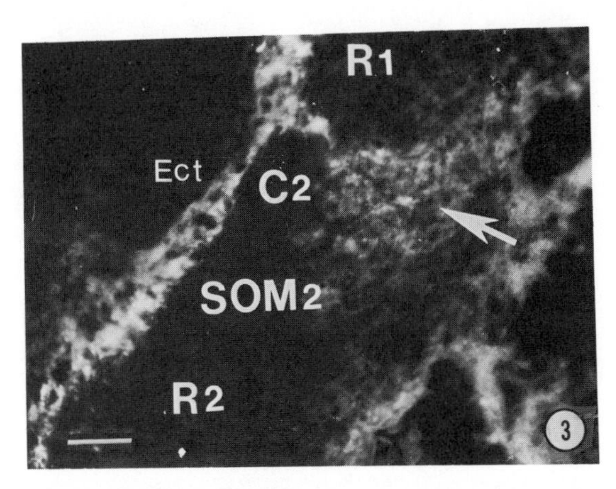


Fig. 3. Coronal section of caudal somites of a HH stage 18 embryo. R1 is the rostral portion of a newly formed somite. The segmental plate is to the right and above R1. SOM2 is the second somite in a rostral direction from the segmental plate. R2 is the rostral portion of SOM2, and C2 is the caudal portion of SOM2. The arrow points to the F22 antibody localization in the ventro-caudal portion of the SOM2. Subectodermal localization (Ect=ectoderm) of the F22 antibody can be seen to the left of the somites, and mesenchymal localization can be seen to the right of the somites. Bar=50 μm.

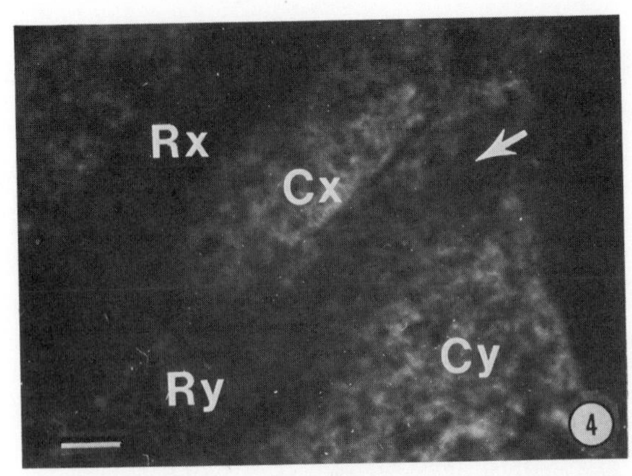


Fig. 4. Coronal section of cervical somites in a HH stage 18 embryo. Rostral (Rx, Ry) regions of somites x and y, and caudal (Cx, Cy) regions of somites x and y are designated. Somite Ry + Cy is the caudal somite. Arrow points to nerve tract in the medial portion of the somites. Bar=50 µm.

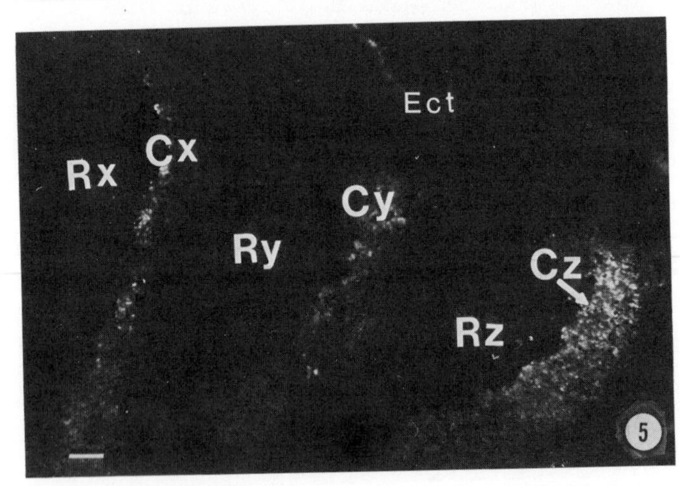


Fig. 5. Coronal section of cervical somites in a HH stage 25 embryo. Rostral (R) and caudal (C) of three somites (x, y and z) are designated. Somite x is the most rostral somite shown. Cz (arrow) is the caudal portion of the most caudal somite shown. Moving from caudal to rostral, the F22 localization diminishes within the somite, as does the subectodermal localization (Ect=ectoderm). Bar=50 µm.

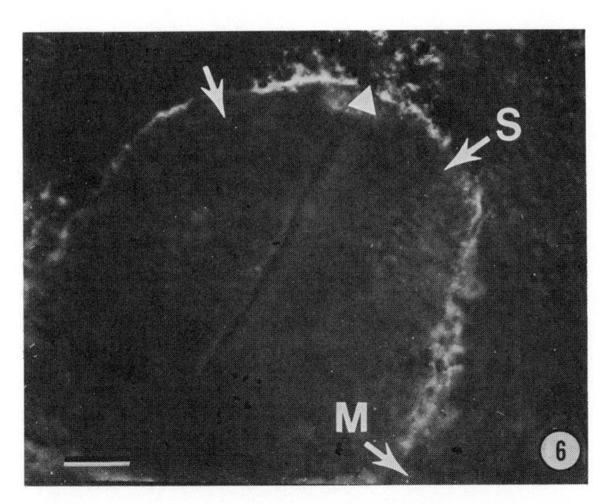


Fig. 6. Cross section in the future thoracic region of the neural tube in a HH stage 18 embryo. F22 localization is prominent in the sheath surrounding the neural tube. Localization is notably absent where sensory fibers enter the neural tube (S, arrow) and where motor fibers (M, arrow) leave the neural tube. Where the neural tube has just completed closure (white triangle) there is no localization. Bar=50 µm.

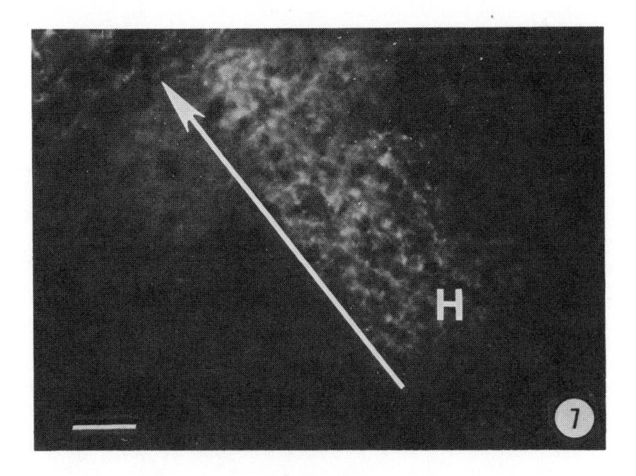


Fig. 7. F22 localization in the chondrogenic area of the forelimb of a HH stage 25 embryo. The arrow points in a distal direction of the humerus (H) anlage. The tip of the arrow is at the future joint. Bar=50 µm.

## SUMMARY AND CONCLUSIONS

Although the source of antigen for the F22 monoclonal antibody came from retinal pigmented epithelium, it has a characteristic and specific localization in other regions of young chicken embryos. As would be expected it reacts with specific regions of the developing eye. It recognizes a component of the retinal interphotoreceptor matrix (IPM). It also displays a specific binding in the cartilaginous plate of the optic sclera, and to a region between the inner limiting membrane and the vitreous. In the optic region, localization is strictly extracellular. There is no evidence of localization in cells.

The extracellular location of the F22 Mab binding is pronounced in the chondrogenic areas of the optic sclera and the limb primordia. Within the somite and the axial neural tube and notochord, localization acquires a specific regional distribution that strongly suggests a developmental role for the F22-reacting antigen. As with other reagents and antibodies (cf. Stern and Keynes, 1986; Norris, et al., 1989; Davies, et al., 1990; Ranscht, and Bronner-Fraser, 1991; Tan, et al., 1987; *inter alia*) there is a growing body of information indicating that there are significant differences between the rostral and the caudal portions of developing somites. Some of these differences are directly correlated with the emigration of neural crest cells or axons (Davies, et al., 1990; Ranscht and Bronner-Fraser, 1991; Cook and Keynes, 1992). The presence of some of these agents appear to inhibit the migration of both neurones and neural crest cells. They thus appear to serve a regulatory function in both the early diversification of the somite, and the subsequent determination of peripheral neuronal pathways. In the early diversification of the somite, the myotomal migration and growth separates the lateral F22-reacting somitic tissue from the more medial tissue. This lateral tissue migrates to the ectoderm and becomes the dermis. Interestingly, Fichard, et al. (1991) have shown that a chondroitin sulfate proteoglycan in the subectodermal region prevents dorsal root ganglia fibers from entering the area. Oakley and Tosney (1991) have shown that chondroitin-6 sulfate is present in tissues that appear to act as barriers to axonal outgrowth. The pattern of barrier tissue, as shown by Oakley and Tosney (loc. cit.) is similar to that demonstrated by the F22 antibody. Previously Cook et al. (1988) had demonstrated the localization of a sulphated polysaccharide in the developing dermatome. Current data is however too scanty to determine whether these similarities indicate a commonality to the molecules given putative regulative qualities for peripheral neuronal outgrowth.

Nevertheless, these new data may yield interesting insights into the regulation of optic cup differentiation, somite diversification, and neuronal outgrowth. How these three major morphogenetic phenomena are related is yet to be determined.

## REFERENCES


Arey, L. B. 1934 "Developmental Anatomy", 593 pp, Saunders, Philadelphia.

Bronner-Fraser, M. and C. Stern 1991 Effects of mesodermal tissues on avian neural crest cell migration. Dev. Biol. 143: 213-217.

Cook, G. M. W., R. J. Keynes, M. Chatterjee, L. Cousens and R. Bellairs 1988 A chick embryo lectin activity towards sulphated\ poly-saccharides. In "Lectins, Biology, Biochemistry, Clinical Biochemistry", vol. 6 (Ed. D. Freed and T. C. Bog-Hansen) Sigma Library.

Cook, G. M. W. and R. J. Keynes 1992 Relations between mesodermal and neural segmentation (this volume)

Davies, D. J., G. M. W. Cook, C. D. Stern and R. J. Keynes 1990 Isolation from chick somites of a glycoprotein fraction that causes collapse of dorsal root ganglion growth cones. Neuron 4: 11-20.

Fichard, A., J-M. Verna, J. Olivares and R. Saxod 1991 Involvement of a chondroitin sulfate proteoglycan in the avoidance of chick epidermis by dorsal root ganglia fibers: A study using b-D-xyloside. Dev. Biol. 148: 1-9.

Jeffs, P. and M. Osmond 1992 Segmentally regulated patterns of cell death during the development of the chick embryo. (this volume)

Keynes, R. J. and C. D. Stern 1988 Mechanisms of vertebrate segmentation. Development 103: 413-429.

Köhler, G. and C. Milstein 1976 Derivation of specific antibody producing tissue culture and tumor cell lines by cell fusion. Eur. J. Immunol. 6: 511-516.

Mackie, E. J., R. P. Tucker, W. Halfter, R. Chiquet-Ehrismann and H. H. Epperlein 1988 The distribution of tenascin coincides with pathways of neural crest cell migration. Development 102: 237-250.

Norris, W. E., C. D. Stern and R. J. Keynes 1989 Molecular differences between the rostral and caudal halves of the sclerotome in the chick embryo. Development 105: 541-548.

Oakley, R. A. and K. W. Tosney 1991 Peanut agglutinin and chondroitin-6-sulfate are molecular markers for tissues that act as barriers to axon advance in the avian embryo. Dev. Biol. 147: 187-206.

Ranscht, B. and M. Bronner-Fraser 1991 T-cadherin expression alternates with migrating neural crest cells in the trunk of the avian embryo. Development 111: 15-22.

Stern, C. D. and R. J. Keynes 1986 Cell lineage and the formation and maintenance of half somites. In "Somites in Developing Embryos (ed. R. Bellairs, D. A. Ede and J. W. Lash), pp. 147-159. New York: Plenum Press.

Stern, C. D., S. M. Sisodiya and R. J. Keynes 1986 Interactions between neurites and somite cells: inhibition and stimulation of nerve growth in the chick embryo. J. Embryol. exp. Morph. 91: 209-226.

Tan, S.-S., K. L. Crossin, S. Hoffman and G. M. Edelman 1987 Asymmetric expression in somites of cytotactin and its proteoglycan ligand is correlated with neural crest cells. Proc. Nat. Acad Sci. U.S.A. 84: 7977-7981.

# RELATIONS BETWEEN MESODERMAL AND NEURAL SEGMENTATION

Geoffrey M. W. Cook and Roger J. Keynes

Department of Anatomy, University of Cambridge
Downing Street, Cambridge CB2 3DY, U.K.

## INTRODUCTION

The importance of the early embryonic mesoderm for peripheral nerve segmentation was amplified some five years ago at the NATO Advanced Workshop on Somite Development (Keynes and Stern, 1986). In the course of normal development, motor and sensory axons emerge in register with the somites, and are arranged in a repeating pattern along the longitudinal axis of the embryo. Earlier this century, studies of spinal nerve development had not been particularly concerned with identifying exactly where they arise in relation to individual somites. More recently, Keynes and Stern (1984), from an examination of whole-mounted embryos stained with zinc iodide - osmium tetroxide, drew attention to the fact that the developing axons cross only the anterior half-sclerotome. This not only confirmed the descriptions of the last century but was important because, by means of a series of surgical manipulations performed on the developing embryo (reviewed by Keynes and Stern, 1986), it was shown that peripheral nerve segmentation results from differences between anterior and posterior half-sclerotome cells. Keynes and Stern (1986) could only speculate, however, on what these differences are likely to be in molecular terms. The purpose of the present contribution is to survey what is now known about molecular differences existing between the two populations of cells that may explain why axons are confined to the anterior half of the somite.

## SPINAL NERVE SEGMENTATION

Does spinal nerve segmentation result from the presence of permissive cues for axon growth in the anterior half-somite, or non-permissive cues in the posterior half, or the operation of both such cues acting in consort? It is now becoming increasingly evident that in certain systems inhibitory interactions play a crucial role in mediating axonal guidance. In fact, one of the earliest hints that patterns of innervation could be controlled by an inhibitory

*Formation and Differentiation of Early Embryonic Mesoderm*
Edited by R. Bellairs *et al.*, Plenum Press, New York, 1992

mechanism came from studies on spinal nerve segmentation. In 1911, Burrows found that when chick embryo neural tubes were grown *in vitro* with attached somites, axons emerged in their normal segmented pattern; in the absence of somites, however, axons emerged along the length of the neural tube. More recently, Tosney (1988) has shown that the deletion of somite tissue results in a continuous row of emerging axons at the operation site while the rest of the embryo shows a normal segmental axonal pattern. These results, together with the demonstration that surgically-constructed "compound anterior somites" (consisting only of cells of the anterior half-somite) permit axons to emerge throughout their length, whereas compound posterior somites are devoid of axons (Stern and Keynes, 1987), strongly suggest that there is an inhibitory activity associated with the posterior half-somite.

## Differences in Glycosylation between Anterior and Posterior Sclerotome

In an attempt to find a molecular difference between the cells of the anterior and posterior half-sclerotome, Stern et al. (1986) examined the staining patterns of chick somites treated with plant lectins as cytochemical reagents. Interestingly, peanut agglutinin (PNA) was found to bind only the cells of the posterior half-sclerotome. The fact that peanuts contain lectin activity has been known for some years (Bird, 1964). Subsequently, several purification schemes have been described, and its specificity for Gal ß1-3 Gal NAc residues has been determined (Lotan et al., 1975).

Of particular interest was the inference (Stern et al., 1986) that, *in vitro*, axons avoid the surface of posterior half-sclerotome cells except in regions subsequently shown to be free of patches of PNA-binding material. Davies et al. (1990) extended these studies using the lectin, jacalin. This plant agglutinin has a similar carbohydrate specificity to PNA, with the exception that binding is not inhibited by sialylation of the disaccharide Gal ß1-3 GalNAc. That Davies et al. (1990) were able to show that, in common with PNA, only posterior half-sclerotome cells reacted with jacalin was important in demonstrating that the differences between the two halves of the somite are not to be found in differential sialylation. This finding also suggested that an attempt to isolate PNA-binding glycoproteins from the somite might help to unravel molecules that are relevant to understanding the inhibitory nature of the posterior half-sclerotome.

## The Assay of Neurite-Inhibitory Molecules

Using the well established technique of lectin affinity chromatography, Davies et al. (1990) isolated a PNA-binding glycoprotein fraction from chick somites. Only relatively small amounts of biological material were available, so it was important to increase the sensitivity of the detection of these glycoproteins, especially as the bulk of such material would be required for testing their biological properties. To this end a small number of somites were metabolically labelled *in vitro* with $^{3}$H-galactose, under conditions which were shown to give optimal labelling of the carbohydrate groups. Radiolabelled somites were then added to the bulk of somites (which had been solubilized in detergent) and subjected to fractionation by lectin affinity chromatography. Following the removal of non-specifically bound radiolabel, specifically-bound radiolabel was eluted by lactose (acting as a competing saccharide), the column fractions being monitored by liquid scintillation counting. A clear peak of lactose-elutable material was detected, and by SDS-PAGE was found to consist of a number of silver-staining protein bands, particularly two components of $M_r$ 48 and 55K. These same bands were found to be missing from manually-dissected anterior half-sclerotomes but were clearly present in posterior half-sclerotome as shown by SDS-PAGE fractionation. The fact that these differences are readily detectable by one dimensional SDS-PAGE would suggest that these two polypeptides are major components of the posterior sclerotome.

Initial attempts to see whether this material might possess any biological activity involved measuring the extent of growth of axons from sensory ganglia cultured on substrates to which test molecules had been adsorbed. In this assay nitrocellulose membranes were coated with bovine serum albumin (BSA) alone or with BSA combined with the glycoprotein fraction obtained by fractionating somite material on immobilized PNA. The median outgrowth on the former substrate was found to be some three-fold greater than on the latter. In this assay it was important to show that the median extent of axon outgrowth was related only to the substrate and not to the size of explanted pieces of ganglia. By explanting pieces of ganglia of various diameters it was established that no obvious correlation existed between ganglion size and the resulting outgrowth, and that the observed inhibition of axon extension reflected the molecular composition of the substrate incorporating the isolated glycoprotein.

At the time that this assay was being examined, Raper and Kapfhammer (1990) were perfecting an elegant growth cone collapse assay for monitoring the presence of neurite inhibitory molecules. Kapfhammer and Raper (1987) had previously observed *in vitro* that when growth cones derived from peripheral neurons make contact with axons from central neurons they collapse, that is, the spread morphology of the growth cones undergoes a dramatic change to a retracted state. The same collapse phenomenon occurs when growth cones from central neurons make contact with peripheral axons. These authors have suggested that neurites have surface membrane-associated labels that initiate growth cone collapse. They further point out that the collapse phenomenon is analogous to the contact inhibition of motility seen between mobile non-neuronal cells in culture. In their assay Raper and Kapfhammer (1990) arranged for neurites growing on planar substrates such as laminin to be challenged with test materials, or with liposomes incorporating them. The reaction of the growth cones was then observed over various periods of time. In these experiments, growing peripheral neurites were confronted with a mixture of molecules obtained from detergent solubized embryonic chicken brain. Within moments of adding liposomes containing these molecules the growth cones collapsed; however, on removing the liposomes the reaction was reversed, showing that collapse was not a result of cellular damage. This assay has been applied to an examination of the glycoprotein fraction obtained from the somite mesoderm by lectin affinity chromatography (Davies et al., 1990) and further details of the experimental strategy used with the somite-derived material have been published (Cook et al., 1991). Briefly, these materials were incorporated into liposomes by dialysis from detergent-solubilized somites and added to cultures of dorsal root ganglia growing on laminin; after one hour the coded cultures were fixed with 4% formaldehyde in phosphate buffered saline containing 10% sucrose. The morphology of the tips of extending neurites was examined by phase-contrast microscropy and the percentage of growth cones that had undergone collapse was then determined (Fig. 1).

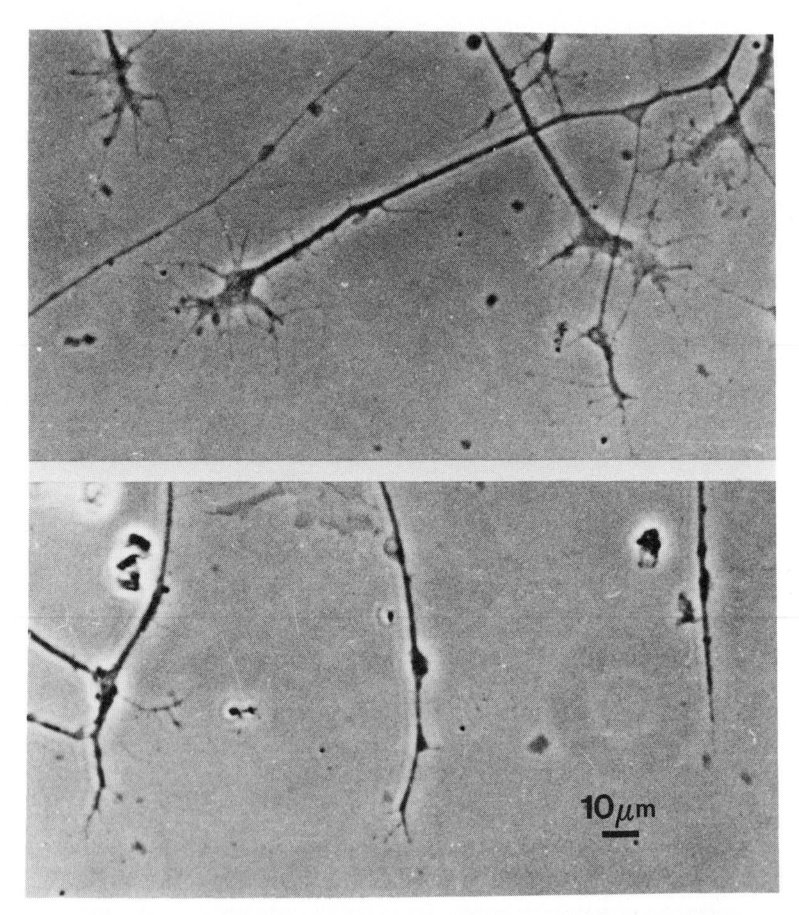


***Fig. 1.*** *Growth Cone Collapse. (Top) A phase contrast micrograph of chick sensory neurons explanted on laminin. Growth cones have a characteristic "spread" morphology. (Bottom) Culture treated for one hour with liposomes incorporating the glycoprotein fraction prepared by PNA affinity chromatography from chick embryo somites. This treatment causes growth cones to collapse. Taken from Davies et al. (1990).*

Though the bulk of such experiments were conducted on cultures growing on laminin, the potent collapse-inducing activity of somite-derived extracts was also detectable using polylysine and Type 1 collagen. With the somite system, the collapse phenomenon is most marked within the first 20 minutes of adding the liposomes and the reaction has reached a plateau within one hour; on removing the liposomes a return to a spread morphology takes place within 4 hours. As with the embryonic brain system described by Raper and Kapfhammer (1990) it appears that the phenomenon is not dependent upon the release of some cellular component that has a general toxic effect. It is reasonable to postulate that the glycoproteins present in the liposomes are responsible for initiating a specific response via receptors located on the growth cone.

The question arises as to the relationship of these inhibitory PNA-binding glycoproteins to the polarity of the somite. The results from the manual dissection of anterior and posterior somite halves and their fractionation by SDS-PAGE provides *prima facie* evidence for the possibility that these inhibitory molecules are confined to the posterior half-sclerotome, from which axons are excluded. Polyclonal antibodies, raised against the 48K and 55K bands and immobilized on agarose beads, are found to remove all the collapse-inducing activity from detergent-solubilized somites. Of particular significance is the finding that this antiserum stains only the posterior half-sclerotome (Fig. 2).

## Structure-Function Relationships

A further question is whether the carbohydrate or polypeptide of the inhibitory glycoproteins is responsible for inducing collapse of the growth cones. Whilst the physiological functions of the carbohydrate groups of most vertebrate glycoproteins remain unknown, there is a working hypothesis that in view of the structural diversity possible in oligosaccharides, glycoconjugates at the cell surface provide the cell with an efficient recognition system (Cook, 1986). Considerable attention has been given to isolating and characterizing lectins from animal cells and, more recently, a new class of adhesion molecules, the selectins or LEC-CAMs, which possess an N-terminal C-type lectin domain, have come to the forefront (Siegelman, 1991). A number of reports have identified the antigen sialyl-$Le^x$ as a minimal carbohydrate structure that binds to this domain. In addition to carbohydrate acting as a ligand or receptor for appropriate lectins, sugar moieties are able to modulate the activity of the protein which carries them; a well known example of this is provided by N-CAM, which is less adhesive when bearing long chains of polysialic acid, as in the embryonic form of this adhesive protein, than it is when shorter sialosyl chains are borne, as in adults (Rutishauser et al., 1985). Carbohydrate groups can also be important for maintaining the gross conformation of a protein, as for example in mucins, where polypeptides are forced into linear strands as a result of these groups (Hill et al., 1977; Rose et al., 1983).

Several approaches are available for investigating the contribution that the carbohydrate of glycoconjugates may make towards their biological properties. Glycosylation of

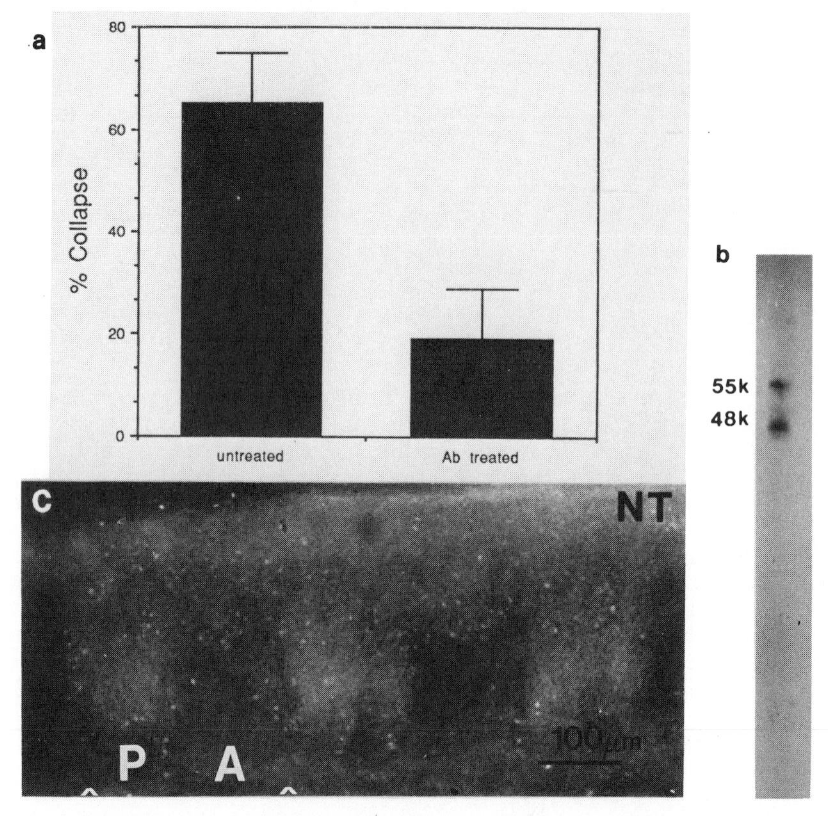


*Fig. 2. Antibodies to the 48K and 55K components from PNA affinity chromatography remove collapse activity. a) CHAPS extracts of stage 17-19 chick trunks, when incubated with immobilized preimmune IgG, retain considerable collapse activity. However, immobilized IgG containing antibodies to the 48K and 55K glycoproteins reduces the collapse activity to control levels. b) A Western blot of trunk proteins, separated by SDS PAGE and incubated with affinity-purified rabbit antibody to the 48K and 55K components. Binding was detected by autoradiography of blots treated with $^{125}I$ protein A. No bands are seen with preimmune serum. c) Frozen saggittal section of stage 19 chick embryo trunk stained with affinity-purified antibody and FITC-conjugated anti-rabbit IgG. Staining is confined to the posterior (P) half-sclerotome, with some staining in the neural tube (NT). A, anterior half-sclerotome; arrowheads denote segment boundaries. Taken from Davies et al. (1990).*

polypeptides can be inhibited *in vivo* by such compounds as tunicamycin, which inhibits N-glycosylation. Enzymatic or chemical deglycosylation may provide an alternative approach. Clearly, the administration of compounds such as tunicamycin to a developing embryo would affect the synthesis of a whole range of glycoproteins in both neural and non-neural cells, marking impossible the identification of the modified glycoprotein that may be responsible for any resultant changes in axon guidance. A more appropriate strategy in the present situation would be to deglycosylate specifically the isolated biologically active molecules, followed by direct measurement of the degree to which the resulting change affects their collapse-inducing properties. It is this approach that we have taken with the glycoproteins isolated from chick somites.

Various purified exoglycosidases, which remove specific sugars from the non-reducing termini of carbohydrate groups, and endoglycosidases, which catalyse the hydrolysis of glycosidic bonds within an oligosaccharide, are now available commercially. To date we have examined the action of the exoglycosidase, neuraminidase (EC 3.2.1.18), which catalyses the release of sialic acid residues, and the endoglycosidases O-glycanase (endo-α-N-acetyl-D-glycosidase: EC 3.2.1.97) and N-glycanase (N -glycosidase F: EC 3.5.1.52), on the collapse-inducing glycoprotein fraction of somites. The enzyme O-glycanase is responsible for the release of the PNA-binding dissacharide Gal ß1-3 GalNAc from serine and threonine residues, provided it is not sialylated (Umemoto et al., 1976). N-glycanase catalyses the hydrolysis of the N-glycosidic linkage between N-acetyl glucosamine and asparagine, and is active on biantennary, triantennary and tetraantennary structures (Tarentino et al., 1985).

Routinely, material isolated from 30 stage 17-19 chick embryo trunks was tested with each glycosidase or with a combination of enzymes. The effects of neuraminidase (50 mU), O-glycanase (5 mU) or N-glycanase (3.1 mU) were examined separately, together with a combination of neuraminidase and O-glycanase. As there have been reports that endoglycosidase may be contaminated with traces of protease, the above incubations were performed in the presence of the metalloprotease inhibitor, 1,10 phenanthroline monohydrate (quantities up to 1.3 mM were tested without any effect on the morphology of the growth cones). Interestingly, treating inhibitory glycoprotein with neuraminidase, O-glycanase and N-glycanase had no significant effect on the ability of this material to induce collapse. However, treating the isolated somite glycoprotein fraction with neuraminidase and O-glycanase eliminates the collapse-inducing activity (Davies, Wajed, Howells, Keynes and Cook, unpublished observations). Neuraminidase treatment might be expected to desialylate any substituted Gal ß1-3 GalNAc core structures, thus making them susceptible to attack by O-glycanase. As this active material binds to PNA it is probable that a significant number of these core structures are already unsubstituted, though interestingly O-glycanase alone has no effect on biological activity.

Whether the effects of glycosidase treatment indicate that carbohydrate is responsible for the collapse-inducing activity awaits further investigation. It is quite possible that a significant degree of deglycosylation, as a result of the combined action of neuraminidase and O-glycanase, might induce steric changes in the isolated glycoprotein with a resulting loss in biological activity. The ganglioside GM1 also carries Gal ß1-3 GalNAc residues, but when this glycolipid, or asialoGM1, is incorporated into liposomes and tested in the growth cone collapse assay, no activity is found (Davies, Keynes and Cook, unpublished observations). At first sight, this might suggest that the carbohydrate groups of the active material isolated from somites are not involved in initiating the collapse reaction. However, the disaccharide Gal ß1-3 GalNAc is likely to be bound to protein via an α-linkage, whilst in GM1 this residue is ß-linked. This structural difference could well affect recognition of the saccharide by a putative carbohydrate-binding protein on the growth cone. An example of how such structural changes can affect binding activity is provided by the exquisite specificity of the plant lectin, jacalin, which binds to Gal ß1-3 GalNAc residues when they are α-linked to the molecule which carries them with 2,000-fold greater affinity than when they are linked in the ß configuration (Surolia et al., 1989). Alternatively, the apparent loss of collapse-inducing activity could result from an impaired uptake of the deglycosylated glycoproteins into the lipid vesicles. To test this hypothesis directly we have radiolabelled the polypeptide portion of the somite glycoprotein fraction with $^{35}$S-methionine and shown that 83% of this labelled material is incorporated into liposomes, and that treatment with neuraminidase and O-glycanase has no effect on this level of incorporation (Howells, Keynes and Cook, unpublished observations). Definitive evidence as to the extent to which carbohydrate groups are involved in the collapse-inducing phenomenon must await further experimentation. Clearly, the isolation of appropriate carbohydrate groups and their linkage to suitable carriers to form a neoglycolipid which could be incorporated into liposomes would provide one route for elucidating the role of sugars in growth cone collapse.

## PNA as a Molecular Marker for Tissues that Act as Barriers to Axon Advance

Oakley and Tosney (1991) have recently examined three tissues that act as barriers for axon advance in avian embryos, namely posterior half-sclerotome, perinotochordal mesenchyme and pelvic girdle precursor mesenchyme. In using the term "barrier" these authors intend it to apply to those areas that have been shown experimentally to provide axon guidance cues that result in growth cone avoidance behaviour. They do not intend that the term should imply a particular mechanism, or that the tissue so identified is necessarily completely impervious to axonal invasion. Interestingly, they find that all three barrier tissues preferentially express both PNA binding sites and chondroitin-6-sulphate immunoreactivity at the time that growth cones avoid these tissues, and that these epitopes are not detected along axon pathways. In the case of the sclerotome they point out that differences in gross structural organization of anterior and posterior cell populations cannot account for the functional differences detected by nerve cells, and that neither scanning nor transmission electron microscopy reveal aligned elements that could channel axons in the anterior half-sclerotome. Oakley and Tosney (1991) are persuaded that functional differences between anterior and posterior half-sclerotome are likely to be related to the differential expression of molecules that can be detected by growth cones. They also note that preferential PNA binding occurs in the posterior half of each epithelial somite, before axon outgrowth and before the onset of neural crest cell migration. This finding extends the results of lectin staining reported by our group (Stern et al., 1986; Davies et al., 1990), which had not detected PNA binding during the earliest stages of somite development. The apparent differences in lectin staining may be attributed to two factors, namely the inclusion of 0.5% cetylpyridinium chloride in the fixative, which markedly enhances the preservation and detection of PNA-binding sites, and also to the use of antibodies to the lectin; in our work peroxidase- or fluorescein-conjugated agglutinins were used in a direct staining procedure. In addition, by making unilateral neural tube deletions, Oakley and Tosney (1991) were able to show that the development of PNA-negative axon pathways is independent of axon outgrowth. This finding rules out the possibility that the lectin staining pattern might result from a modification of the cellular environment by the axons, for example by proteases released from the growth cones. While Oakley and Tosney (1991) affirm that PNA-binding molecules act as mediators of barrier function they also point out that the particular PNA-binding sugar moieties carried by these proteins are not necessarily implicated, as binding sites for this lectin are absent in the posterior half-sclerotome of the quail. They rightly point out that it is currently unknown if differentially glycosylated forms of the glycoproteins identified by Davies et al. (1990) in chick are expressed in the quail posterior half-sclerotome. The studies described here on the effect of glycosidases on collapse-inducing glycoproteins are relevant to this latter consideration: that O-glycanase treatment alone has no effect on the growth cone collapse-inducing properties of the somite glycoprotein fraction is in full agreement with the view that the PNA-binding disaccharide Gal ß1-3 GalNAc is not essential for barrier function.

## CONCLUSIONS AND FUTURE APPROACHES

It is now possible to implicate a sclerotome glycoprotein fraction as playing a crucial role in spinal nerve segmentation. Because of this material, no-go areas are repeated along the length of the embryonic axis and force developing spinal nerves to traverse only the anterior half-sclerotomes. Fine molecular details still have to be elucidated. What, for example, is the relation of 48K material to the 55K band? Are they both subunits of the native glycoprotein responsible for growth cone collapse? Are PNA-binding sites present on both components? Whilst these sugar residues may not be directly involved in initiating collapse the role of other carbohydrate residues has still be elucidated. It can be confidently predicted, however, that the PNA-binding glycoprotein fraction of the somite will prove to be an important element in any explanation of the relationship between mesodermal and neural segmentation.

ACKNOWLEDGEMENTS

This work is supported by grants from the Wellcome Trust, the Medical Research Council and Action Research for the Crippled Child. G.M.W.C. is a Member of the M.R.C. External Scientific Staff.

REFERENCES

Bird, G. W. G., 1964, "Anti-T in peanuts". Vox Sang. 9: 748-749.

Burrows, M. T., 1911, The growth of tissue of the chick embryo outside the animal body with special reference to the nervous system, J. Exp. Zool. 115: 44-55.

Cook, G. M. W., 1986, Cell surface carbohydrates: molecules in search of a function? J. Cell Sci. Suppl. 4: 45-70.

Cook, G. M. W., Davies, J. A. and Keynes, R. J., 1991, Growth cone collapse: a simple assay for monitoring cell-cell repulsion, in: Cell Signalling: Experimental strategies, E. Reid, G. M. W. Cook and J. P. Luzio, eds. pp 359-366. Royal Society of Chemistry, Cambridge.

Davies, J. A., Cook, G. M. W., Stern, C. D. and Keynes, R.J., 1990, Isolation from chick somites of a glycoprotein fraction that causes collapse of dorsal root ganglion growth cones, Neuron 4: 11-20.

Hill, H. D., Reynolds, J. A. and Hill, R. L., 1977, Purification, composition, molecular weight, and subunit structure of ovine submaxillary mucin, J. Biol. Chem. 252: 3791-3798.

Kapfhammer, J. P. and Raper, J. A., 1987, Interactions between growth cones and neurites from different neuronal tissues in culture, J. Neurosci. 7: 1595-1600.

Keynes, R. J. and Stern, C. D., 1984, Segmentation in the vertebrate nervous system, Nature 310: 786-789.

Keynes, R. J. and Stern, C. D., 1986, Somites and neural development, in: Somites in Developing Embryos, R. Bellairs, D.A. Ede and J.W. Lash, eds. pp 289-299, Plenum, New York.

Lotan, R., Skutelsky, E., Danon, D. and Sharon, N., 1975, The purification, composition and specificity of the anti-T lectin from peanut (Arachis hypogaea), J. Biol. Chem. 250: 8518-8523.

Oakley, R. A. and Tosney, K. W., 1991, Peanut agglutinin and chondroitin-6-sulfate are molecular markers for tissues that act as barriers to axon advance in the avian embryo, Dev. Biol. 147: 187-206.

Raper, J. A. and Kapfhammer, J. P., 1990, The enrichment of a neuronal growth cone collapsing activity from embryonic chick brain, Neuron 4: 21-29.

Rose, M. C., Voter, W. A., Sage, H., Brown, C. F. and Kaufman, B., 1983, Effects of deglycosylation on the architecture of ovine submaxillary mucin glycoprotein, J. Biol.Chem. 259: 3167-3172.

Rutishauser, U., Watanabe, M., Silver, J., Troy, F. A. and Vimr, E.R., 1985, Specific alteration of N CAM-mediated cell adhesion by an endoneuraminidase, J. Cell Biol. 101: 1842-1849.

Siegelman, M., 1991, Sweetening the selectin pot, Current Biology 1: 125-128.

Stern, C.D., Sisodiya, S.M. and Keynes, R.J., 1986, Interactions between neurites and somite cells: inhibition and stimulation of nerve growth in the chick embyro, J. Embryol exp. Morph. 91: 209-226.

Stern, C.D. and Keynes, R.J., 1987, Interactions between somite cells: the formation and maintenance of segment boundaries in the chick embryo, Development, 99: 261-272.

Surolia, A., Mahanta, S.K. and Sastry, M.V.K., 1989, Thermodynamics of saccharide binding of Artocarpus integrifolia reveals its exquisite specificity for Thomsen-Friedenreich antigen, p394, Proceedings X$^{th}$ International Symposium on Glycoconjugates, Jerusalem, Israel.

Tarentino, A.L., Gomez, C.M. and Plummer, T.H., Jr., 1985, Deglycosylation of asparagine-linked glycans by peptide: N-glycosidase F, Biochemistry, 24: 4665-4671.

Tosney, K.W., 1988, Proximal tissues and patterned neurite outgrowth at the lumbosacral level of the chick embryo: partial and complete deletion of the somite, Dev. Biol. 127: 266-286.

Umemoto, J., Bhavanandan, V.P. and Davidson, E.A., 1976, Purification and properties of an endo-α-N-acetyl-D-galactosaminidase from Diplococcus pneumoniae, J. Biol. Chem. 252: 8609-8614.

# SOMITIC MESODERM: MODULATION OF CHONDROGENIC EXPRESSION BY RETINOIC ACID

Nagaswami S. Vasan

Department of Anatomy
UMDNJ-New Jersey Medical School
Newark, New Jersey 07103-2714

## INTRODUCTION

Somites are mesodermal structures that appear transiently along both sides of the neural tube in a repetitive manner. Embryologically, somites are the first segmented structures to form, and possibly these are aggregates of cells (mesodermal) in part arising from the primitive streak ("pre-somite clusters"-Bellairs, 1985). There is now extensive evidence that somitomeres in the segmental plate exist in all the vertebrate groups and that these presomitic collections of cells are arranged in a characteristic rosette-like pattern (see Jacobson and Meier, 1986 and references therein). When first formed from the unsegmented mesoderm, somites appear as epithelial vesicles consisting of one cell type (Trelstad, et al., 1967). After segmentation, the medial portion of the somite soon becomes mesenchymal. This is the sclerotome, the cartilage forming part of the somite. The rest of the somite, the dermamyotome remains a compact mass for a longer time. Subsequently, the sclerotome cells disperse and migrate medially towards the notochord and neural tube.

It has been well documented that the chondrogenic pathway of somite (sclerotome) differentiation is influenced by the inducer tissues, notochord and spinal cord (for complete references see review, Vasan, 1987). In vitro studies indicate that extracellular matrix components (collagen and proteoglycans) are involved in somite expression of the chondrogenic phenotype (Lash and Vasan, 1978; Vasan, 1987). Enzymatic removal (Kosher and Lash, 1975) or alteration of the peri-notochordal matrix (Vasan, 1981; 1983) reduces the inductive capacity of the notochord. Another study observed that, when cultured in suitable medium, somites formed cartilage-specific matrix even in the absence of inducer tissues (Vasan, 1983). Such a spontaneous cartilage formation shows that sclerotomes possess a chondrogenic bias, and that ECM of the notochord participates in the enacting of a program already set. Notochords enhance and accelerate the rate of chondrogenesis in somite explants. This conclusion agrees with an earlier study in which whole somites formed cartilage when cultured in suitable medium without notochord (Ellison and Lash, 1971).

A complete understanding of the molecular basis of interactive events would involve the chemical characterization of signal substances and, at a molecular level, of their mode of action on target, all resulting in the expression of a new phenotype. The search for the 'inducer,' which began in the early 1950's, has identified a number of tissues and substances (see Review by Hall, 1977; Vasan, 1987).

In the early 1980's it was discovered that retinoic acid (RA) induces pattern duplication in chick wing buds, and thus could be a new morphogenic factor (Tickle et al., 1982; Summerbell, 1983). Retinoic acid and its analogs are essential for normal growth and a variety of cellular functions. In excess, however, retinoids induce dedifferentiation of chondrocytes (Solursh and Meier, 1973; Vasan and Lash, 1975; Vasan, 1981; Horton et al., 1987) and are also a potent teratogen (Rosa et al., 1986). On the other hand, very low concentrations of RA promote proliferation of chondrocytes (Enomoto et al., 1990) as well as differentiation (Ide and Aono, 1988; Takishita et al., 1990).

In a continued search for the 'inducer' substances, the present study was undertaken to examine the influence of low levels of RA on the chondrogenic expression of somites. In these experiments we have utilized a medium that contain bovine fetal calf serum depleted of vitamin A. Our results indicate that RA induces cell proliferation, synthesis of sulfated glycosaminoglycans [GAG], and large cartilage specific proteoglycan aggregates. These effects were primarily on the sclerotomal portion of the somites.

## MATERIALS AND METHODS

Fertile eggs of White Leghorn chickens were incubated at 38°C to HH stages 17/18. The embryos were freed from their membranes, cut at the level of the wing bud and stripped of all external tissues. The resulting trunks consisted of notochord, spinal cord and somites. Somites and notochord used in these studies were isolated by microdissection without the aid of trypsin. Explants of either 30 random somites in a cluster or one 2mm notochord piece surrounded by 16 random somites were placed on Millipore filters which supported them at the air-medium interface. The liquid nutrient feeding medium consisted of Simms' balanced salt solution (SBSS), fetal calf serum and nutrient supplement F12X (Gibco, Grand Island, NY) in the proportions of 2:2:1 (Lash and Vasan, 1978). All cultures were maintained at 38°C in a humidified atmosphere of 95% air/5% $CO_2$ for up to 7 days. One or two drops of appropriate fresh medium was added daily. For isolation of sclerotome and dermamyotome blocks of somites stripped from the trunks were briefly trypsinized (1.25% trypsin, 60 sec at 37°C) prior to micro-dissection.

VITAMIN A DEPLETED FETAL CALF SERUM: Bovine fetal calf serum was treated with activated charcoal (Sigma Chemicals) in a 1:10 ($^{v}/_{v}$) suspension at 4°C for 24-36 hours. The suspension was then passed through a 0.22 micron filter to remove the charcoal particles and to sterilize the serum. Charcoal treatment was found to remove as much as 90% of the vitamin A from serum (Enomoto et al., 1990). Retinoic acid (all trans type XX, Sigma Chemicals) dissolved in a small volume of ethanol was added to the feeding medium. The same volume of ethanol was added to control cultures (final concentration of 0.01%).

CELL PROLIFERATION AND DNA SYNTHESIS: Filters containing explants were dropped into a test tube containing cold Ca/Mg free phosphate buffered saline and were washed three times. Single cell suspension was obtained by adding 0.25% trypsin and agitating the tubes for 10 minutes on an orbital shaker. Washed cells were then counted in a haemocytometer. At random, cell samples were judged for viability using the Trypan Blue exclusion method. For determination of DNA synthesis, cultures were labelled for 6 hours with 10μCi/ml of $^{3}$H-thymidine. Thymidine incorporation into the TCA precipitable material was determined.

DNA QUANTITATION: Samples to be analyzed were diluted in water to 50 microliters. The analysis was done in a Beckman DU50 spectrophotometer. A patented soft pack module was employed for the determination of nucleic acid by a fixed parameter analysis.

ANALYSIS OF SULFATED GLYCOSAMINOGLYCAN SYNTHESIS: Eighteen hours prior to harvest, explants were exposed to medium containing 50μCi/ml of $Na_2{}^{35}SO_4$. To extract the glycosaminoglycan, tissues were sonicated (aliquot removed for DNA analysis), digested with pronase (Calbiochem-Behring, San Diego, CA), treated with cold TCA, and the supernatant dialyzed against 0.04 M sodium sulfate and distilled water. An aliquot of known volume was counted in a Beckman Model LS-8000 Liquid scintillation counter.

PROTEOGLYCAN EXTRACTION: Proteoglycan was extracted from the tissues with 4.0 M guanidine hydrochloride containing protease inhibitors (Oegema et al., 1975). The extraction was done at 4°C for 36 hours on an orbital shaker with gentle agitation.

GEL CHROMATOGRAPHY: Prior to chromatography on a Sepharose CL-2B gel column (95cm X 0.8cm), the proteoglycan extract was dialyzed briefly against distilled water, followed by 24 hours against elution buffer (0.5 M sodium acetate buffer, pH 5.8 containing 0.02% sodium azide and protease inhibitors). The sample in 0.5ml elution buffer was applied onto the column. The eluate was collected in 1.0ml fractions and the radioactivity was measured.

RETINOIC ACID EFFECT ON SCLEROTOME CELL AGGREGATION: Sclerotome was isolated as described earlier, and cells were isolated by trypsin treatment (Vasan and Lash, 1975). Washed cells were plated in low density (i.e., 24 chamber multi well plates). Twenty-four hours after the initial plating, control cells in medium containing vitamin A depleted serum received equal volumes of ethanol, and the experimental cells received 30 ng/ml of RA in ethanol. Cells were cultured for several days, and at fixed time intervals, labelled with radioactive sulfate as described earlier. Collected cells were sonicated and aliquot removed for DNA analysis and the remaining sample processed for sulfated GAG determination. Aggregation of cells were recorded photographically.

HEPARAN SULFATE ANALYSIS: An aliquot from the above $^{35}S$-sulfate labelled GAG sample was subjected to heparitinase digestion (Sigma Chemicals). Subsequent to digestion, 100 μg of heparan sulfate was added as carrier to the sample. The undigested material was precipitated with ethnol-potassium acetate (1.3% potassium acetate in ethanol). The supernatant which contains heparan sulfate was counted. We also used control samples which were untreated with enzyme but processed similarly.

## RESULTS AND DISCUSSION

MORPHOLOGY: Somite explants cultured in the presence of 30 ng/ml of RA assumed a well rounded compact shape (Fig. 1a), and by day 6 the presence of cartilge nodules was observed. The explants cultured in medium containing vitamin A depleted serum, however, were less compact and exhibited nodules only sporadically (Fig. 1b). When explants were cultured in medium containing higher concentrations of RA (150 and 300 ng/ml) cell debris both on the filters and in the medium, probably arising from cell death was seen. These cultures also exhibited excessive tissue spreading at times even difficult to visualize (Fig. 1c). Is there a correlation between compact morphology of the explants and chondrogenic expression? Earlier it was reported that somite explants are compact and rounded when cultured in medium containing exogenous proteoglycans (Lash and Vasan, 1978). Somite explants also exhibit different morphologies depending upon the type of additives present in the nutrient medium or coated onto the millipore filters on which they were grown. At that time, it was not possible to draw a relationship between tissue compactness and synthetic activities with respect to chondrogenesis (Lash and Vasan, 1978). As seen in earlier results (Lash and Vasan, 1978) the RA-treated explants, observed as compact somite clusters showed enhanced proliferation, survival rate, and GAG synthesis (Figs. 2-5, Table 1). Hence, to answer the above question, somite tissue compactness is indirectly related to chondrogenic expression via cell proliferation, and to explant survival, resulting in an abundance of matrix synthesis.

EFFECT OF RA ON CELL PROLIFERATION AND DNA SYNTHESIS: The cell number and the DNA content in the explants and their [$^{3}H$]-thymidine incorporation were measured. The results indicate that RA increased the DNA content of the somite explants in a dose-dependent manner (Fig. 2), the DNA content markedly increasing at 30 ng/ml. It was reported that RA at low concentration increased the DNA content in cultured chondrocytes (Takishita, et al., 1990), and that removal of RA on the fourth day of culture decreased DNA accumulation significantly compared to cultures treated continuously for 7

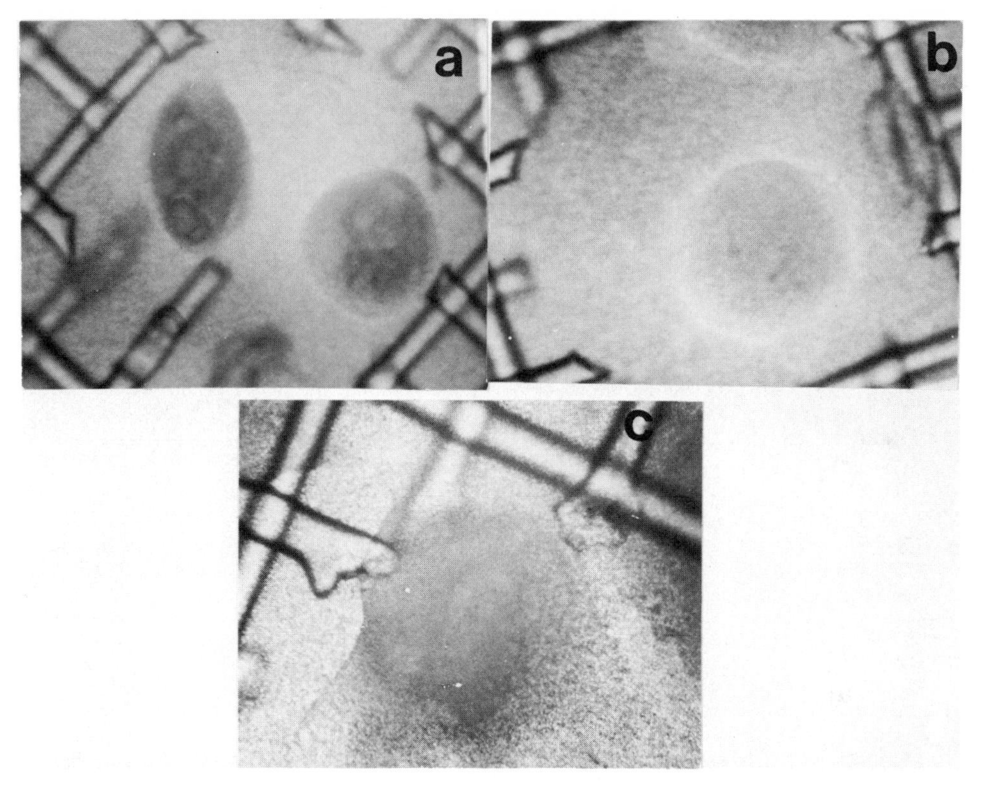


**Figure 1.** Morphology of somite explants at the end of 6 days in culture in the presence of 30ng/ml RA (A), medium containing vitamin A depleted serum (B), and 150-300ng/ml RA (C).

days. At higher concentrations (150-300 ng/ml) the DNA content of the somite explants dramatically decreased. This is probably due to inhibition of DNA synthesis as well as to increased cell death. Takishita, et al (1990) also made a similar observation of the effect of high RA level on cultured chondrocytes. Earlier studies from our laboratory also showed that vitamin A at higher concentrations decreased the DNA content and cell number of cultured sternal chondrocytes (Vasan and Lash, 1975; Vasan, 1981). Reduction in growth, malformation, and fetal death are commonly associated with higher (teratogenic) levels of vitamin A (Rosa, et al., 1986. When somite explants were cultured in the continued presence of 30ng/ml of RA, a time course study showed that the DNA content steadily increased (Fig. 3). The level of DNA on day 6 was two-fold higher than that of control explants cultured in medium containing vitamin A depleted serum. Inclusion of notochord with these control explants caused only a 33% increase in DNA level (Fig. 3). Interestingly, addition of 30ng/ml of RA to these somite + notochord control culture did not cause an additive effect (DNA level on day 6 was 468ng/explant).

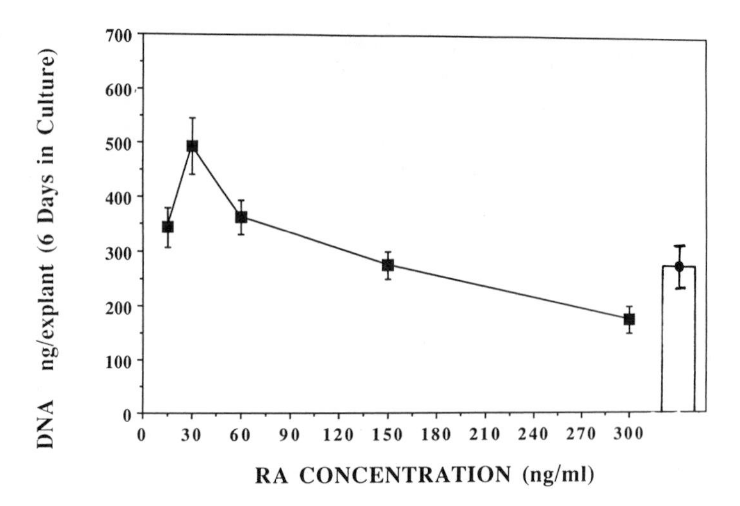


**Figure 2.** Effect of various concentrations of RA on somite explants. Results expressed in ng of DNA/explant are the average of five experiments. The bar on the graph indicates the DNA level in control cultures.

Retinoic acid at 30ng/ml also promoted cell proliferation in somite explants (Fig. 4). Earlier, it was reported that low levels of RA induced cell proliferation in cultured avian chondrocytes (Takishita, et al., 1990), rabbit costal growth cartilage cells (Enomoto, et al., 1990), and distal mesodermal limb bud cells (Ide and Aono, 1988). Somites at stage 17/18 consist of predominantly mesenchymal cells. The cell proliferation induced by RA correlates well with the observation made on embryonic mesodermal cells (Ide and Aono, 1988). It was also reported that, in high density whole limb bud cell culture, low concentrations of RA slightly stimulated DNA synthesis, but it was inferred that the stimulation reflect a nutrition supplementation (Paulsen, et al., 1988). The growth promotion by low levels of RA that we have observed in somite explants could also be due to diminished cell death (i.e., increased cell survival) as seen in developing avian wing buds (Tickle, et al., 1985). Retinoic acid has also increased cell proliferation in other cells like rat tracheal epithelial cell lines (Klann and Marchok, 1982).

[$^3$H]-thymidine incorporation was measured to determine how the continuous presence of RA affected DNA synthesis in somite explants. The results indicate that somite explants cultured in RA containing medium synthesized 68% more DNA than control explants (Fig. 5). From the figure, it is also clear that explants cultured in serum free medium also synthesized DNA but at a reduced level. Control explants cultured in medium containing vitamin A depleted serum also synthesized DNA which was 30% higher than the explants cultured in serum free medium. This incorporation was considerably lower than

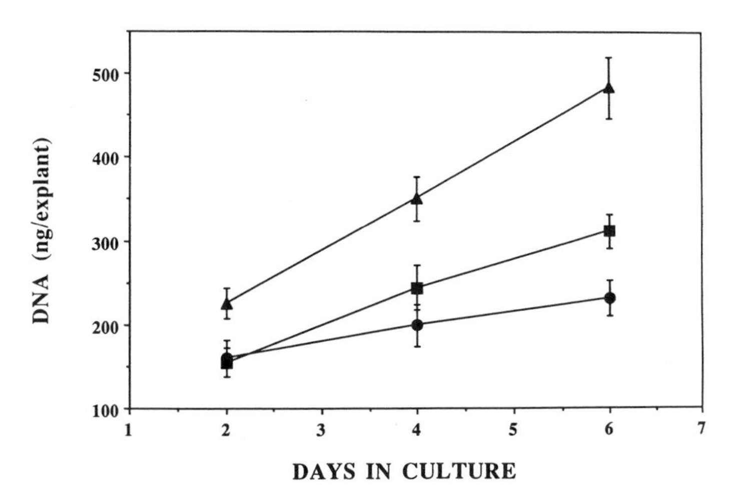


**Figure 3.** Effect of RA at 30ng/ml on DNA levels. Somites cultured in medium containing vitamin A depleted serum (o), and in the presence of 30ng/ml RA (▲). DNA of normal somite-notochord explants (■). Average of five experiments.

the RA supplemented cultures. Takishita, et al. (1990) observed that for chondrocytes in both growth phase and confluency, pretreatment with 0.1$\mu$M RA increased DNA synthesis significantly, though DNA synthesis in cultured chondrocytes without pretreatment was not affected. They explained their result by pointing out the fact that serum contains many kinds of vitamins and growth factors. It is worthwhile to indicate that the requirement of RA may be different for differentiated chondrocytes (studied by Takishita, et al., 1990) and for limb mesodermal cells (Ide and Aono, 1988). Proximal cells obtained from stage 23 limb did not respond to RA, whereas distal cells proliferated significantly (Ide and Aono, 1988). Furthermore, it has recently been reported that RA applied into proximal mesoderm of chick limb bud induces local deletion of limb structures, while RA applied onto anterior distal mesoderm induces duplicate limb element formation (Tickle and Crawley, 1988). This study also supports an earlier <u>in vitro</u> observation that distal mesodermal cells are different from proximal mesodermal cells in their responses to RA (Zimmerman and Tsambaos, 1985).

The observation of the effect of RA on cell proliferation and DNA synthesis can be summarized as follows: Retinoic acid (1) at low concentration stimulates DNA synthesis in a dose dependent manner; (2) in excess causes decreased DNA level (synthesis) and increased cell death; (3) at 30ng/ml stimulates cell proliferation; and may also augment cell survival/decrease cell death; (4) increases DNA synthesis, either by itself or interacting with other serum factors; and (5) has different effect on somites, distal and proximal mesodermal cells, and differentiated chondrocytes.

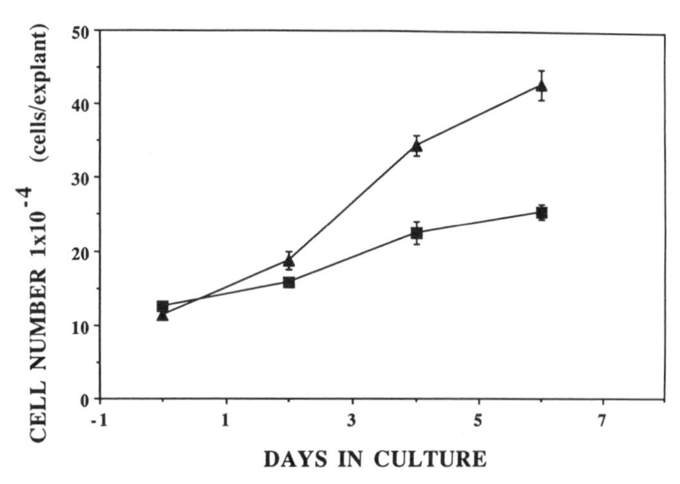


**Figure 4**. Effect of RA on cell number in the presence of vitamin A depleted serum (■) and 30ng/ml RA (▲). Average of three experiments.

EFFECT OF RA ON GLYCOSAMINOGLYCAN SYNTHESIS: Synthesis and accumulation of type II collagen and sulfated GAGs are two of the extracellular matrix components frequently studied as markers of chondrogenic expression. In the present investigation, we have studied the synthesis of sulfated GAGs by somite explants in the presence of RA. We have observed that with time, and under various experimental conditions, somite explants continued to synthesize sulfated GAG (Table 1). We have also found that on day 6 somite explants cultured in medium containing vitamin A depleted serum incorporated 1744 dpm of $^{35}$S-sulfate per ng DNA, whereas somites cultured in medium containing untreated serum incorporated 30% more radioactive sulfate (result not shown in Table 1). When notochord was subsequently included with the somite explants cultured in vitamin A depleted serum, the rate of incorporation was 2860 dpm/ng DNA (Table 1), a rate approximately 60% higher than the somite control under identical culture conditions. This increase may be due to the notochord's enhancement of the somite survival rate, as can also be seen above in the increase in DNA content of the explants (Fig.3). Furthermore, in earlier studies, when somite and notochord explants were co-cultured in normal (untreated) fetal calf serum, the notochord increased the (35S)-sulfate incorporation by 2-to-3 fold (Lash and Vasan,1978; Vasan, et al, 1986a). However, in the present study (Table 1) somites cultured in vitamin A depleted serum increased their incorporation rate by only 60% after the addition of notochord. This significant decline may be due to the lack of vitamin A in the serum which could interact/influence the serum sulfation factor (Salmon and Daughaday, 1957). This possibility is supported by the results of an experiment where

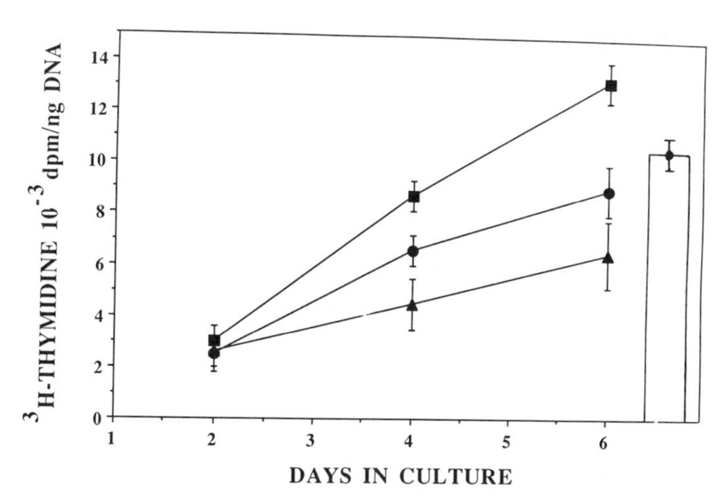


**Figure 5.** Somite explants cultured in medium containing vitamin A depleted serum (o), 30ng/ml RA (■), and serum-free medium (▲). Bar indicates [$^3$H]thymidine incorporation in normal serum.

**TABLE 1. RADIOACTIVE SULFATE INCORPORATION INTO SULFATED GLYCOSAMINOGLYCANS**

| Days in Culture | Somite | Somite + Notochord | a% | Somite + RA | b% |
|---|---|---|---|---|---|
| 2 | 1193 ± 74 | 1620 ± 86 | 36 | 1595 ± 91 | 33 |
| 4 | 1458 ± 113 | 2208 ± 173 | 51 | 2736 ± 62 | 87 |
| 6 | 1745 ± 120 | 2860 ± 307 | 63 | 3733 ± 104 | 114 |

Cultures were exposed to 25μCi/ml of radioactive sulfate for a period of 18-20 hours. Results expressed as dpm/μg DNA were from five experiments. "a" and "b" in percent shows the value over control cultures. All cultures were grown in medium containing vitamin A depleted serum. RA was added to one set of somite and notochord cultures. The result indicated a 30% increase, showing an additive effect of RA in these cultures.

**TABLE 2. EFFECT OF RETINOIC ACID ON DNA LEVEL OF SCLEROTOME AND DERMAMYOTOME**

| Days in Culture | Sclerotome | Sclerotome + RA | Dermamyotome | Dermamyotome + RA |
|---|---|---|---|---|
| 0 | 290 ± 25 | 281 ± 26 | 194 ± 20 | 171 ± 13 |
| 3 | 356 ± 19 | 230 ± 18 | 230 ± 18 | 201 ± 24 |
| 6 | 410 ± 20 | 586 ± 29 | 241 ± 18 | 267 ± 20 |

Microdissection of sclerotome and dermamyotome is described in Methods. The results expressed as ng of DNA per explant is the average from three experiments.

somite explants, cultured in medium containing vitamin A depleted serum and 30ng/ml of RA, incorporated 114% more $^{35}$S-sulfate. Earlier studies which also lend support to our findings that low levels of RA promote chondrogenesis were done in limb bud cell cultures (Paulsen and Solursh, 1987; Ide and Aono, 1988).

A study (results not included) was done on only one set of experimental tissues (due to lack of material), but nevertheless support the earlier results (Figs. 2-5 and Table 1). Somite explants were cultured for 6 days in serum free and serum containing medium to which 30ng/ml of RA was added. The results show that RA in serum free medium stimulates both cell proliferation (i.e., [$^3$H]-thymidine incorporation) and sulfated GAG synthesis (i.e., [$^{35}$S]-sulfate incorporation). This increase however, was significantly smaller when compared to tissues which were cultured in serum containing medium to which RA was added.

Taken together the results on DNA synthesis and sulfated GAG synthesis indicate that low levels of RA may act in the following manner. First, RA interacts with growth factors to stimulate cell proliferation, thereby increasing the cell density. Secondly, a transient increase in the synthesis of heparan sulfate in the presence of RA facilitates cell aggregation thus improving cell viability (Lash and Vasan, 1978) or decreasing cell death (Tickle et al., 1985). Thirdly, RA interacts with serum sulfation factor to stimulate increased sulfated GAG synthesis.

EFFECT OF RA ON PROTEOGLYCAN SYNTHESIS: Synthesis and accumulation of large size cartilage-specific proteoglycan is considered one of the phenotypic expressions in chondrogenic tissues. The proteoglycan synthesized by differentiating chondrocytes has been shown to elute as three major populations (Palmoski and Goetinck, 1972; Vasan, 1982; Vasan et al., 1986a). In this experiment, total proteoglycan was extracted from the tissues and applied onto a Sepharose CL-2B gel column. For comparative purposes, somite explants were

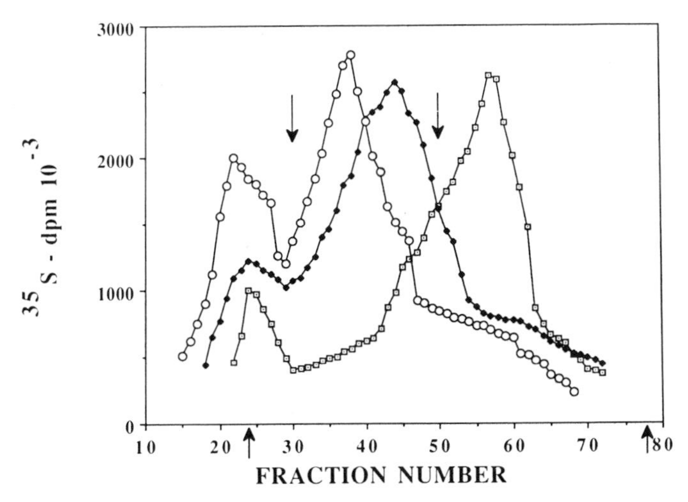


**Figure 6.** Percent of various peaks separated on Sepharose CL-2B. Data from Figure 6.

co-cultured with notochord in addition to the RA treatment. In the presence of RA, somite explants synthesized and accumulated 26% of the total proteoglycan in the form of large aggregates that excluded from the column (Figs. 6 & 7). The proportion of the intermediate size proteoglycan (55%) formed a majority of the total population. Induction of the somites by the notochord slightly increased the proportion of large size molecules but increased significantly the proportion of the intermediate (57%) size population. Comparatively, the control somites cultured in medium containing vitamin A depleted serum synthesized more than 50% of the proteoglycan as small size molecules. Such small size proteoglycan synthesis is commonly observed in mesenchymal cells (Vasan and Lash, 1979; Vasan, 1982). Retinoic acid at 150ng/ml resulted in the loss of the synthesis of large and intermediate size proteoglycan molecules, but in the significant increase of smaller mesenchymal type molecules (data not shown). This is in agreement with earlier reports that sternal chondrocytes perturbed with vitamin A dedifferentiated and synthesized only smaller size proteoglycans (Vasan, 1981), and stopped synthesizing the chondrocyte-specific pro$\alpha$1(II) chain of collagen II and a 370-kDa precursor protein of cartilage-specific proteoglycans (Horten et al., 1987).

Our earlier study showed that the number and length of chondroitin sulfate chains attached to the core protein of the proteoglycan caused the difference in their size, i.e., the variations in the glycosylation of the core protein (Vasan, 1982). Thus in the present study (Fig. 6), the increase in the proportion of large and intermediate size proteoglycan may be due to the increased glycosylation of the core protein in the presence of RA. Our results support earlier findings that vitamin A stimulates the synthesis of specific glycoproteins when added to incubated corneas from vitamin A deficient rats (Kiorpes, et al., 1979). Another study of mouse limb

**TABLE 3. EFFECT OF RETINOIC ACID ON SULFATED GAG SYNTHESIS OF SCLEROTOME AND DERMAMYOTOME**

| Days in Culture | Sclerotome | Sclerotome + RA | Dermamytome | Dermamyotome + RA |
|---|---|---|---|---|
| 3 | 876±42 | 1364±95 | 329±32 | 469±50 |
| 6 | 1237±83 | 2952±167 | 461±46 | 642±38 |

Radioactive sulfate 25 $\mu$Ci/ml was added to the medium 18-20 hours prior to sample collection. Results expressed as dpm per $\mu$g of DNA is the average from three experiments.

mesenchymal cells also showed that vitamin A stimulates the mannsylation of certain fractions of glycopeptides (Pennypacker, et al., 1978). A more recent study reported that RA caused a general increase in protein synthesis of rabbit costal growth cartilage (Enomoto et al., 1990).

EFFECT OF RA ON SCLEROTOME AND DERMAMYOTOME: Somite tissue consists of two components, sclerotome and dermamyotome. In HH stage 17/18 embryos when the sclerotomal cells undergo a mesenchymal transition, the dermamyotomal portion still remains intact and epithelial (Ede and El-Gadi, 1986). The effect of RA on somite explants observed in the present study could be due to: (1) selective stimulation of sclerotomal cells, (2) selective inhibition and/or death of dermamyotomal cells, and (3) transformation of some of the epithelial dermamyotomal cells into sclerotomal mesenchyme. The effect of RA on epithelial cells has been well established in a variety of tissues and cell lines (see Review by Sporn and Roberts, 1990). It is noteworthy that an earlier investigation has clearly demonstrated that at HH stage 17/18 it is the sclerotome, and not the dermamyotome, that respond to notochordal induction expressing chondrogenesis (Cheney and Lash, 1981).

The present study utilized sclerotome and dermamyotome that were separated by a quick trypsin treatment followed by microdissection. A time course study was done over a period of 6 days to investigate the effect of RA DNA level. During the 6 day period the level of DNA increased steadily in both tissues regardless of added RA, thus showing a healthy tissue growth (Table 2). At the end of the 6 day period however, RA treated sclerotomal cultures contained 108% more DNA compared to day zero, while explants grown in vitamin A depleted medium showed only a 40% increase during the same period of time. Dermamyotome, on the other hand, showed an increase of 56% due to RA enrichment. Retinoic acid thus promoted cell proliferation in both tissues, but its effect was two times greater on sclerotomal cells than on dermamyotome. This selective effect may be due to the fact that the sclerotomal cells at this stage of embryonic development have more RA receptors. Recent studies have demonstrated a spatial and temporal expression of the RA receptors in developing avian

**TABLE 4. HEPARAN SULFATE SYNTHESIS BY SCLEROTOMAL CELLS IN CULTURE**

| Retinoic Acid | Days in Culture | | |
|---|---|---|---|
| | 2 | 3 | 6 |
| None | 414 ± 83 | 537 ± 102 | 431 ± 70 |
| 30ng/ml | 595 ± 112 | 1176 ± 267 | 628 ± 96 |

Sclerotomal cells were exposed to 25 μCi/ml of radioactive sulfate 18 hours prior to harvest. Glycosaminoglycan was isolated as described in methods, and an aliquot used for digestion with heparitinase. Results are expressed as dpm per μg of DNA is the average of three experiments.

(Maden, et al., 1988) and regenerating amphibian limbs (Giguere, et al., 1989). It has been proposed that RA acts by binding to a cellular RA-binding protein and entering the nucleus, thereby altering the pattern of gene activity (Chytil and Ong, 1984; Takase, et al., 1986).

EFFECT OF RA ON THE SULFATED GAG SYNTHESIS IN SCLEROTOME AND DERMAMYOTOME: Results of the stimulatory effect of RA on sulfated GAG synthesis indicate that the RA effect was more pronounced on sclerotome than on dermamyotome (Table 3). After 6 days of culture, RA-treated sclerotome incorporated 2952 dpm of [$^{35}$S]-sulfate per ng DNA. This is about 138% higher than the control culture for the same time period. Retinoic acid treated dermamyotome, however, incorporated only 39% more radioactive sulfate than the untreated control. The marked increase in the DNA content and sulfated GAG synthesis of the sclerotomal tissue suggests that a sclerotomal specific cellular response occurs in the presence of RA (speculation number 1 above). Furthermore, no selective cell death occurs in dermamyotomal tissues (speculation number 2), and that some dermamyotomal cells which are still epithelial may possibly be transformed into sclerotomal cells (speculation number 3). Although these findings are based on limited experimentations and require further study, it is important to note that (1) RA has been shown to influence the synthetic mechanisms in a variety of epithelial tissues, and (2) sclerotomal and dermamyotome respond differently to the inductive notochordal tissues (Cheney and Lash, 1981).

In vivo, sclerotomal cell migrate medially towards the spinal cord and notochord and, in part influenced by the perinotochordal matrix, undergo cellular condensation and differentiate into chondrocytes (see Vasan, 1987 and references therein). In an unrelated RA experiment, isolated single cells of sclerotome in suspension culture formed aggregates (within 48 hours) prior to chondrogenic expression. During this period, cellular synthesis of heparan sulfate ceased after 24 hours, while the synthesis of chondroitin 4-sulfate continued to increase (unpublished results).

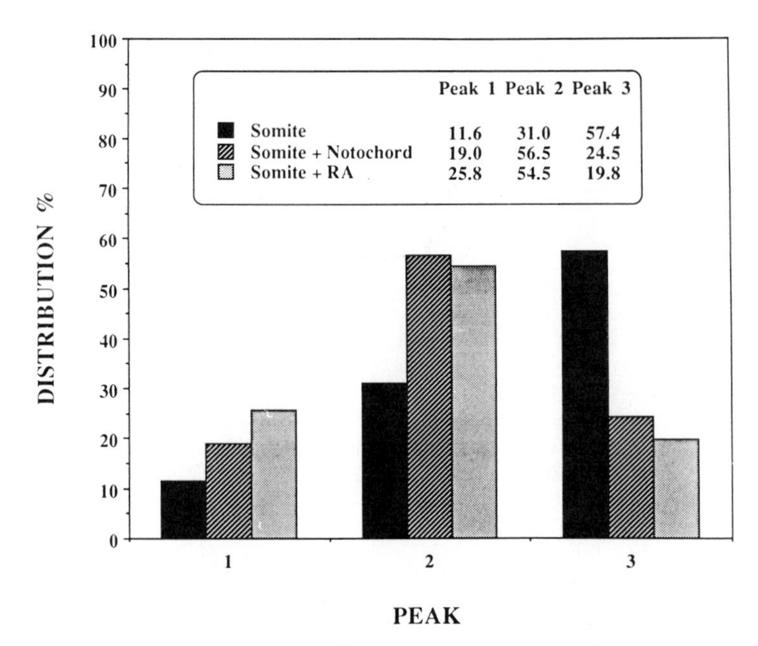


**Figure 7.** Percent of various peaks on Sepharose CL-2B. Data from Figure 6.

In the present study it was found that 48 hours after RA addition, sclerotomal cells show maximum aggregation compared to controls cultured in medium containing vitamin A depleted serum. Sulfated GAG analysis at this time (day 3) revealed that RA had caused a two-fold increase in the heparan sulfate in these cells (Table 4). Further culturing of these cells to 6 days, we found that the heparan sulfate level had decreased (Table 4) and that the level of chondroitin 4-sulfate, a marker of chondrogenic expression, had increased (result not included in the Table). Solursh et al. (1990) reported that in 9-day mouse embryonic limb, heparan sulfate proteoglycan (Syndecan) was present throughout the limb region. By day 13, the immunostaining for syndecan was lost in regions destined for chondrogenesis but persisted in distal mesenchyme. In an earlier experiment, heparan sulfate was transiently expressed during mesenchymal condensation around stage 23/24 of avian limb development. By stage 25, the heparan sulfate decreased significantly with an increase in chondroitin 4-sulfate (onset of chondrogenesis) (Vasan, 1986c; Manuscript in preparation).

POSSIBLE MOLECULAR BASIS OF RA EFFECT: The effects of RA observed in this and other studies reflect an up-regulation of other genes. It has been recently reported that RA regulates the growth hormone genes of human GH1 pituitary cells by acting either independently, or synergistically with glucocorticoids or thyroid hormone (Bedo, et al., 1989). An earlier study on somite explants showed that growth hormone supplementation in the medium increased the radioactive sulfate incorporation into sulfated GAG by two-fold (Lash and

Whitehouse, 1960). A recent study also showed that the expression of epidermal growth factor receptor gene in normal rat kidney fibroblasts is regulated by RA (Thompson and Rosner, 1989). Reports are also available of other gene sequences being transcriptionally regulated by RA (Wang, et al., 1985). From molecular studies of the steroid receptor superfamily, it has been recognized that there are at least five different molecular species of RA receptors, and they all appear to be nuclear transcription factors, serving to regulate transcription target genes by binding to specific regulatory sequences known as RA response elements (Sporn and Roberts, 1991). Furthermore, various subtypes of these RA receptors are expressed in distinct patterns during embryonic development and during adulthood, indicating that they regulate different functions (Kurst, et al., 1989; Mangelsdorf, et al., 1990). Since the receptors of RA are transcriptional regulators, it has been suggested that RA affects pattern formation through the regulation of one or more key genes during early development (Smith, et al., 1989). While it is not clear which genes are directly regulated, RA has been shown to influence a number of homeobox genes (see Smith, et al., 1989 for references).The molecular mechanism of the action of RA is being understood in various cell lines, but studies on mesenchymal cells and chondrocytes are just at the beginning stage.

## ACKNOWLEDGEMENTS

The author wishes to thank Marcus Meyenhofer for photographic assistance. Some of the tedious dissection was done by Ms. Heather Bost. This investigation was supported by grants from NIH 5SO7RR-05393 and the Foundation of the University of Medicine and Dentistry of New Jersey.

## REFERENCES

Bedo, G., Santisteban, P., and Aranda, A., 1989, Retinoic acid regulates growth hormone gene expression. Nature (London), 339:231-234.

Belsky, E., Vasan, N.S., and Lash, J.W., 1980, Extracellular matrix components and somite chondrogenesis: A Microscopical Analysis. Develop. Biol., 79:159-180.

Chytil, F. and Ong, D.E., 1984, Cellular retinoid-binding proteins. in The Retinoids', Vol. 2 (eds M.B. Sporn, A.B. Roberts, and D.S. Goodman), PP. 90-123, Academic Press, New York.

Ede, D.A. and El-Gadi, A.O., 1986, Genetic modifications of devel-opmental acts in chick and mouse somite development. in, Somites om Developing Embryos (eds. R. Bellairs, D.A. Ede and J.W. Lash) PP 209-224, Plenum Press, New York.

Eichele, G., Tickle, C., and Alberts, B.M., 1985, Studies on the mechanism of retinoid - induced pattern duplication in the early chick limb bud: temporal and spatial aspects. J. Cell Biol., 101:1913-1920.

Enomoto, M., Pan, H., Suzuki, F., and Takigawa, M., 1990, Physiolo-gical role of vitamin A in growth cartilage cells: Low concen-tration of retinoic acid strongly promote the proliferation of rabbit costal growth cartilage cells in culture. J. Biochem., 107:743-748.

Giguere, V., Ong, E.S., Evans, R.M., and Tabin, C.J., 1989, Spatial and temporal expression of the retinoic acid receptor in the regenerating amphibian limb. Nature (London), 337:566-569.

Hall, B.K., 1977, Chondrogenesis of the somitic mesoderm. Adv. Anat. Embryol. Cell Biol., 53:1-50.

Hamburger, V. and Hamilton, H.L., 1951, A series of normal stages in the development of the chick embryo. J. Morphol., 88:48-92.

Horton, W.E., Yamada, Y., and Hassell, J.R., 1987, Retinoic acid rapidly reduces cartilage matrix synthesis by altering gene transcription in chondrocytes. Develop. Biol., 123:508-516.

Ide, H. and Aono, H., 1988, Retinoic acid promotes proliferation and chondrogenesis in the distal mesodermal cells of chick limb bud. Develop. Biol., 130:767-773.

Kiopes, T.C., Kim, C.L.-Y., and Wolf, G., 1979, Stimulation of the synthesis of specific glycoproteins in corneal epithelium by vitamin A. Exp. Eye Res., 28:23-35.

Kosher, R.A., Lash, J.W., and Minor, R., 1973, Environmental enhancement of in vitro chondrogenesis. IV. Stimulation of somite chondrogenesis by exogenous chondromucoprotein. Develop. Biol., 35:210-220.

Kosher, R.A. and Church, R.L., 1975, Stimulation of in vitro somite chondrogenesis by procollagen and collagen. Nature (London), 228:327-329.

Kosher, R.A. and Lash, J.W., 1975, Notochordal stimulation of in vitro somite chondrogenesis before and after enzymatic removal of perinotochordal materials. Develop. Biol., 42:362-368.

Krust, A., Kastner, P.H., Petkovich, M., Zelent, A., and Chambon, P., 1989, A third human retinoic acid receptor, hRAR-y. Proc. Natl. Acad. Aci. USA, 86:5310-5314.

Lash, J.W., 1967, Differential behavior of anterior and posterior embryonic chick somites in vitro. J. Exp. Zool., 165:47-56.

Lash, J.W. and Whitehouse, M.W., 1960, The in vitro effect of hormones amd drugs upon the synthesis of acid mucopolysaccharides. Abst. First International Congress of Endocrinology, 125.

Lash, J.W. and Vasan, N.S., 1978, Somite chondrogenesis in vitro: stimulation by exogenous extracellular matrix components. Develop. Biol. 66:151-171.

Maden, M., Ong, D.E., Summerbell, D., and Chytil, F., 1988, Spatial distribution of cellular protein binding to retinoic acid in the chick limb bud. Nature (London). 335:733-735.

Mangeldorf, D.J., Ong, E.S., Dyck, J.A., and Evans, R.M., 1990, Nuclear receptor that identifies a noval retinoic acid response pathway. Nature (London). 354:224-229.

Nevo, Z. and Dorfman, A., 1972, Stimulation of chondromucoprotein synthesis in chondrocytes by extracellular chondromucoprotein. Proc. Natl. Acad. Sci. USA. 69:2069-2071.

Oegema, T.R., Jr., Hascall, V.C., and Dziewiatkowski, D.D., 1975, Isolation and characterization of proteoglycans from the rat chondrosarcoma. J. Biol. Chem. 250:6151-6159.

Palmoski, M.J. and Goetinck, P.F., 1972, Synthesis of proteochon-droitin sulfate by normal, nanomelic, and 5-bromodeoxyuridine-chondrocytes in cell culture. Proc. Natl. Acad. Sci. USA 69:3385-3388.

Paulsen, D.F., Langille, R.M., Dress, V., and Solursh, M., 1988, Selective stimulation of in vitro limb-bud chondrogenesis by retinoic acid. Differentiation 39:123-130.

Pennypacker, J.P., Lewis, C.A., and Hassell, J.R., 1978, Altered proteoglycan metabolism in mouse limb mesenchyme cell cultures treated with vitamin A. Arch. Biochem. biophys., 186:351-358.

Rosa, F.W., Wilk, A.L., and Kelsey, F.O., 1986, Teratogen update: Vitamin A Congeners. Teratology 33:355-364.

Salmon, W.D. and Daughaday, W.H., 1957, A hormonally controlled serum factor which stimulates sulfate incorporation by carti-lage in vitro. J. Lab. Clin. Med., 49:825-836.

Solursh, M. and Meier, S., 1973, The selective inhibition of muco-polysaccharide synthesis by vitamin A treatment of cultured chick embryo chondrocytes. Calcif. Tiss. Res. 13:131-142.

Solursh, M., Reiter, R.S., Jensen, K.L., Kato, M., and Bernfield, M., 1990, Transient expression of a cell surface heparan sulfate proteoglycan (Syndecan) during limb development. Develop. Biol. 140:83-92.

Sporns, M.B. and Roberts, A.B., 1991, Interactions of retinoids and transforming growth factor-$\beta$ in regulation of cell differentiation and proliferation. Mole. Endocrinol. 5:3-7.

Summerbell, D., 1983, The effect of local application of retinoic acid to the anterior margin of the developing chick limb. J. Emrbyol. exp. Morph. 78:269-289.

Takase, S., Ong, D.E., and Chytil, F., 1986, Transfer of retinoic acid from its complex with cellular retinoic acid-binding protein in the nucleus. Arch. Biochem. biophys. 247:328-334.

Thaller, C. and Eichele, G., 1987, Identification and spatial distribution of retinoids in the developing chick limb bud. Nature (London) 327:625-628.

Thompson, K.L. and Rosner, M.R., 1989, Regulation of epidermal growth factor receptor gene expression by retinoic acid and epidermal growth factor. J. Biol. Chem. 264:3230-3234.

Tickle, C., Summerbell, D., and Wolpert, L., 1975, Positional signalling and specification of digits in chick limb morphogenesis. Nature (London) 254:199-202.

Tickle, C., Lee, J., and Eichele, G., 1985, A quantitative analysis of the effects of all-trans-retinoic acid on the pattern of chick wing development. Develop. Biol. 109:82-95.

Tickle, C., Alberts, B.M., Wolpert, L., and Lee, J., 1982, Local application of retinoic acid to the limb bud mimics the action of the polarizing region. Nature (London 296:564-565.

Takishita, Y., Hiraiwa, K., and Nagayama, M., 1990, Effects of retinoic acid on proliferation and differentiation of cultured chondrocytes in terminal differentiation. J. Biochem., 107:592-596.
Vasan, N.S., 1981, Analysis of perinotochordal materials: 1. Studies on proteoglycan synthesis. J. Exp. Zool. 215:229-223.
Vasan, N.S., 1981, Proteoglycan synthesis by sternal chondrocytes perturbed with vitamin A. J. Embryol. exp. Morph. 63:181-191.
Vasan, N.S., 1982, Analysis of intermediate size proteoglycans from the developing chick limb buds. J. Embryol. exp. Morph. 70:61-74.
Vasan, N.S., 1983, Analysis of perinotochordal materials: 2. Studies on the influence of proteoglycans in somite chondrogenesis. J. Embryol. exp. Morph. 73:263-274.
Vasan, N.S., 1986c, Analysis of extracellular matrix material during mesenchymal cell condensation in limbs development. J. Cell Biol. 103:97a.
Vasan, N.S., 1987, Somite chondrogenesis: The role of microenvironment. Cell Differ. 21:147-159.
Vasan, N.S. and Lash, J.W., 1975, Chondrocyte metabolism as affected by vitamin A. Calcif. Tiss. Res. 19:99-107.
Vasan, N.S. and Lash, J.W., 1979, Monomeric and aggregate proteoglycans in the chondrogenic differentiation of embryonic chick limb buds. J. Embryol. exp. Morph. 49:47-59.
Vasan, N.S. and Miller, E., 1985, Somite chondrogenesis in vitro differential induction by modified matrix - a biochemical and morphological study. Dev. Growth Diff. 27:405-417.
Vasan, N.S., Lamb, K.M., and LaManna, O., 1986a, Somite chondrogenesis in vitro: 1. Alterations in proteoglycan synthesis. Cell Differ. 18:79-90.
Vasan, N.S., Lamb, K.M., and LaManna, O., 1986b, Somite chondrogenesis in vitro: 2. Changes in the hyaluronic acid synthesis. Cell Differ. 18:91-99.
Wang, S.Y., LaRosa, G.J., and Gudas, L.J., 1985, Molecular cloning of gene sequences transcriptionally regulated by retinoic acid and dibutyryl cyclic AMP in cultured mouse teratocarcinoma cells. Develop. Biol. 107:75-86.
Zimmerman, B. and Tsambaos, D., 1985, Evaluation of the sensitive step of inhibition of chondrogenesis by retinoids in limb mesenchymal cells in vitro. Cell Differ. 17:95-103.

# EXPRESSION OF MYOGENIC FACTORS IN SOMITES AND LIMB BUDS DURING MOUSE EMBRYOGENESIS

Marie-Odile Ott and Margaret Buckingham

Department of Molecular Biology, UA 1148 CNRS
Pasteur Institute
25, rue du Dr. Roux, 75724 Paris Cedex, France

## INTRODUCTION

Our main interest is focused on the study of myogenesis in the mouse. Both cardiac and skeletal striated muscle are mesodermal derivatives. It was shown a number of years ago that one of the earliest indications of muscle formation in response to mesoderm induction is the expression of α-actin which precedes somitogenesis and skeletal muscle formation in *Xenopus*. In the mouse, gastrulation takes place from 6.5 d post coitum (p.c.) and mesodermal derivatives are clearly present by the 7th day. The heart muscle is the first to be formed followed by somitogenesis and skeletal myogenesis both characterized by early cardiac actin expression (Sassoon et al., 1988). The identification of regulatory myogenic genes of the MyoD family has provided an opportunity to analyse the molecular mechanisms of myoblast lineage determination and differentiation. Myogenesis can be schematized in the following steps: 1) determination of the mesodermal lineage during gastrulation, 2) determination of the myoblast lineage among the mesodermal derivatives, 3) maintenance of the determined state during the proliferating period, 4) muscle differentiation, characterized by the cessation of cell division and expression of different muscle protein genes together with cell fusion and finally 5) skeletal muscle maturation which results in different muscle fiber types. We have developed specific probes which permit us to define by *in situ* hybridization when and where different muscle structural genes are expressed in the embryo. We have also used probes for the detection of different members of the MyoD1 family of genes in order to examine the potential role of these regulatory sequences during myogenesis *in vivo*.

## THE HEART

The first striated muscle to form in the mammalian embryo is the heart (Rugh, 1990) which is evident in the mouse as a cardiac tube between 7-8 days p.c., resulting from the fusion of the two cardiac primordia, which are formed from lateral mesoderm. By 9 days p.c. the heart is beating and acquires a more adult morphology with distinguishable atrial and ventricular compartments. In the cardiac tube from the earliest stages that we have examined (circa 7.5 days p.c.) all the major myosin sequences are already accumulated (Lyons et al., 1990a). Cardiac actin, and to a

*Formation and Differentiation of Early Embryonic Mesoderm*
Edited by R. Bellairs *et al.*, Plenum Press, New York, 1992

lesser extent, skeletal actin transcripts and protein are also present (Sassoon et al., 1988; Lyons et al., 1991a). These genes are among the first characteristic of an adult phenotype to be expressed during embryogenesis, and cardiac actin, for example, is a potentially interesting early marker for tracing cardiac cells as they emerge from the precardiac mesoderm. Subsequently, as the heart acquires its mature morphology, different myosin sequences become restricted to atrial or ventricular compartments. This process takes place asynchronously; the ß-MHC (myosin heavy chain), for example, is only present in the ventricle by 10.5 days p.c., whereas the ventricular MLC1V (myosin alkali light chain) form is still present in the atria at this time, and is confined to the ventricular compartment only by 17.5 days p.c. No family of regulatory genes equivalent to the MyoD1 family has yet been characterized in the heart, and the molecular regulation of muscle genes, many of which are expressed in both skeletal and cardiac muscle, is less well understood in the latter.

## SKELETAL MUSCLE

In higher vertebrates our understanding of the origin of skeletal muscle is mostly based on experiments with chick/quail chimaeras which demonstrated that muscle precursor cells are present in the somites (e.g. Chevallier et al., 1977; Christ et al., 1977; Bellairs et al., 1986; Ott et al., 1990) or in the case of certain head muscles in the pre-chordal plate which precedes the first somite (Wachtler et al., 1986). With the possible exception of the latter where an additional neural crest component has been suggested (Le Douarin, 1982), no other source of skeletal muscle cells has been identified. Initially somites consist of balls of epithelial-like cells which form in a rostro-caudal gradient by segmentation of paraxial mesoderm on either side of the neural tube. In the mouse, the first somites are formed from 7.8 d p.c. (Theiler, 1989), pairs of somites are added, following the regression of the primitive streak. By 15 days of development, about 65 pairs of somites have been formed. The early somite rapidly differentiates into two cellular compartments, the dermamyotome which contains precursor cells for muscle or skin, and the sclerotome which contributes to the formation of skeletal structures. Cells migrate out from the dermamyotome mainly from the ventro-lateral edge (Jacob et al., 1978) to form muscle masses elsewhere in the embryo; somites at the level of the limbs contribute the cells for formation of limb musculature (Wachtler et al., 1982). The first skeletal muscle to form is the myotome, in the central region of the somite, as a result of migration of precursor cells from the cranio-medial edge of the dermamyotome (Kaehn et al., 1988). At later stages myotomes contribute to the formation of trunk musculature, such as the intercostal muscles. The myotome is clearly visible in the more rostral somites from 8.5 d p.c. Expression of muscle markers such as cardiac actin is observed in the early myotome, before it becomes a functional muscle, 2 to 3 days later.

## MYOGENIC REGULATORY FACTORS

During the last five years four myogenic regulatory sequences have been isolated and characterized in higher vertebrates. This has been made possible by the existence of cell culture systems which reproduce the process of myogenic differentiation *in vivo* (for review see Emerson, 1990).

Mononucleated myoblasts, usually obtained from late foetal or new born muscle, will undergo cell fusion to form muscle fibres and this morphological differentiation is accompanied by the onset of muscle protein synthesis. This process takes place spontaneously when the myoblasts reach a certain cell density. In addition to primary cultures a number of mammalian muscle cell lines exist, again mostly derived from late foetal myoblasts or indeed even adult muscle satellite cells. Such cell lines or primary cultures therefore provide model systems for examining the regulation of muscle differentiation. Using muscle cell cultures, W. Wright cloned the myogenin regulatory sequence (Wright et al 1989) by subtractive hybridization of sequences from differentiating cells against those from myoblasts. The C3H 10T1/2 cell line represents a model system for examining earlier stages of myogenesis (see Konieczny and Emerson, 1984). These cells resemble embryonic fibroblasts and have the notable property of becoming myogenic, when treated with the drug 5'aza-cytidine which prevents DNA methylation. This cell system was used to clone the MyoD1 regulatory sequence by subtractive hybridization of sequences from myoblasts against those from the untreated 10T1/2 cells (Davis et al 1987). The other two members of the MyoD family have been identified by cross-hybridization to the MyoD sequence: myf-5 (Braun et al., 1989) and MRF4, also called myf-6 or herculin (Rhodes and Konieczny, 1989; Braun et al., 1990; Miner and Wold, 1990). These sequences all contain a basic DNA binding domain and a helix-loop-helix motif involved in their dimerization. For a number of muscle structural genes, it has been shown that they act as muscle specific transcriptional factors. In addition, however, they all demonstrate the striking property of myogenic conversion. When transfected into the 10T1/2 cell line or indeed into many other cell types including primary cultures of diverse origin, the cells begin to express muscle specific genes and in many cases form muscle fibres. It is on this basis that the MyoD family of genes have been called myogenic determination factors (see Weintraub et al., 1991).

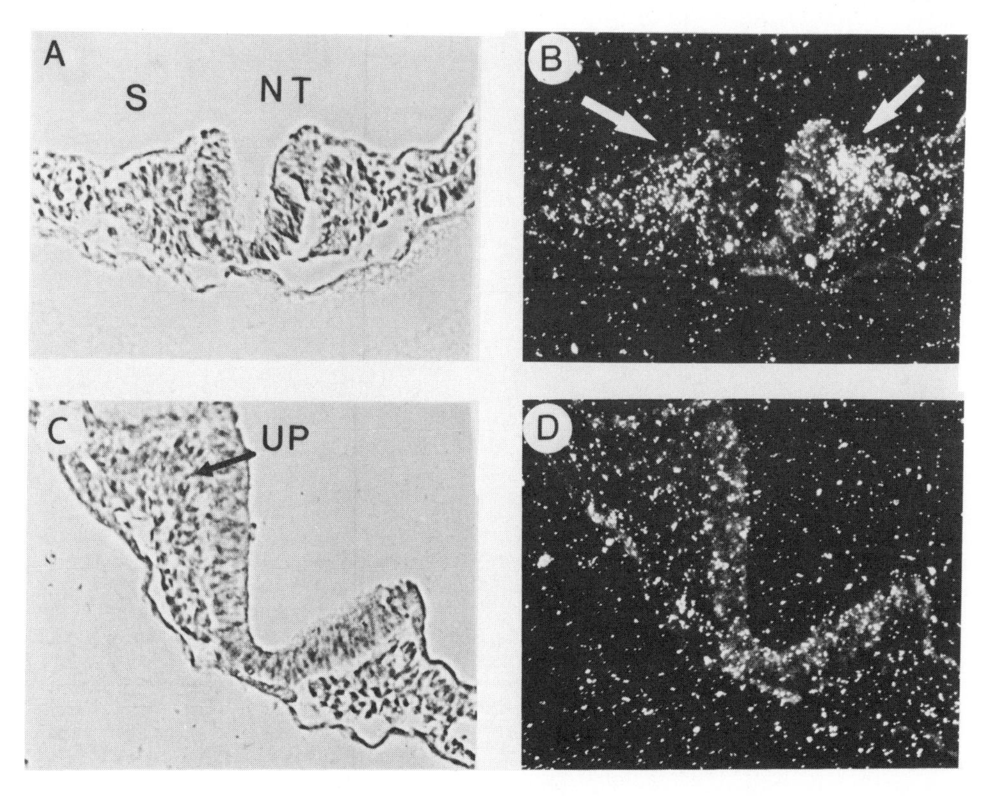


Fig.1 Transverse sections through an early 8 d mouse embryo. (A) and (C) phase contrast photomicrographs of sections at the levels of early somites: (A) where the neural tube is not yet formed and (C) through the unsegmented plate in a more posterior section. (B) and (D) dark field micrographs of (A) and (C) showing localization of myf-5 transcripts only in the early somite (arrows in B). No specific signal of hybridization is detectable before segmentation (D).
S = somite; TN = neural tube; UP = unsegmented mesoderm

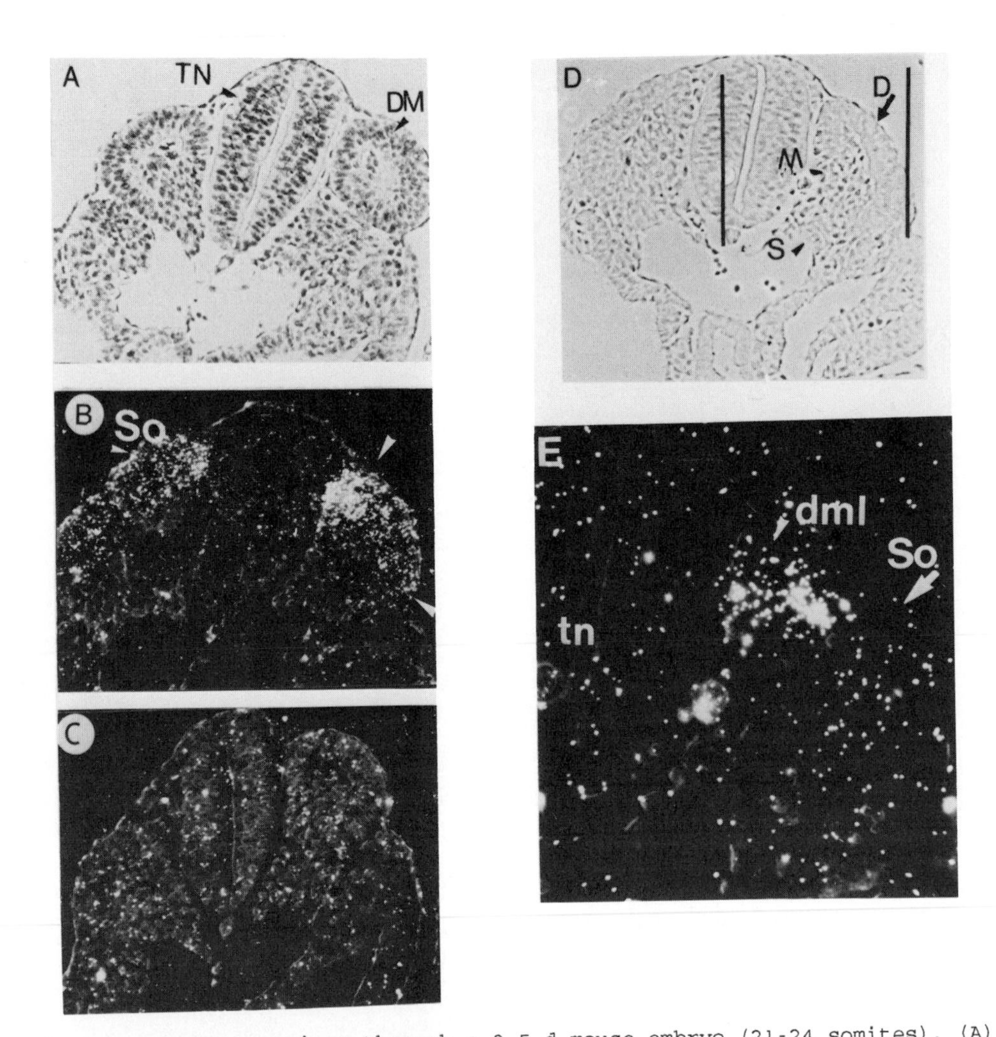


Fig.2.Transverse sections through a 9.5 d mouse embryo (21-24 somites). (A) phase contrast picture of a section in the caudal region of the embryo showing the neural tube flanked by two somites. No myotome is yet formed. (B) dark field micrographs of a serial section hybridized with myf-5. Transcripts are accumulated throughout the dermamyotome but mainly in the upper dorsal part close to the neural tube (arrows). No signal is detectable with the myogenin probe (C). (D) shows a transverse section from a slightly more rostral part of the same embryo. The signal obtained after hybridization with myf-5 is now localized under the dorso medial lip of the dermatome, in the forming myotome (higher magnification in E).
TN or tn = neural tube; DM = dermamyotome; D = dermatome; S = sclerotome; SO = somite; M = myotome; dml = dorso medial lip.

We address here the question of the role during myogenesis *in vivo* of these four different factors, which share common properties *in vitro*. Is there a hierarchy or a cascade in correlation with specific events during embryonic development? Or are the four sequences redundant and interchangeable?

If the myogenic regulatory genes play a role in muscle determination *in vivo* one might expect them to be expressed in the dermamyotome before myotome formation, and also in precursor muscle cells before and during their migration to other sites of muscle formation in the embryo.

In fact the only myogenic regulatory gene to be expressed as early as this is myf-5 (Ott et al., 1991). Myf-5 transcripts are not detectable in the embryo prior to somite formation (Fig.1, C and D) but are present at low levels in the first somites as they form (from 8 days). Even at this stage the distribution of transcripts does not appear to be uniform, but is rather higher in cells adjacent to the neural tube (Fig.1, A and B). This would suggest that there may already be some distinction between muscle precursors and other cells. At the next stage of somite maturation this becomes evident. There is no labelling of cells in the sclerotome, whereas cells in the dermamyotome particularly in the dorsal lip region adjacent to the neural tube are strongly positive when hybridized to a myf-5 specific probe (Fig. 2B). In transverse section this corresponds to the region where myotomal precursor cells would be expected to be present (Fig.2, D and E). Interestingly, at least in the chick, the cells in this region are dividing rapidly before moving into the myotomal compartment (Kaehn et al., 1988). These observations demonstrate that the myf-5 gene is activated prior to muscle formation in the mouse embryo, preceding any sign of muscle differentiation. The location of the transcripts would suggest that myf-5 may be implicated in the commitment of muscle precursor cells which will differentiate in the myotome. As the myotome forms myf-5 transcripts are localized in the myotomal cells (Fig.3a). Once the myotome has formed, myogenin transcripts are also detected. They are concentrated in the central region, whereas myf-5 transcripts are more widely distributed throughout the myotome suggesting that all myotomal cells are not identical at this stage.

Myf-5 transcripts are detectable at other sites in the embryo. They accumulate in one of the visceral arches and the precursor cells of extraoculo-motor muscles in the head, before any other muscle marker examined (Ott et al., 1991). Precursor muscle cells at these sites originate, at least in part, from the prechordal plate. The earliest time at which we obtain a positive signal in the limb is at 10 days in the forelimbs or 10.5 days in the hindlimbs. Here again myf-5 expression precedes that of other muscle markers. It is later than the time when myogenic precursor cells would be expected to have migrated from the somites to sites in the limb, based on chick/quail experiments. Some cells throughout the dermamyotome are myf-5 positive although we have not detected any peculiar concentration of label in the ventro-lateral edge. It is quite possible that this migration occurs relatively later in the mouse. The *in situ* hybridization technique with $^{35}$S-labelled probes on mouse embryo sections does not permit resolution at the single cell level and it is perhaps not surprising that signal is not detectable in cells as they migrate out from the somite. However we would expect to detect positive cells concentrated in the pre-muscle masses of the limb buds, as soon as they have migrated from the somites, given the numbers of cells probably involved (e.g. see Wachtler et al., 1982; Milaire, 1976). It is also possible that cells which have been myf-5 positive in the dermamyotome cease to express this gene during their migration from the somites, and then re-express it again at a slightly
later stage. Alternatively the muscle precursor cells, which form extra-myotomal muscle masses, and which are clearly committed to myogenesis in that they never contribute to another tissue type, may not be initially programmed by myf-5, but by another type of determination factor, acting upstream of the MyoD family.

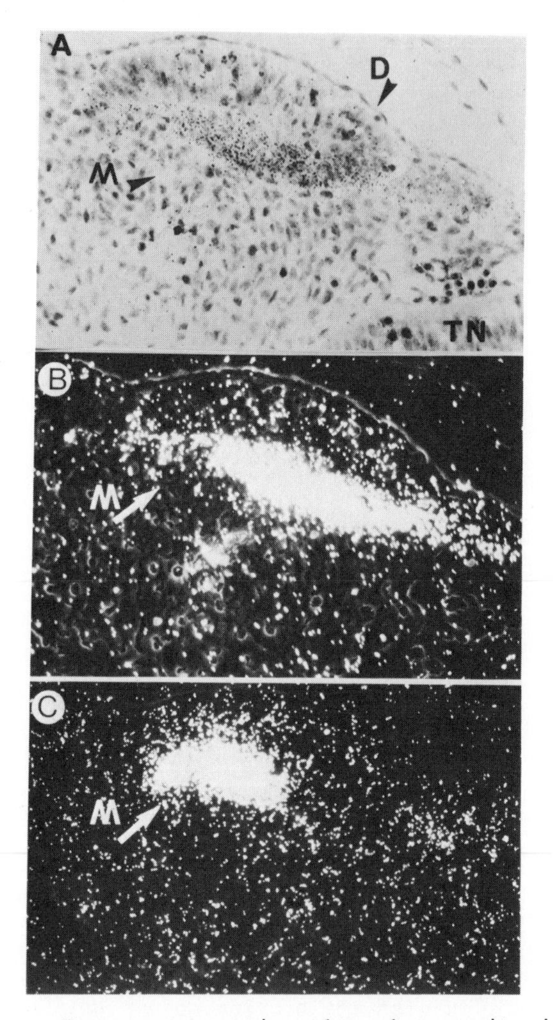


Fig.3.11 d embryo. Transverse section through a somite in the region of the mid back. The phase contrast picture (A) shows the myotome formed underneath the dermatome. The grains are readily visible. (B) dark field of the same section hybridized with the myf-5 probe. A strong signal covers the myotome adjacent to the neighbouring somites, while transcripts for myogenin (C) are concentrated in the central part of the myotome at this stage.
D = dermatome; M = myotome; TN = neural tube

**Table 1**. **Expression of myogenic regulatory genes and muscle markers in somites, myotomes and skeletal muscle during mouse embryogenesis**

| Days post coitum | 7.8-8 | 8.75 | 9 | 9.5 | 10.5 | 11.5 | 12.5 | 13.5 | 16.5 |
|---|---|---|---|---|---|---|---|---|---|
| myf-5 | + | ++ | +++ | ++ | +++ | ++ | + | - | - |
| myogenin | - | ++ | +++ | +++ | +++ | +++ | +++ | +++ | +++ |
| MyoD | - | - | - | - | ++ | +++ | +++ | +++ | +++ |
| myf-6 | - | - | + | + | ++ | - | - | - | ++ |
| α-cardiac actin | - | + | ++ | +++ | +++ | +++ | +++ | +++ | +++ |
| α-skeletal actin | - | +/- | + | + | ++ | ++ | +++ | +++ | +++ |
| $MLC1_A$ | - | - | - | + | +++ | +++ | +++ | +++ | +++ |
| $MLC1_V$ | - | - | - | - | - | - | - | - | + |
| M-CK | - | - | - | - | - | - | - | - | + |

MLC1A= myosin light chain gene 1a; MLC1V= myosin light chain 1V; M-CK= muscle creatine kinase
These results are based on *in situ* hybridization experiments cited in the text.

## MYOGENIC REGULATORY SEQUENCES IN DIFFERENTIATING MUSCLE

Apart from myf-5, the other mammalian myogenic regulatory sequences are only detectable in skeletal muscle. Each sequence has a distinct temporal pattern of expression (Bober et al., 1991; Ott et al., 1991; Sassoon et al., 1989). Myogenin transcripts accumulate in the newly formed myotome, from 8.5 d.p.c. Myogenin expression appears in the somites along the antero-posterior axis, with some delay compared to myf-5 transcripts, which also become restricted to the myotome as it forms. Myf-6 transcripts accumulate slightly later from 9 to 11 d.p.c. in the myotome. As indicated in Table 1, there is a two day time lag between myotome formation and MyoD expression. Some muscle protein genes are already expressed two days before MyoD is present, and are therefore transcribed in the absence of this factor. In the limb, too, myf-5 expression precedes that of myogenin and MyoD although it is very transient. MyoD and myogenin transcripts co-accumulate, whereas myf-6 is not detectable in embryonic limb muscle. This is in contrast to what happens in the myotome, and may reflect the fact that the cells which migrate out from the somite correspond to a different myogenic lineage present in the dermamyotome. Alternatively different environmental cues may be responsible for this striking difference in the myogenic regulatory programme. In the myotome myf-6 expression decreases rapidly from 10.5 days (Bober et al., 1991). Myf-5 expression also decreases and transcripts are no longer detectable by *in situ* hybridization by about 15 days anywhere in the foetus (Ott et al., 1991). Myf-6, unlike myf-5, begins to be re-expressed in all skeletal muscles from about this time and is a major transcript post-natally when MyoD and myogenin decline (Bober et al., 1991; Rhodes and Konieczny, 1989). There is no indication that these myogenic factors are preferentially associated with a particular type of muscle fibre, as foetal muscles mature. At earlier stages there is no indication that secondary versus primary fibres have a distinct myogenic factor composition.

The data summarized here are based on levels of transcripts. Antibodies for myf-5 and myf-6 which work on mouse sections are not yet available. However in the case of MyoD1 and myogenin it has been possible to look at the equivalent proteins (Cusella-de Angelis et al., 1991). In the limb the proteins are detectable together at the same time as the corresponding mRNAs. In the myotome, on the other hand, the MyoD protein is present from the same time as its mRNA but myogenin protein is not detectable prior to this time, while the transcript is clearly present earlier. At the protein level, therefore, MyoD and myogenin appear to be co-expressed in the somite, only from a later stage. Prior to this time in the myotome, myf-5 and myf-6 are the candidate myogenic factors. Indeed, the expression of this pair of genes during the early period of myotome maturation, and not in the differentiating muscle masses of the limb, is probably related to the absence of myogenin and MyoD proteins in the former.

## EXPRESSION OF MUSCLE STRUCTURAL GENES DURING MYOGENESIS *IN VIVO*

The myogenic regulatory genes code for DNA binding proteins, which have been shown to act as positive regulators of the transcription of a number of muscle structural genes (see Weintraub et al., 1991). In order to examine more closely possible differences in their function, it is of interest to compare the onset of expression of muscle structural genes with the myogenic regulatory sequences present. We have examined in detail the expression of genes of the actin (Sassoon et al., 1988) and myosin (Lyons et al., 1990b) multigene families in the embryo. Each of these genes has a distinct pattern of expression and the onset of muscle gene activation in the myotome is asynchronous; cardiac actin transcripts and protein precede those of the embryonic myosin heavy chain by a day. Different developmental isoforms both between gene families and within a family appear to be independently regulated. These conclusions are based on the detection of transcripts by *in situ* hybridization but where it has been possible to look at transcription rather than RNA accumulation the result is similar (Cox and Buckingham, 1991). There is no correlation between the

**Table 2. Expression of myogenic factors in forelimb buds and visceral arches during mouse embryogenesis**

| Days post coitum | 9.5 | 10 | 10.5 | 11 | 11.5 | 14 |
|---|---|---|---|---|---|---|
| myf-5 | - | - | + | ++ | + | - |
| myf-6 | - | - | - | - | - | +++ |
| myogenin | - | - | - | ++ | ++ | +++ |
| MyoD | - | - | - | ++ | ++ | +++ |

N.B. The respective timing of onset of expression of the different factors in the hindlimbs is the same, with a delay in time due to their later formation.

The results are based on in situ hybridization experiments cited in the text.

onset of expression of a myogenic regulatory sequence and the activation of structural genes except perhaps in a few cases (Table 1). A striking example is that of transcripts encoding the M-isoform of creatine phosphokinase which are only detectable from 13 days in the myotome, several days after the accumulation of myogenin and MyoD proteins (Cusella-de Angelis et al., 1991). The myosin light chain MLC1F is expressed early in the myotome (Lyons et al., 1990b) before the appearance of MyoD which has been shown to regulate the enhancer of this gene, the activity of which is necessary for MLC1F expression. In this context the gene is probably activated by myf-5 and or myf-6 in the myotome. Most of these genes are also expressed in the limb where MLC1F is probably activated by MyoD and/or myogenin. We would suggest that different muscle structural genes have different threshold requirements for activation by myogenic factors. A certain level of myf-5 and/or myf-6 is sufficient to activate the cardiac actin gene, but not the embryonic myosin heavy chain gene initially. Other factors may also be required at different threshold levels. It is also possible that the levels of myf-5 or myf-6 required to activate these genes in the myotome differ from those of MyoD or myogenin which are necessary in the limb. The overall pattern of expression of muscle structural genes is similar but not identical in the myotome and limb (M. Ontell, M. Buckingham et al., in preparation).

## THE MYOGENIC REGULATORY GENES IN DIFFERENT SPECIES

The myogenic factors are members of a superfamily of helix-loop-helix regulatory proteins. Genes related to the myogenic subclass have been found in many species, in vertebrates such as birds or Xenopus as well as in invertebrates, such as Drosophila or *Caenorabditis elegans*. In all species examined to date, the major site of expression of these genes is skeletal muscle or its precursors. No expression in cardiac muscle has ever been observed.

### Vertebrates

It is probable that all four myogenic factors are present in birds, as well as mammals (Y. Nabeshima and A. Fujisawa, personal communication). If each sequence is strictly equivalent to the mammalian counterpart - the sequence homologies are only of the order of 65-75% - then the relative timing of activation during myogenesis is different. In the chick, myogenin transcripts are only detected in the somite after myotome formation, whereas CMD1 (MyoD) transcripts are present earlier (Lyons et al., 1991b). In the quail the timing of expression of the factors qmf-1 (MyoD) qmf-2 (myogenin) and qmf-3 (myf-5) has been examined in detail. Qmf-1 is already expressed in early somites, prior to compartmentation, followed by qmf-3. Qmf-2 and α-actin accumulate together with qmf-1 and qmf-3 in the myotome (C. Emerson, personal communication).

Chick somitic cells are derived from distinct precursors. Somitic precursors from Hensen's node would contribute to the medial halves of the somites, while the lateral part of somites and notochord originate from progenitor cells localized more caudally, in the primitive streak (Selleck and Stern, 1991). The medial and lateral halves of the somites not only differ in their origin but also in their developmental fates. C. Ordahl and N. Le Douarin recently showed, using the classical chick-quail chimaera system, that the medial part contributes quasi-exclusively to the trunk musculature while the lateral half contributes only to the limb muscle masses (Ordahl and Le Douarin, in press), as already suggested by classical work on migration of muscle precursor cells from the dermamyotome (Christ et al., 1977). By making different combinations of the two halves of the newly formed somites, however they showed that they retain developmental flexibility in this respect. Earlier experiments indicated that the somite can also subdivide into rostral and caudal territories with respect to neural crest cell migration and motor nerve outgrowth (Keynes and Stern, 1984, Kalcheim and Teillet, 1989).

This said, it is worth noticing that in the mouse myf-5 transcripts in the early somite are mainly localized in the dorsomedial lip of the dermamyotome, but a non negligible accumulation is also observed over the whole dermamyotome, including the ventrolateral edge, from which cells are supposed to migrate to colonize the limb muscle masses, according to the classical work of Christ and his colleagues.

As the embryological manipulation of murine somites is still in its infancy we can only make some assumptions from the avian experiments. However, somite transplantations in 8 to 9 d.p.c. mouse embryos have been carried out with some success, although somite orientation has not yet been monitored (R. Beddington et al., in press). These experiments suggest that either the developmental fates of the somites are fixed later in embryogenesis, or that murine somites may stay "plastic" slightly later, keeping a potential for cellular re-localization.

The Xenopus embryo presents a different situation. Muscle formation differs from that in higher vertebrates in that a muscle specific gene such as the muscle actin gene is expressed several hours before the formation of the first somites. For some years now expression of $\alpha$-actin has been used as a marker of mesoderm induction.

Two myogenic factors have been reported in Xenopus, XMyoD and Xmyf-5 (Hopwood et al., 1989, 1991) and it is probable that Xenopus, too, has the equivalent of all four myogenic sequences.

Xmyf-5 transcripts are detected at the early gastrula stage, slightly before the onset of XMyoD expression and about two hours before accumulation of cardiac actin gene transcripts (Hopwood et al., 1991). Xmyf-5 always precedes XMyoD in somites as they form along the anteroposterior axis, as observed for myf-5 and myogenin in the murine embryo. Decline of its expression starts soon after neurulation and also precedes that of XMyoD. The main accumulation of XMyoD, and Xmyf-5 transcripts, is in muscle or mesodermal muscle precursor cells. However maternal MyoD transcripts are present in Xenopus before expression of the zygotic MyoD following mesoderm induction (Harvey, 1990). The two zygotic MyoD genes are weakly activated at the midblastula stage (Rupp and Weintraub, 1991; Harvey, 1991) prior to a specific activation in response to mesoderm induction. MyoD protein is detected from the midgastrula stage in cells surrounding the blastopore, most of which will form muscle later (Hopwood et al., 1991)

Invertebrates

Expression of a myogenic regulatory sequence prior to muscle formation is also observed in *C.elegans* where a single MyoD like sequence has been isolated. Endogenous MyoD protein is detected in the embryo at the 100 cell stage exactly at the time of clonal commitment of precursors of body wall musculature. Interestingly, the *C.elegans* embryo has the ability to express MyoD earlier than this (Krause et al., 1990). Expression of Ce-MyoD-ß galactosidase fusion protein can be observed in some blastomeres as early as the 28 cell stage, before the clonal commitment of muscle precursors, suggesting that a phenomenon of restriction rather than induction is important in the subsequent expression of MyoD in muscle.

A single MyoD homologue has also been isolated in Drosophila: nautilus (Michelson et al., 1990) or DMyd (Paterson et al., 1991). In Drosophila, muscle formation differs from that in vertebrates in that muscle masses appear to be founded by the clustering of pioneer cells after gastrulation, with subsequent recruitment of mesodermal cells. Expression of nautilus precedes that of the marker sequences of muscle differentiation. In the embryo, its expression is first detected in somatic mesoderm and then is restricted to a subset of muscle precursor cells. Subsequent expression in the adult is not detectable in many muscles. However in a "chase" experiment a transgene consisting of ß-galactosidase under the control of the Drosophila MyoD promoter, shows galactosidase expression in all muscles, suggesting that the precursors of these muscles expressed DMyoD1 (Paterson et al., 1991).

Expression of nautilus or Dmyd follows gastrulation in Drosophila. Several genes expressed before nautilus are known to be crucial for mesoderm determination and development. twist in Drosophila is expressed in muscle precursor cells and its expression disappears with the onset of muscle differentiation (Bate et al., 1991). twist is probably required for the activation of downstream mesodermal genes, for maintenance of its own expression and for full expression of snail which prevents expression in the mesoderm of genes destined to be active elsewhere (Leptin, 1991). Both these genes are transcriptional regulators. The great potential for genetic analysis in Drosophila should permit definition of the relationships between genes such as these and myogenesis.

## CONCLUDING REMARKS

In the mouse myf-5 expression is detectable at the time of somite formation (Montarras et al., 1991; Ott et al, 1991). From this stage, it seems probable that myf-5 expression marks muscle precursor cells and that subsequently myf-5 and the other three myogenic factors are directly involved in the transcriptional activation of muscle specific genes. The sequence of regulatory events acting upstream of myf-5 expression is not yet well defined. The murine twist gene is expressed only after formation of the somite and therefore does not play the same role as the twist gene in Drosophila (Wolf et al., 1991). The gene responsible for the brachyury phenotype, the T gene, which is expressed in the primitive streak during gastrulation and later in paraxial and lateral plate mesoderm, has been recently cloned (Hermann et al., 1990). Loss of function mutations in brachyury embryos lead to severe defects in the notochord, and to defective mesoderm (Hermann, 1991). This is therefore an important regulatory gene for mesoderm formation. Other genes such as FGF5, int2 or TGFß2 (Wilkinson et al., 1988; Hebert et al., 1991) are also expressed during gastrulation. It will be interesting to determine the putative relationships of such genes with the subsequent onset of myf-5 expression.

A key question raised by the similar properties of the myogenic factors when analysed *in vitro* is that of their interchangeability. The existence of specific patterns of expression during myogenesis *in vivo* argues against this hypothesis and suggests that each of them probably specifies different events in response to distinct upstream regulatory signals and in different combinations with other regulatory factors, ubiquitous or not. The use of transgenic mice, with appropriate reporter sequences under the regulatory elements of these myogenic factors should help to clarify the question of cell migration from the early somite. Targeted mutations of these genes by homologous recombination in the mouse is a way of creating loss of function mutants. If the phenotype obtained is not lethal at early stages this approach should help to clarify their respective roles, and the rules and regulations governing their participation in myogenesis.

## ACKNOWLEDGEMENTS

The laboratory is supported by grants from the Pasteur Institute, the Centre National de la Recherche Scientifique, the Institut National de la Santé et de la Recherche Médicale and the Association Française contre les Myopathies.

REFERENCES


Bate, M., Rushton, E. and Currie, D. A., 1991, Cells with persistent *twist* expression are the embryonic precursors of adult muscles in *Drosophila*, Development., 113: 79-89.

Beddington, R., Puschel, A. W. and Rashbass, P., 1991, Ciba Symposium 165. In press.

Bellairs, R., Ede, D. A. and Lash, J. W., 1986, "Somites in developing embryos", in : "Somites in developing embryos"., Plenum Press, New York and London.

Bober, E., Lyons, G., Braun, T., Cossu, G., Buckingham, M. and Arnold, H. H., 1991, The muscle regulatory gene, myf-6, has a biphasic pattern of expression during early mouse development, J. Cell Biol., 113: 1255-1265.

Braun, T., Bober, E., Winter, B., Rosenthal, N. and Arnold, H. H., 1990, Myf-6 a new member of the human gene family of myogenic determination factors: evidence for a gene cluster on chromosome 12, EMBO J., 9: 821-831.

Braun, T., Buschhausen-Denker, G., Bober, E., Tannich, E. and Arnold, H. H., 1989, A novel human muscle factor related to but distinct from MyoD1 induces myogenic conversion in 10T1/2 fibroblasts, EMBO J., 8: 701-709.

Charles de la Brousse, F. and Emerson, C. P., 1990, Localized expression of a myogenic regulatory gene, *qmf1*, in the somite dermatome of avian embryos, Genes and Development., 4: 567-581.

Chevallier, A., Kieny, M. and Mauger, A., 1977, Limb-somite relationship: origin of the limb musculature, J. Embryol. Exp. Morph., 41: 245-258.

Christ, B., Jacob, H. J. and Jacob, M., 1977, Experimental analysis of the origin of the wing musculature in avian embryos, Anat. Embryol., 150: 171-186.

Cox, R. and Buckingham, M., 1991, Actin and myosin genes are transcriptionally regulated during mouse skeletal muscle development, Dev. Biol., In press.

Cusella-De Angelis, M. G., Lyons, G., Sonnino, C., De Angelis, L., Vivarelli, E., Farmer, K., Wright, W. E., Molinaro, M., Bouché, M., Buckingham, M. and Cossu, G., 1991, MyoD1, myogenin independent differentiation of primordial myoblasts in mouse somites., J. Cell Biol., In press.

Davis, R. L., Weintraub, H. and Lassar, A. B., 1987, Expression of a single transfected cDNA converts fibroblasts to myoblasts., Cell., 51: 987-1000.

Ede, D. A. and El-Gadi, A. O. A., 1986, Genetic modifications of developmental acts in chick and mouse somite development, in : "Somites in developing embryos"., Plenum Press, New York and London.

Emerson, C. P., 1990, Myogenesis and developmental control genes, Current Opinion in Cell Biology., 2: 1065-1075.

Harvey, R. P., 1991, Widespread expression of MyoD genes in Xenopus embryos is amplified in presumptive muscle as a delayed response to mesoderm induction, Proc. Natl. Acad. Sci. USA., 88: 9198-9202.

Hebert, J. M., Boyle, M. and Martin, G. R., 1991, mRNA localization studies suggest that murine FGF-5 plays a role in gastrulation, Development., 112: 407-416.

Herrmann, B. G., 1991, Expression pattern of the brachyury gene in whole mount $T^{wis}/T^{wis}$ mutant embryos., Development., 113: 913-918.

Herrmann, B. G., Labeit, S., Poustka, A., King, T. R. and Lehrach, H., 1990, Cloning of the T gene required in mesoderm formation in the mouse, Nature., 343: 617-622.

Hopwood, N. D., Pluck, A. and Gurdon, J. B., 1989, MyoD expression in the forming somites is an early response to mesoderm induction in *Xenopus* embryos, EMBO J., 8: 3409-3417.

Hopwood, N. D., Pluck, A. and Gurdon, J. B., 1991, Xenopus myf-6 marks early muscle cells and can activate muscle genes ectopically in early embryos, Development., 111: 551-560.

Hopwood, N. D., Pluck, A., Gurdon, J. B. and Dilworth, S. M., 1991, Expression of XMyoD protein in early Xenopus laevis embryos, Development, In press.

Jacob, M., Christ, B. and Jacob, H. J., 1978, On the migration of myogenic stem cells into the prospective wing region of chick embryos. A scanning and transmission electron microscopic study., Anat. Embryol., 153: 179-193.

Kaehn, K., Jacob, H. J., Christ, B., Hinricksen, K. and Poelmann, R. E., 1988, The onset of myotome formation in the chick, Anat. Embryol., 177: 191-201.

Kalcheim, C. and Douarin, N. L., 1986, Requirement of a neural tube signal for the differentiation of neural crest cells into dorsal root ganglia, Dev. Biol., 116: 451-466.

Keynes, R. J. and Stern, C. D., 1988, Mechanisms of vertebrate segmentation, Development., 103: 413-429.

Krause, M., Fire, A., White-Harrison, S., Priess, J. and Weintraub, H., 1990, CeMyoD accumulation defines the body wall muscle cell fate during C.elegans embryogenesis, Cell., 63: 907-919.

Le Douarin, N., 1982, "The neural crest," Cambridge University Press, Cambridge

Leptin, M., 1991, *twist* and *snail* as positive and negative regulators during *Drosophila* mesoderm development, Genes Dev., 5: 1568-1576.

Lyons, G. E., Schiaffino, S., Sassoon, D., Barton, P. and Buckingham, M., 1990a, Developmental regulation of myosin gene expression in mouse cardiac muscle., J. Cell Biol., 111: 2427-2436.

Lyons, G. E., Ontell, M., Cox, R., Sassoon, D. and Buckingham, M., 1990b, The expression of myosin genes in developing skeletal muscle in the mouse embryo, J. Cell Biol., 111: 1465-1476.

Lyons, G. E., Buckingham, M. E. and Mannhertz, H., 1991a, α-actin proteins and gene transcripts are colocalized in embryonic mouse muscle, Development., 111: 451-454.

Lyons, G. E., Mühlebach, S., Moser, A., Masood, R., Paterson, B. M., Buckingham, M. E. and Perriard, J. C., 1991b, Developmental regulation of creatine kinase gene expression by myogenic factors in embryonic mouse and chick skeletal muscle., Development., In press.

Michelson, A. M., Abmayr, S. M., Bate, M., Martinez-Arias, A. and Maniatis, T., 1990, Expression of a MyoD family member prefigures muscle pattern in *Drosophila* embryos, Genes Dev., 4: 2086-2097.

Milaire, J., 1976, Contribution cellulaire des somites à la genèse des bourgeons de membres postérieurs chez la souris, Arch. Biol., 87: 315-343.

Miner, J. H. and Wold, B., 1990, Herculin a fourth member of the MyoD family of myogenic regulatory genes, Proc. Natl. Acad. Sci. USA., 87: 1089-1093.

Montarras, D., Chelly, J., Bober, E., Arnold, H., Ott, M. O., Gros, F. and Pinset, C., 1991, Developmental patterns in the expression of myf-5, MyoD, myogenin and MRF4 during myogenesis., The New Biologist., 3: 592-600.

Ordahl, C. and LeDouarin, N., Two myogenic lineages within the developing somite, Development., In press.

Ott, M.-O., Bober, E., Lyons, G. E., Arnold, H. H. and Buckingham, M. E., 1991, Early expression of the myogenic regulatory gene, myf5, in precursor cells of skeletal muscle in the mouse embryo, Development., 111: 1097-1107.

Ott, M. O., Robert, B. and Buckingham, M., 1990, Le muscle d'où vient-il?, Médecine/Science., 6: 653-663.

Paterson, B. M., Walldorf, U., Eldridge, J., Dübendorfer, A., Frasch, M. and Gehring, W. J., 1991, The *Drosophila* homologue of vertebrate myogenic-determination genes encodes a transiently expressed nuclear protein marking primary myogenic cells, Proc. Natl. Acad. Sci. USA., 88: 3782-3786.

Rhodes, S. J. and Konieczny, S. F., 1989, Identification of MFR4, a new member of the muscle regulatory factor gene family., Genes Dev., 3: 2050-2061.

Rugh, R., 1990, "The Mouse: its Reproduction and Development," Burgess Publishing Company, Minneapolis

Rupp, R. A. W. and Weintraub, H., 1991, Ubiquitous MyoD transcription at the midblastula transition precedes induction-dependent MyoD expression in presumptive mesoderm of X. Laevis, Cell., 65: 927-937.

Sassoon, D., Garner, I. and Buckingham, M., 1988, Transcripts of α-cardiac and α-skeletal actins are early markers for myogenesis in the mouse embryo, Development., 104: 155-164.

Sassoon, D., Lyons, G. E., Wright, W., Lin, V., Lassar, A., Weintraub, H. and Buckingham, M., 1989, Expression of two myogenic regulatory factors: myogenin and MyoD1 during mouse embryogenesis, Nature., 341: 303-307.

Scales, J. B., Olson, E. N. and Perry, M., 1990, Two distinct *Xenopus* genes with homology to MyoD1 are expressed before somite formation in early embryogenesis., Mol. Cell. Biol., 10: 1516-1524.

Selleck, M. and Stern, C., 1991, Fate mapping and cell lineage analysis of Hensen's node in the chick embryo., Development., 112: 615-626.

Taylor, M. V., Gurdon, J. B., Hopwood, N. D., Towers, N. and Mohun, T. J., 1991, *Xenopus* embryos contain a somite-specific, MyoD-like protein that binds to a promoter site required for muscle actin expression, Genes Dev., 5: 1149-1160.

Theiler, K., 1989, "The house mouse: atlas of embryonic development," Springer-Verlag, New York.

Wachtler, F., Christ, B. and Jacob, H. J., 1982, Grafting experiments on determination and migratory behaviour of presomatic and splanchnopleural cells in avian embryos., Anat. embryol., 164: 369.

Wachtler, F. and Jacob, M., 1986, Origin and development of the cranial skeletal muscles, Biblthca anat., 29: 24-46.

Weintraub, H., Davis, R., Tapscott, S., Thayer, M., Krause, M., Benezra, R., Blackwell, K., Turner, D., Rupp, R., Hollenberg, S., Zhuang, Y. and Lassar, A., 1991, The MyoD gene family: nodal point during specification of the muscle cell lineage., Science., 251: 607-617.

Wilkinson, D. G., Peters, G., Dickson, C. and McMahon, A. P., 1988, Expression of the FGF-related proto-oncogene *int-2* during gastrulation and neurulation in the mouse, EMBO J., 7: 691-695.

Wolf, C., Thisse, C., Stoetzel, C., Thisse, B., Gerlinger, P. and Perrin-Schmitt, F., 1991, The *M-twist* gene of *Mus* is expressed in subsets of mesodermal cells and is closely related to the *Xenopus X-twi* and the *Drosophila twist* genes, Dev. Biol., 143: 363-373.

Wright, W. E., Sassoon, D. and Lin, V. K., 1989, Myogenin, a factor regulating myogenesis has a domain homologous to MyoD., Cell., 56: 607-617.

# ROLE OF GROWTH AND MOTILITY FACTORS IN EARLY AVIAN DEVELOPMENT

Sarah E. Herrick[1], Heather Taylor[2] and
Grenham W. Ireland[1]

[1]Department of Cell and Structural Biology
University of Manchester, Coupland 3 Bldg.
Manchester, M13 9PL, UK

[2]Newnham College, University of Cambridge, UK

## INTRODUCTION

Mesoderm formation in avian embryos and higher vertebrates involves the formation of a primitive streak (Vakaet, 1984; Bellairs, 1986; Harrisson, 1989). This axial thickening arises by de-epithelialisation of the upper layer of the embryo, the epiblast, and the ingression and coalescence of cells. The primitive streak represents the first overt sign of the cephalo-caudal axis. Later, when the primitive streak regresses, the body plan becomes laid out as paired somites either side of the midline notochord and developing neural tube. From the primitive streak arise both the mesoderm and the definitive endoderm. The former will form the musculo-skeletal system, circulatory system and other connective tissues, and the latter will form the gut lining as well as contributing to components of internal organs. The formation of early mesoderm thus involves both morphogenesis and cytodifferentiation.

The formation of the primitive streak involves changes in cell-cell and cell-matrix adhesion, increased motility and the development of an invasive phenotype. The signals responsible in the chick are largely unknown, but considerable progress has been made in the amphibian embryo by studying two main families of mesoderm inducing factors (Slack *et al*, 1987; Smith, 1987; Rosa *et al*, 1988). The first are heparin binding growth factors which include basic fibroblast growth factor (bFGF) and the second are related to transforming growth factors beta, including Activin A, transforming growth factor beta 1 (TGFß1) and beta 2 (TGFß2).

Many 'growth' factors are known to have other effects besides stimulating proliferation. These include influencing cell motility, receptor expression and extracellular matrix biosynthesis. Thus it is possible that these factors could mediate most of the observed changes seen at gastrulation. Some of these factors have been implicated in chick gastrulation but usually indirectly. However, activin A has been reported to restore axis formation in isolated epiblasts (Mitrani *et al*, 1990) or to reverse axis formation in whole embryos (Cooke and Wong, 1991). It has also been suggested that bFGF can reverse

*Formation and Differentiation of Early Embryonic Mesoderm*
Edited by R. Bellairs *et al*., Plenum Press, New York, 1992

axis formation (Cooke and Wong, 1991) but substances known to inhibit FGF activity have been reported to block mesoderm formation (Mitrani *et al*, 1990).

Other factors which are presently less well characterised could also be involved. Scatter factor is an interesting protein whose effects include a paracrine influence on the motility of epithelial cells (Stoker *et al*, 1987; Stoker, 1989; Rosen *et al*, 1989) and promotion of epithelial cell invasion (Weidner *et al*, 1990). Scatter factor is produced only by certain embryonic fibroblasts, smooth muscle cells, and some cell lines but it only influences the motility of epithelial cells and thus might be a good candidate for an inducer molecule. When we grafted scatter factor-producing cells into primitive streak or pre-primitive streak stage chick embryos, a variety of embryonic anomalies resulted (Ireland *et al*, 1987; Stern *et al*, 1990). In particular, these embryos showed characteristic changes in the their axial structures: at one extreme secondary streaks were produced, and at the other bent streaks. These results were not seen with embryos grafted with non-producing cells. Subsequently, the experiments were repeated using beads loaded with purified scatter factor (Stern *et al*, 1990). Again embryos with anomalies were produced which were significantly different from those grafted with control beads.

We concluded that one explanation of the results was that scatter factor might be acting as a mesoderm inducing factor. However, there was no evidence that scatter factor was present in chick embryos and no *in vitro* assay system in which to test the activity of putative mesoderm inducing factors. The work reported here is an attempt to rectify these deficiencies.

## METHODS

### *Cell culture and collection of conditioned media*

Cell lines were grown in 9 cm. dishes in Dulbecco's modification of Eagles medium (DMEM) with 5 or 10% foetal calf serum at $37^0$C in an incubator gassed with 5% $CO_2$. Conditioned medium was obtained by taking just subconfluent dishes and replacing the normal growth medium with serum free medium, which was then harvested after 4 days.

### *Treatment of conditioned media*

Samples of conditioned media were heated to either 40, 60 or $100^0$C for 30min. Other samples were dialysed against a 100x volume of serum free medium for 24h. at $4^0$C. Acid or alkaline tolerance was tested by adjusting to pH 2 or 12 with 1M HCl or NaOH leaving for 2h. and then readjusting. Sensitivity to trypsin was tested by treatment with 100µg/ml for 45min. at $37^0$C before or after treatment with 200µg/ml soyabean trypsin inhibitor (SBTI). The protein content of all samples was determined by optical density measurements using protein assay reagent (Pierce) and BSA standards.

### *Early embryos*

Fertile hens' eggs were obtained from the Northern Biological Supply Company, Lytham St. Annes, and incubated until the desired stage. Embryos were removed from eggs aseptically into Tyrode's saline and staged according to Eyal-Giladi and Kochav (1976) and Hamburger and Hamilton (1951). Explants were dissected using iridectomy knives and entomological pins. Multiple explants used for producing conditioned media were grown in 4 well plates (Nunc).

For dissociated and reconstituted streak pieces ('pseudoexplants'), the method of Stern *et al* (1990) was used with minor modifications. For each preparation at least 25 primitive streaks of stage $3^+$ were dissected and gently washed in calcium and magnesium free (CMF) Tyrode's saline in a sterile Eppendorf tube. They were fully dissociated by incubation at $37^0$C for 20 min. and gentle pipetting using a Gilson with a blue tip. The Eppendorf was centrifuged at 6500rpm. in a microcentrifuge (MSE) for 20 s. The supernatant was discarded and fresh Tyrode's saline (with calcium and magnesium) gently added. The pellet was loosened from the tube with a fine pin and transferred to a dish where it was cut into explant sized pieces.

### *Late embryos*

7 and 10 day old embryos were transferred to CMF Tyrode's saline. The heads were cut from the bodies and the eyes and brain removed under the dissecting microscope using fine forceps. Each tissue was washed and placed in separate 2ml. syringes. The plunger was replaced and the tissue squirted into separate universals containing 0.25% trypsin and 0.02% versene. The tubes were incubated at $37^0$C for 20 minutes with pipetting to break up the lumps. The trypsin was stopped by the addition of medium containing serum, large lumps were allowed to settle, the cell suspension was then pelleted and plated at $2 \times 10^6$ cells per dish. For whole embryo cultures, the embryos were placed straight into the syringe. For cultures of chick skin fibroblasts, the skin was peeled off the back of 11 day embryos taking care to avoid underlying muscle. The tissue was dissociated and plated as above except that the feather germs were allowed to settle out and were discarded. Chick limb bud mesenchyme cells were produced according to the method of Ahrens *et al* (1977).

### *Scatter factor assay*

The method of Stoker and Perryman (1985) was used with minor modifications. Briefly the medium to be tested was added to the first well of a 96 well tissue culture plate and doubling dilutions performed in normal medium. 3000 MDCK cells were added to each well in a final volume of 200µl containing 5% serum. After overnight incubation, the plate was fixed in formol saline and stained with Harris's haematoxylin. After washing in running tap water, rinsing in distilled water and air drying the plate was examined under the microscope and scored for 'scattering' activity as previously described (Stoker and Perryman, 1985). The titre was the last well in which activity was detectable.

### *Proliferation measurements*

Subconfluent MDCK cells, deprived of serum for 24h., were trypsinised and replicate dishes set up with 20,000 cells per 35mm dish. At each of the subsequent times, four dishes were

Table I. Scatter factor activity in conditioned media of established cells and cell lines using MDCK assay.

| CELL LINE/STRAIN | TITRE | ACTIVITY |
|---|---|---|
| ras-NIH3T3 (mouse) | 128 | 640 |
| MRC-5 (human embryo lung) | 32 | 160 |
| F110 (human foetal skin) | 2 | 10 |
| FSF-37 (human adult foreskin) | 2 | 10 |
| Pigmented retinal epithelium (human) | <2 | <10 |
| Swiss 3T3K (mouse) | <2 | <10 |

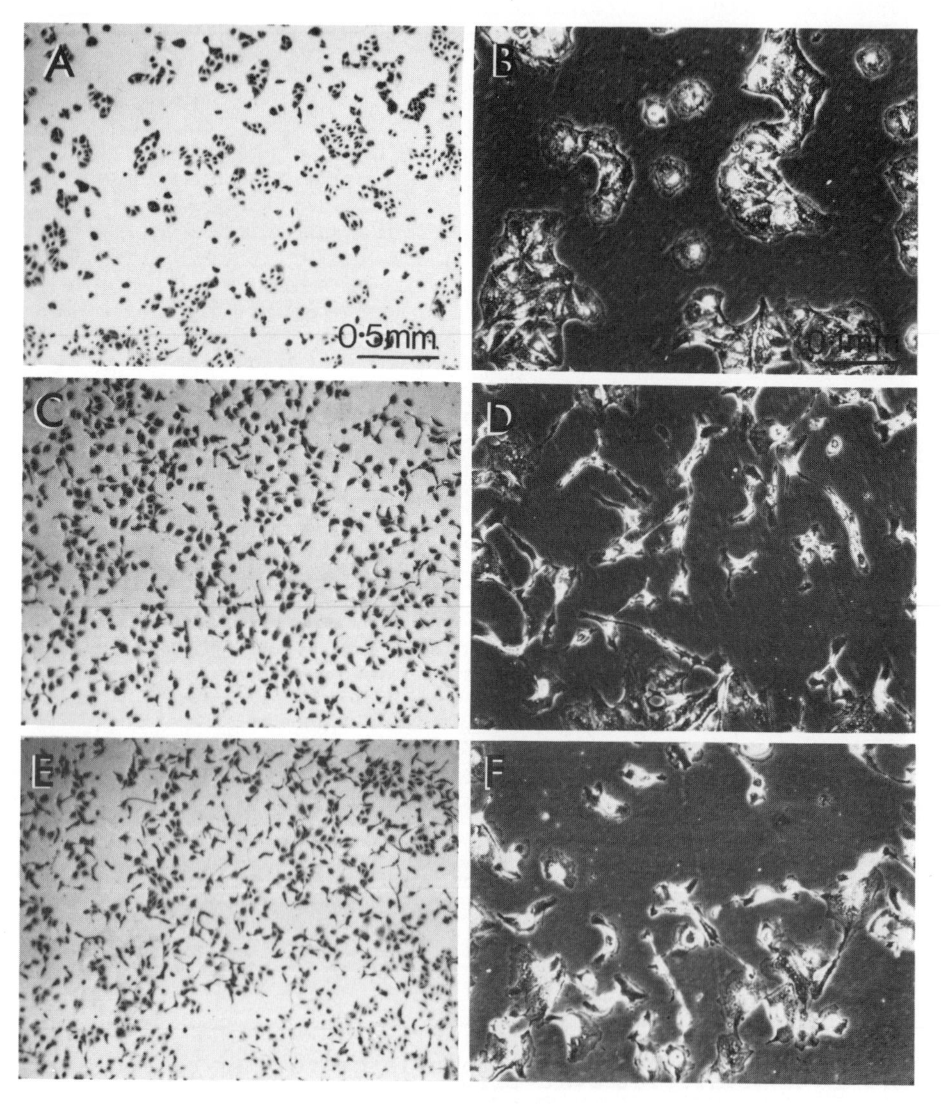


Fig 1. Scatter factor assays showing negative control (A & B), chick head fibroblast conditioned medium (C & D) and ras-NIH3T3 conditioned medium (E & F).

trypsinised and washed, and the pooled cells diluted in Isoton and counted in a Coulter counter and a final average produced.

*Collagen Gels*

The method of Schor and Court (1979) was used. Briefly, tails were obtained from young adult rats and the tendons removed aseptically using bone forceps and washed in 70% ethanol. The tendons were extracted in 3% acetic acid for 24h. at $4^0$C, centrifuged and the supernatant dialysed against distilled water for 48h. Optical density measurements were made and the concentration adjusted to 2mg/ml. The collagen gels were prepared by mixing 1ml. of 10x Ham's F12 medium, 0.5ml. of 7.5% sodium bicarbonate and 8.5ml. of collagen on ice. 300µl of solution was pipetted into each well of a 24 well plate (Nunc) and allowed to set by placing in the incubator. Later 200µl of solution was added and immediately the chick tissue was added using a Gilson P20. After returning to the incubator to set the gel was then overlaid with DMEM with or without bFGF (British Biotechnology) to a final concentration of 10µg/ml.

*Scoring*

At daily intervals explants in collagen gels were examined using an Olympus inverted (IMT-2) microscope equipped with hot stage. Explants were scored for outgrowth, presence of beating tissue and other features. The pooled results for three experiments are shown.

**RESULTS**

*Scatter factor activity is produced by some embryonic chick cells*

Following observations on the effect of scatter factor-producing cells and purified scatter factor on chick embryos, it became important to establish whether scatter factor was present in chick embryos. Whilst no immunological assay is available, Stoker and colleagues have developed a biological assay using the changes in the morphology of MDCK cells (Stoker and Perryman, 1985) which is distinctive since no other growth factors have been shown to provoke a similar response (but see Discussion). Normally MDCK cells seeded at low density and allowed to spread for 18h. form a mixture of single rounded cells and groups of spread cells in tight patches (Fig 1A & B). However, in the presence of scatter factor single cells become elongated, the tight patches become more spread but looser, and are dispersed or 'scattered' (Fig 1E & F).

Conditioned medium collected from established cells and cell lines was tested in the scatter factor assay. Only the known producers *ras*-NIH3T3 (mouse) and MRC-5 (human embryonic lung) fibroblasts had activity but other established cells had weak or no detectible activity (Table I).

Conditioned media collected from cultures of cells from primitive streak stage and pre-streak embryos tested negative (Table II). In most cases we obtained nearly confluent wells but, in our hands, it was difficult to propagate cells from these embryonic stages and obtain large volumes of media which could be concentrated. For this reason we decided to examine older embryos. Fibroblastic cells, presumably of mesenchymal origin, were established in culture from whole 7 or 10 day old chick embryos. When these were assayed scatter factor activity

Table II. Scatter factor activity in conditioned media of explant cultures from early chick embryos and primary cultures of fibroblastic cells from older embryos.

| | TITRE | ACTIVITY |
|---|---|---|
| **REGION OF EARLY CHICK EMBRYO** | | |
| Area opaca (stage 3) | <2 | <10 |
| Primitive streak (stage 3/4) | <2 | <10 |
| Hensen's node (stage 4) | 2 | 10 |
| Epiblast (stage XIV) | <2 | <10 |
| Hypoblast (stage XIV) | <2 | <10 |
| **ORIGIN OF FIBROBLASTIC CELLS** | | |
| Whole chick embryo (7 or 10 day) | 8 | 40 |
| Chick embryo head (7 or 10 day) | 16 | 80 |
| Chick embryo body (7 or 10 day) | 8 | 40 |
| Chick embryo brain (10 day) | 4 | 20 |
| Chick embryo eye (10 day) | 32 | 160 |
| Chick embryo head remains (10 day) | 16 | 80 |
| Chick skin (11 day) | <2 | <10 |
| Chick limb bud (4 day) | <2 | <10 |

Table III. Change in scatter factor activity with time of medium conditioned by either *ras*-NIH3T3 or chick embryonic eye-derived fibroblasts.

| | ras-NIH3T3 | | CE-SFA | |
|---|---|---|---|---|
| | TITRE | ACTIVITY | TITRE | ACTIVITY |
| **Day 0 (6h)** | 8 | 40 | <2 | <10 |
| **Day 1** | 32 | 160 | 4 | 20 |
| **Day 2** | 64 | 320 | 8 | 40 |
| **Day 3** | 128 | 640 | 8 | 40 |
| **Day 4** | 128 | 640 | 16 | 80 |
| **Day 5** | 128 | 640 | 32 | 160 |

Table IV. Comparison of various treatments on the scatter factor activity of *ras*-NIH3T3 and chick embryonic eye derived-fibroblast conditioned media.

| | ras-NIH3T3 | | CE-SFA | |
|---|---|---|---|---|
| **TREATMENT** | **TITRE** | **ACTIVITY** | **TITRE** | **ACTIVITY** |
| Undiluted | 128 | 640 | 32 | 160 |
| Dialysed | 64 | 320 | 8 | 40 |
| $4^0$C | 128 | 640 | 32 | 160 |
| $40^0$C | 64 | 320 | 16 | 80 |
| $60^0$C | 4 | 20 | 4 | 20 |
| $100^0$C | <2 | <10 | <2 | <10 |
| Trypsin/SBTI | <2 | <10 | <2 | <10 |
| SBTI | 32 | 160 | 16 | 80 |
| pH2 | 128 | 640 | 32 | 160 |
| pH12 | 128 | 640 | 32 | 160 |

Table V. Scatter factor activity (/ml) and measured protein concentration (µg/ml) of fractions recovered following fractionation of conditioned media.

| **AMMONIUM SULPHATE PPTN.** | **ras-NIH3T3** | | **CH-SFA** | | **CE-SFA** | |
|---|---|---|---|---|---|---|
| | **ACT.** | **PROT.** | **ACT.** | **PROT.** | **ACT.** | **PROT.** |
| Undiluted | 640 | 53 | 80 | 45 | 160 | 47 |
| 0-10% | <10 | 4 | <10 | 1 | Not Done | |
| 11-20% | 20 | 3 | <10 | 3 | Not Done | |
| 21-30% | 10 | 9 | <10 | 5 | Not Done | |
| 31-50% | 40 | 30 | 20 | 8 | Not Done | |
| 51-80% | 40 | 53 | 40 | 15 | Not Done | |
| >81% | <10 | 12 | <10 | 3 | Not Done | |
| **HEPARIN AGAROSE CHROMAT.** | **ras-NIH3T3** | | **CH-SFA** | | **CE-SFA** | |
| | **ACT.** | **PROT.** | **ACT.** | **PROT.** | **ACT.** | **PROT.** |
| 0.1M | <10 | 22 | <10 | 18 | <10 | 37 |
| 0.3M | <10 | 7 | <10 | 7 | 10 | 23 |
| 0.6M | 10 | 9 | 20 | 2 | 10 | 5 |
| 1.0M | 160 | 3 | 40 | <1 | 20 | 3 |
| 1.5M | <10 | 2 | <10 | 2 | <10 | 1 |
| 3.0M | <10 | <1 | <10 | 4 | <10 | 1 |

was detected (Table II). In an attempt to localise the origin of the cells producing this activity, pieces from different regions of embryos were isolated, dissociated and cultures established. Scatter factor activity was detected in many cultures, except skin and limb bud mesenchyme, but higher activities were consistently obtained from cultures derived from embryonic heads (chick head scatter factor activity, CH-SFA) and the highest titres were obtained from eye-derived fibroblasts (chick eye scatter factor activity, CE-SFA) (Table II). Whilst these activities were always lower than the known standard producer line *ras*-NIH3T3 they were comparable to the other producer MRC-5 and showed similar morphological changes to both (Fig 1C & D).

In order to exclude the possibility that the activity was carried over in the isolated material, conditioned medium was assayed at different times after culture (Table III). The activity of the medium was found to increase with increasing time in culture although the experiment could not be continued because the cells would not survive longer than 5 days in serum free medium.

It was important to establish whether this activity was the same as scatter factor. A comparison with *ras*-NIH3T3 conditioned medium following various treatments showed that both activities were stable at acid and alkaline pH, and were non-dialysable, heat and trypsin sensitive (Table IV). A partial fractionation was carried out showing that both activities were in the 30-80% ammonium sulphate fractions and both eluted off a heparin agarose column with 1.0M NaCl (Table V).

The morphological changes produced on MDCK cells by scatter factor are known to be caused without effect on their proliferation (Gherardi *et al*, 1989). When 4 x $10^4$ cells were plated in different media and 48 h. later counted, the following means +/- standard deviations were obtained, 5% serum (2.85 x $10^5$ +/- 1.7 x $10^4$), 5% serum + ras-NIH3T3 conditioned medium diluted 1:4 (2.60 x $10^5$ +/- 1.6 x $10^4$) and 5% serum + chick eye fibroblast conditioned medium diluted 1:2 (2.27 x $10^5$ +/- 3.5 x $10^4$). Therefore we could detect no stimulatory effects of CE-SFA on proliferation.

CH-SFA or CE-SFA has not been characterised further due to poor yields and loss of activity during purification.

### *An assay for testing the effect of soluble factors on differentiation*

Explants of chick tissue, whilst attaching and outgrowing in conventional tissue culture, often show poor rates of proliferation and differentiation. Moreover, the type of results obtained with different explants can differ depending on their speed of attachment and size. Explants of primitive streak attach well but pieces of epiblast often roll up and form spherical vesicles. In seeking a reproducible method, we have cultured tissue in standard type I three-dimensional collagen gels.

We examined the ability of explants of various regions to grow out into the gel. Explants of primitive streak (stage 3 and 4) grew most vigorously into the gel (Fig 2). Some of the cell processes were extremely long and resembled neurites. After 3-4 days, the central regions of the primitive streak

Table VI. Scoring of explants, grown in high or low serum with or without fibroblast growth factor, for 'muscle' or 'blood'.

| | 0.5% serum | | 0.5% serum + FGF (10µg/ml) | |
|---|---|---|---|---|
| 'Muscle' | 8/13 | 62% | 10/13 | 77% |
| 'Blood' | 1/13 | 8% | 0/13 | 0% |
| Neither | 4/13 | 30% | 3/13 | 23% |
| | **5% serum** | | **5% serum + FGF (10µg/ml)** | |
| 'Muscle' | 0/12 | 0% | 9/13 | 69% |
| 'Blood' | 11/12 | 92% | 0/13 | 0% |
| Neither | 1/12 | 8% | 4/13 | 31% |

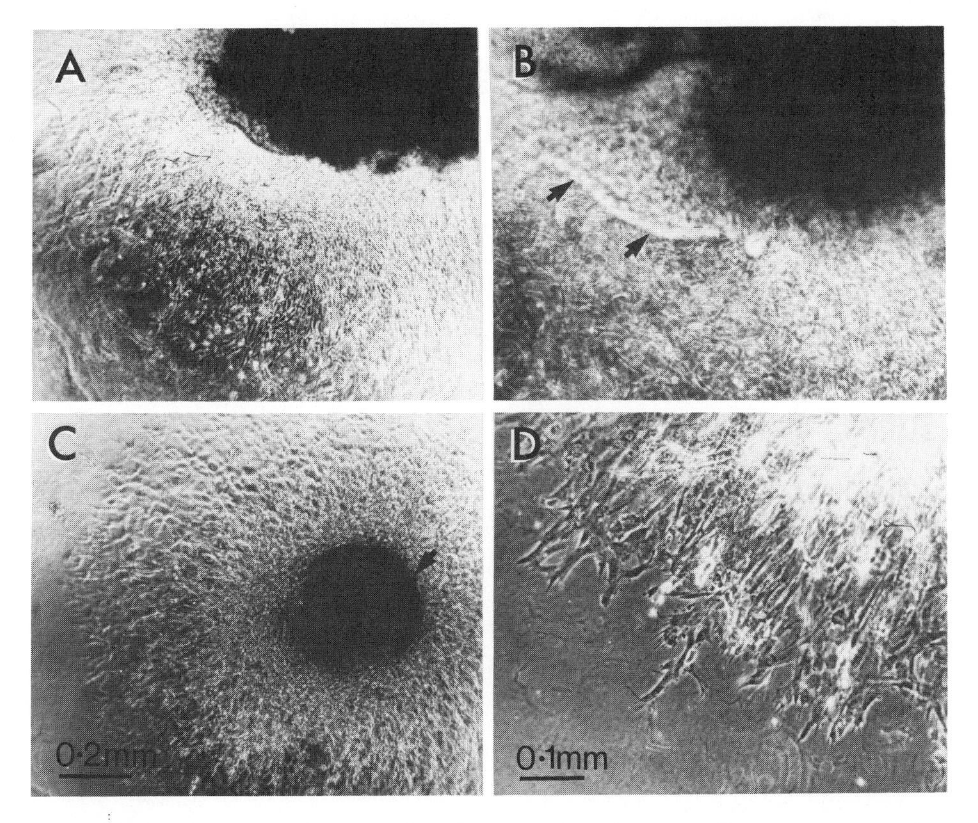


Fig 2. Explants of streak tissue showing profuse outgrowth into collagen gel and spontaneously beating region (A & B). Streak explant displaying central vesicle (C) and cells migrating into collagen gel (D).

explants appeared to show various forms of differentiation. Some explants had regions where groups of cells showed spontaneous contraction, 'muscle' (Fig 2A & B). In other explants, vesicles of different sizes formed (Fig 2C). Some of these vesicles and some central portions of explants developed a red coloration, 'blood'. At the periphery of all explants cells invaded the collagen gel (Fig 2D).

The type of morphology shown by the explants varied considerably and in order to try and obtain more reproducible results we examined the results obtained from different regions of the primitive streak. We found that nodes grew and differentiated poorly compared to posterior regions of the streak. However, even with similar sized pieces of posterior streak, the results obtained were not consistent. One problem appeared to be due to batches of collagen used. Another problem could be that, despite accurate staging, cells with different commitment will be taken in each explant because of the dynamic nature of the population of cells in the streak. Therefore, we decided to use dissociated primitive streaks.

Full thickness streaks from over 20 embryos were dissociated and then re-associated and small pieces used as 'pseudoexplants' in the collagen gel as before. Now we obtained reproducible results (Table VI). Over 60% of 'pseudoexplants' set up in medium containing 0.5% serum formed muscle whereas none formed muscle in medium containing 5% serum ($X^2 = 9.7$, $P = < 0.01$, when tested as a 2 x 2 contingency table with 1 df).

We wanted to test the effect of a factor known to induce mesoderm in *Xenopus*. We chose to examine the effect of bFGF in this system. The addition of bFGF (10µg/ml) caused an increase in 'muscle' in 5% serum containing medium ($X^2 = 12.0$, $P = < 0.001$, when tested as a 2 x 2 contingency table with 1df), but little effect in 0.5% serum containing medium. The appearance of explants with 'blood' was found to increase if medium containing 5% serum was used but decrease when bFGF was included.

**DISCUSSION**

Our results imply that a 'scatter factor activity' is produced by cultured mesenchymal cells from 7-10 day chick embryos. While we cannot be sure that it is identical to scatter factor, the activity is similar in all respects we have tested to the activity in conditioned media from *ras*-NIH3T3 cells from which scatter factor has been purified by others (Gherardi *et al*, 1989). It scatters MDCK cells, and is precipitated by 30-50% ammonium sulphate, eluted from a heparin agarose column with 1.0M NaCl and fails to stimulate MDCK cell proliferation.

Three growth factor activities have been isolated from the bovine eye termed eye-derived growth factors (EDGF) I, II & III (Courty *et al*, 1985). EDGF-I and EDGF-II have been identified as bFGF and acidic FGF, respectively, whilst EDGF-III does not bind heparin (reviewed McAvoy and Chamberlain, 1990). bFGF, which binds heparin, has been purified from chick embryos (Kimura *et al*, 1989) but FGF is negative in scatter factor assays (personal observations) so none of these factors are likely to be scatter factor.

Recently it has been observed that the N-terminal sequence of the B subunit of scatter factor is highly related (if not identical) to hepatocyte growth factor (HGF)/hepatopoetin A (Gherardi and Stoker, 1990; Weidner *et al*, 1991). This is a potent mitogen for rat hepatocytes and is thought to be involved in liver regeneration. HGF mRNA is distributed in rat lung, liver, brain and kidney (Tashiro *et al*, 1990), and therefore its presence in 7 day chick embryos would not be surprising. However, our results suggested that head or eye-derived fibroblasts provided the best source of the activity, and it will be interesting to examine the distribution of HGF mRNA in the early chick. It may be that a family of similar factors are involved. A factor produced by human placenta also has 'scatter factor activity' but may be a different protein (Rosen *et al*, 1990).

We have not been able to detect scatter factor activity in cells derived from any region of the early chick embryos. However, it remains a distinct possibility that the methods we have employed are not sensitive enough to detect the production of a small amount of factor. With the recent appearance of sequence data on scatter factor, biochemical methods of greater sensitivity will soon be available to test this possibility more fully. Other factors are being identified in the early chick embryo although their role is still not clear. TGFs have been implicated using a bioassay (Macintyre *et al*, 1988) and located using immunological methods (Sanders and Prasad, 1991). Insulin-like growth factors and their receptors are present (Ralphs *et al*, 1990; Scavo *et al*, 1991) and FGF is found in the yolk and albumin of unfertilised eggs (Seed *et al*, 1988) developing striated muscle (Joseph-Silverstein *et al*, 1989) and the developing nervous system (Kalcheim and Neufeld, 1990).

We have explored the use of collagen gels to examine the overt differentiation of streak tissue. We found that consistent results could be obtained by dissociating streak cells and re-associating them to form 'pseudoexplants'. In our hands spontaneously contracting muscle was only formed when 'pseudoexplants' were grown in 0.5% serum and not in 5% serum. However, addition of bFGF to explants cultured in 5% serum caused the reappearance of spontaneously contracting tissue.

bFGF is known to be able to induce mesoderm in the animal cap assay in *Xenopus*. Whilst it seems to induce more ventral mesodermal structures than activin it does induces muscle at high concentrations (Slack *et al*, 1987). Therefore our finding that bFGF increases 'muscle' in explants grown in 5% serum is perhaps not too surprising. Vesicles similar to those we have observed were attributed by Zagris (1980) to blood formation so it is possible that bFGF shifts the differentiation pathway from blood to muscle formation. Alternativley it may counteract another factor present in serum. However, we need to use appropriate biochemical markers to establish which differentiated tissues are formed and whether the contracting tissue is skeletal or cardiac muscle.

We believe that the effect of growth factors, individually and in combination, might be tested using this assay. Ideally an equivalent to the animal cap assay in *Xenopus* is needed but further work on manipulating chick epiblast without damaging cells will be required.

**ACKNOWLEDGMENTS**

We thank the Medical Research Council for a grant supporting S.H. and the Wellcome Trust for a vacation scholarship for H.T. We particularly thank Claudio Stern (University of Oxford) and Anne-Marie Grey for helpful tips and useful discussions and other members of the department for providing conditioned media.

## REFERENCES

Ahrens, P. B., Solursh, M. and Reiter, R. S., 1977, Stage-related capacity for limb chrondrogenesis in cell culture, Devl. Biol., 60:69-82.

Bellairs, R., 1986, The primitive streak, Anat. Embryol., 174:1-14.

Cooke, J. and Wong, A., 1991, Growth factor-related proteins that are inducers in early amphibian development may mediate similar steps in amniote (bird) development, Development., 111:197-212.

Courty, J., Loret, C., Moenner, M., Chevallier, B., Lagente, O., Courtois, Y. and Barritault, D., 1985, Bovine retina contains three growth factor activites with different affinity to heparin: eye derived growth factor I, II, III. Biochemie 67: 265-269.

Eyal-Giladi, H. and Kochav, S., 1976, From cleavage to primitive streak formation: a complementary normal table and a new look at the first stages of the development of the chick, Devl. Biol., 49:321-337.

Gherardi, E. and Stoker, M., 1990, Hepatocytes and scatter factor, Nature, 346:228.

Gherardi, E., Gray, J., Stoker, M., Perryman, M. and Furlong, R., 1989, Purification of scatter factor, a fibroblast-derived basic protein which modulates epithelial interactions and movement, Proc. Natl, Acad. Sci. USA, 86:5844-5848.

Hamburger, V. and Hamilton, H. L., 1951, A series of normal stages in the development of the chick, J. Morph., 88:49-92.

Harrisson, F., 1989, The extracellular matrix and the cell surface, mediators of cell interactions in chicken gastrulation, Int. J. Dev. Biol., 33:417-438.

Ireland, G. W., Stern, C. D. and Stoker, M., 1987, Human MRC-5 cells induce a secondary primitive streak when grafted into chick embryos, J. Anat., 152:223-224.

Joseph-Silverstein, J., Consigli, S. A., Lyser, K. M. and ver Pault, C., 1989, Basic fibroblast growth factor in the chick embryo: immunolocalisation to striated muscles and their precursors, J. Cell Biol. 108:2459-2466.

Kalcheim, C. and Neufeld, G., 1990, Expression of basic fibroblast growth factor in the nervous system of early avian embryos, Development, 109:203-215.

Kimura, I., Gotoh, Y. and Ozawa, E., 1989, Further purification of a fibroblast growth factor-like factor from chick embryo extract by heparin-affinity chromatography, In Vitro, 25:236-242.

Macintyre, J., Hume, D. D., Smith, J. and McLachlan, J., 1988, A microwell assay for anchorage independent cell growth, Tiss. Cell 20:331-338.

McAvoy, J.W. and Chamberlain, C.G., 1990, Growth factors in the eye, Prog. Growth Factor Res. 2:29-43.

Mitrani, E., Grunebaum, Y., Shohat, H. and Ziv, T., 1990, Fibroblast growth factor during mesoderm induction in the early embryo, Development, 109:387-393.
Ralphs, J. R., Wylie, L. and Hill, D. J., 1990, Distribution of insulin-like growth factor peptides in the developing chick embryo, Development, 109:51-58.
Rosa, F., Roberts, A. B., Danielpour, D., Dart, L. L., Sporn, M. B., and Dawid, I. B., 1988, Mesoderm induction in amphibians: the role of TGFß2 like factors, Science 239:783-785.
Rosen E. M., Goldberg, I. D., Kacinski, B. M., Buckholz, T. and Vinter, D. W., 1989, Smooth muscle releases an epithelial cell scatter factor which binds to heparin, In Vitro, 25:163-173.
Rosen, E. M., Meromsky, L., Romero, R., Setter, E. and Goldberg, I., 1990, Human placenta contains an epithelial scatter protein, Biochem. Biophys. Res. Comm., 168:1082-1088.
Sanders, E. J., and Prasad, S., 1991, Possible roles for TGFß1 in the gastrulating chick embryo, J. Cell Sci., 99:617-626.
Scavo, L. M., Serrano, J., Roth, J. and de Pablo, F., 1991, Genes for the insulin receptor and the insulin-like growth factor I recptor are expressed in the chicken embryo blastoderm and throughout organogenesis, Biochem. Biophys. Res. Comm. 176:1393-1401.
Schor, S. and Court, J., 1979, Different mechanisms involved in the attachment of cells to native and denatured collagen. J. Cell Sci. 38:267-281.
Seed, J., Olwin, B. B. and Hauschka, S., 1988, FGF levels in the whole embryo and limb bud during chick development, Devl. Biol., 128:50-58.
Slack, J. M. W., Darlington, B. G., Heath, J. K. and Godsave, S. F., 1987, Mesoderm induction in the early *Xenopus* embryo by heparin-binding growth factors, Nature, 326:197-200.
Smith, J. C., 1987, A mesoderm inducing factor is produced by a *Xenopus* cell line, Development, 29:109-115.
Stern, C. D., Ireland, G. W., Herrick, S. E., Gherardi, E., Gray, J., Perryman, M. and Stoker, M., 1990, Epithelial scatter factor and development of the chick embryonic axis, Development, 110:1271-1284.
Stoker, M., 1989, Effect of scatter factor on motility of epithelial cells and fibroblasts, J. Cell Physiol., 139:565-569.
Stoker, M. and Perryman, M., 1985, An epithelial scatter factor released by embryo fibroblasts, J. Cell Sci., 77:209-223.
Stoker, M., Gherardi, E., Perryman, M. and Gray, J., 1987, Scatter factor is a fibroblast-derived modulator of epithelial cell mobility. Nature, 327:239-242.
Tashiro, K., Hagiya, M., Nishizawa, T., Seki, T., Shimonishi, M., Shimizu, S. and Nakamura, T., 1990, Deduced primary structure of rat hepatocyte growth factor and expression of the mRNA in rat tissues, Proc. Natl. Acad. Sci. USA., 87:3200-3204.
Vakaet, L., 1984, The initiation of gastrular ingression in the chick blastoderm, Am. Zool., 24:555-562.

Weidner, K. M., Behrens, J., Vanderkerkhove, J. and Birchmeier, W., 1991, Scatter factor: molecular characteristics and effect on the invasiveness of epithelial cells, J. Cell Biol., 111:2097-2108.
Zagris, N., 1980, Erythroid cell differentiation in unicubated chick blastoderm in culture, J. Emb. exp. Morph., 58:209-216.

# THE ROLE OF THRESHOLDS AND MESODERM INDUCING FACTORS IN AXIS PATTERNING IN XENOPUS

Jeremy B.A. Green

Laboratory of Developmental Biology
National Institute for Medical Research
The Ridgeway, Mill Hill
London NW7 1AA, UK

## ABSTRACT

Mesoderm of amphibia such as the frog *Xenopus* is formed via inductive interactions in which cells of the embryo's vegetal hemisphere act on cells of the overlying animal hemisphere. Two groups of molecules mimic this inductive action, namely members of the fibroblast growth factor (FGF) and transforming growth factor beta families. The factors yield tissue types characteristic of ventro-posterior and dorso-anterior parts of the tadpole body respectively. Experiments with disaggregated animal pole cells reveal behaviour reminiscent of theoretically postulated morphogens: at least three different tissue types are elicited by activin dose ranges separated by sharp thresholds. Relationships between factor doses and genes induced are consistent with quite simple gradients of concentration being responsible *in vivo* for patterning of the mesoderm at early stages.

## INTRODUCTION: SPEMANN, NIEUWKOOP AND INDUCTION

Embryonic induction was first described in detail and, effectively, defined at the beginning of this century by Hans Spemann in his work on the amphibian eye . He showed how underlying optic vesicle can induce lens differentiation in overlying ectoderm (see Saha, Spann et al., 1989) and this has remained a "paradigm for induction in general" ever since (Hamburger, 1988). It is no surprise, therefore, that Spemann saw the results of the famous Organiser experiment as another example of the phenomenon, that is as an (albeit complex) induction. In the Organiser Experiment a dorsal blastopore lip transplanted into the ventral side of a host embryo was able to recruit host tissues to form an entire second body axis in mirror-symmetry with the host's. The Organiser was able both to induce new tissue types and organise long-range pattern. Thus, Spemann made no distinction between the processes of induction and pattern organisation. Nowadays it is possible to make a distinction: induction can be defined as an all-or-nothing switch in which one part of the embryo diverts the specification of a neighbouring part, while the Organiser

*Formation and Differentiation of Early Embryonic Mesoderm*
Edited by R. Bellairs *et al.*, Plenum Press, New York, 1992

effect is an example of Positional Information in which a whole sequence of parts is patterned by relatively long range influences (see Wolpert, 1989 for a detailed comparison of these two processes).

In an important series of experiments in the 1970s, Nieuwkoop and Nakamura showed that, in amphibians including the clawed frog Xenopus, mesoderm arises through the action of cells of the embryo's vegetal hemisphere on overlying animal cells (see Nieuwkoop, Johnen et al., 1985). This interaction is known as mesoderm induction, analogous to other inductions. In the last few years, great leaps have been made in our understanding of mesoderm induction thanks to the discovery and characterisation of two types of mesoderm inducing factor (MIF). It is perhaps ironic that one conclusion that is now emerging from this work is that mesoderm induction is not, in fact, an induction at all. Instead, mesoderm "induction" is a process of transmission of positional information. In this chapter I will present evidence that MIFs can act like the much-theorised but rarely demonstrated morphogens of simple positional information models. Furthermore, I will describe specific characteristics of each of the two principal classes of MIF suggesting that their distributions and local concentrations constitute a two dimensional coordinate system that specifies the pattern of mesoderm as early as the blastula stage.

## ACTIVIN AND FGF: THE TWO MIFS

Nieuwkoop showed that Xenopus animal cap explants cultured in isolation become atypical epidermis but when juxtaposed with vegetal pole explants form significant amounts of muscle and other mesodermal tissue. Applying various substances to isolated animal caps in an effort to mimic this mesodermalisation, Smith discovered mesoderm inducing activity or factor (Smith, 1987). Smith used a medium conditioned by a Xenopus cell line (XTC) from which he purified the active fraction, activin A (Smith, Price et al., 1990). Activin A is a member of the transforming growth factor beta superfamily, previously identified and named for its action in the adult reproductive cycle and differentiation of red blood cells (erythrocyte differentiation factor). At about the same time, Slack and Kimelman both screened a large number of "likely" candidate MIFs including growth factors and found that bovine basic fibroblast growth factor (FGF) is active in the animal cap assay (Slack, Darlington et al., 1987; Kimelman and Kirschner, 1987).

From the beginning, there were hints that the two MIFs have slightly different effects. Activin makes cap cells to converge and extend forming a long narrow protuberance. Histological analysis confirms the impression that this protuberance is like the dorsal axis of the embryo: it often contains notochord cells and segmented somitic muscle. Activin-induced caps bear a resemblance to explanted dorsal marginal zones. FGF, on the other hand, generates caps that resemble ventrolateral marginal zone explants: they differentiate as fluid-filled vesicles and contain little muscle which is generally unsegmented. These early impressions were strengthened by a side-by-side comparison of Xenopus activin A (purified from XTC medium) and recombinant Xenopus bFGF. This study (Green, Howes et al., 1990) showed that increasing doses of activin yielded tissue types of an increasingly dorsal character, while FGF gave only non-dorsal cell types.

## DOSE-RESPONSE EXPERIMENTS REVEAL MORPHOGEN PROPERTIES OF ACTIVIN

The increasingly dorsal character of induced tissue type with increasing dose of activin is a weak trend, requiring a relatively large number of animal caps to be observed at a statistically significant level (Green, Howes et al., 1990). Nonetheless, the very fact of a dose-dependent specification of cell type means that activin might be capable of generating positional information, that is it could be a morphogen. To understand this property better it was necessary to develop an assay system in which the effects would be more clear cut. One obvious limitation of the animal cap assay is that the cap is a thick pad of tissue that "rounds up" during incubation in normal amphibian culture medium, as shown in fig.1. The outside of such rounded-up caps (the original outer layer of the blastula) is impervious to factors (Cooke, Smith et al., 1987). It is thus clear that cells on the inner surface of a freshly dissected cap are more exposed to factors in the medium than either cells within the cap or cells that form the outer layer. To avoid the

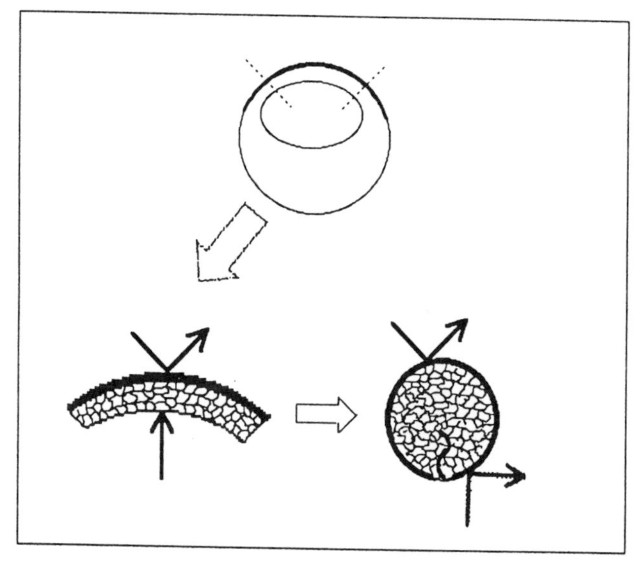

Fig. 1. Rounding up of animal cap explants: Animal caps are excised from embryos (top) as a two- or several-layer pad of cells (lower left). Inner layer but not outer layer cells are exposed to factors (solid arrows) but incubation in standard amphibian media causes the cells to round up with time and exclude inducing factors (Cooke et al., 1987)

rounding-up problem, caps were disaggregated into single cells by incubation in medium lacking calcium and magnesium ions. Cells thus treated are viable and can be reaggregated by restoration of calcium and magnesium to the medium. Outer layer cells of the cap were removed in this way since these are less responsive to factors than inner layer cells even when their inner surface is exposed (unpublished observations). Dispersed cells were then treated for an hour with varying doses of activin, washed by gentle centrifugation-resuspension and reaggregated by restoration of

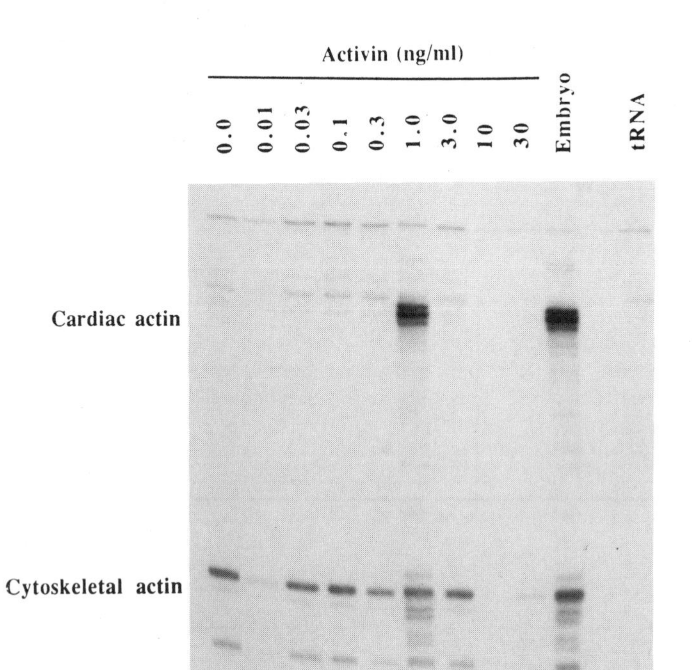


Fig. 2. Induction of muscle actin in a single cell aggregate experiment: Aggregates were incubated until control stage 18/19 (tailbud) and RNA extracted. Expression of actin genes was analysed using RNAse protection with a probe that recognises both a muscle-specific "cardiac" actin and a cytoskeleton actin which acts as an internal loading control. Note that the cardiac actin is found in all tadpole muscle, not just the heart. (See Green & Smith, 1990)

calcium and magnesium ions. Single cell aggregates were then incubated to various stages and analysed for gene expression (generally by RNAse protection) or histological differentiation (using antibody staining). Fig.2 shows the striking result of one such experiment: muscle, the dominant mesodermal cell type in tadpoles, is expressed only within a narrow dose-range or window. This range is bounded by sharp thresholds (fig.2 and Green and Smith, 1990). Below the lower threshold, aggregates express epidermal-specific keratin, as do untreated cells. Above the upper threshold cells express an unknown cell type that has the ability to induce neural gene expression when mixed with untreated cells. In our early experiments we found that notochord was induced in the same dose range as muscle (Green and Smith, 1990) but more careful recent work has shown that the notochord-inducing range overlaps with but is, in fact, somewhat higher than the muscle-inducing dose range (Green & Smith, manuscript in preparation).

This sequence of tissue types, skin-muscle-notochord-neuroinductive, is important for two reasons. Firstly it shows that activin can act like a morphogen in specifying tissue-type with dose and being able to make the difference between more than two such tissue types (see Green and Smith, 1991). The sharp dose-thresholds mean that a smooth distribution of activin can give rise to a sharply-defined pattern, as shown in fig. 3. The

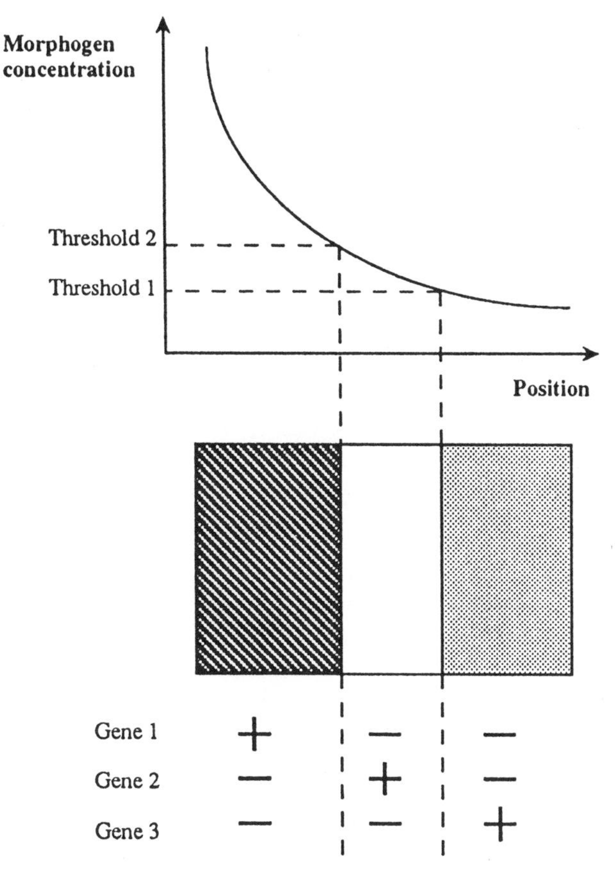


Fig. 3. Thresholds convert a smooth morphogen distribution into a sharply defined pattern: A graded distribution of morphogen (top) is applied to cells. Cells interpret local concentration according to preset thresholds and particular genes become activated or repressed (bottom) giving rise to a pattern with sharp boundaries between its elements (centre). If the morphogen is activin then genes, 1, 2 and 3 correspond to neural inducer, muscle actin and epidermal keratin respectively.

sequence of induced tissue types is also important because it confirms the increasing dorsal-specific character of induced mesoderm with increased activin dose.

## ACTIVIN VERSUS FGF: WHERE ARE THE AXES?

The dorsoventral distinction between the effects of activin and FGF was confused somewhat by the finding that FGF is a strong inducer of posteriorly expressed genes such as Xhox3 and XlHbox6 while activin is a weak inducer of these genes (Ruiz i Altaba and Melton, 1989; Cho and DeRobertis, 1990). The implication of this result, considered in isolation, was that the activin-FGF difference is in the anteroposterior axis rather that the dorsoventral one. In fact, the idea that FGF is a posterior inducer had also emerged from our histological analysis in which we observed induction by FGF of structures possessing a striking resemblance to tadpole anal pores (Green, Howes et al., 1990). How could this confusion between AP and DV axes be resolved?

In the tadpole it is, of course, quite easy to distinguish the anteroposterior axis from the dorsoventral one since they are, by definition, at right angles. However, an important point has been underestimated in efforts to explain the role of the two factors in pattern formation. The fact is that at the time when activin and FGF are acting, namely during blastula stages, the anteroposterior and dorsoventral axes are not clearly defined. In makes sense, therefore, to consider not axes in the tadpole but asymmetries at earlier stages as the basis for understanding the operational axes for MIFs. If one does this things become much clearer. Two asymmetries in the blastula/gastrula, though not at right angles, are evident by simple inspection. One asymmetry is along the animal-vegetal axis of the egg. The animal hemisphere is pigmented while the vegetal hemisphere is unpigmented and much richer in yolk. The other asymmetry, although just visible from the first cell cycle, becomes obvious with the first signs of gastrulation. This is the asymmetry between so-called dorsal and ventral sides. The pole of this asymmetry is the dorsal lip - Spemann's Organiser. As long ago as 1938, Albert Dalcq proposed that "physicochemical gradients" corresponding to these two asymmetries are responsible for patterning the embryo (Dalcq, 1938; see also Needham, 1942). Could these gradients consist of activin and FGF? Certainly, the fate of cells near the dorsal lip is to be dorsoanterior and those further away are increasingly ventral and posterior. Along the animal-vegetal axis the cells of the prospective mesoderm that are nearest to the animal pole are fated to be the most posterior ones, those nearer the vegetal pole have increasingly anterior fates. It is hard to avoid the suggestion that the "Organiser-aborganiser" and animal-vegetal differences correspond to high-low concentration differences of activin and FGF respectively.

## THE ROLE OF FGF: SYNERGY OR MODULATION?

It is now clear that both FGF and at least one type of activin are present in the embryo (Kimelman, Abraham et al., 1988; Slack and Isaacs, 1989; Asashima, Nakano et al., 1991) and one therefore has to consider the interactions of the two factors as well as their effects when applied singly. Experiments using two MIFs are inevitably more complicated to do and to analyse than single-factor experiments, which may explain why so little has been published on this subject. Kimelman & Kirschner (1987) reported that FGF and TGF-beta 1 are synergistic in induction of caps to make muscle. TGF-beta 1 on its own is unable to induced muscle while FGF induction of muscle is only weak. When the two are applied together, however, large amounts of muscle are formed (Kimelman and Kirschner, 1987). One interpretation of this result is that the two factors are needed on the same cells for "optimum" muscle formation. However, given that the two factors can act independently and given the possiblitity of spatial differences, another interpretation is possible. The TGF-beta/FGF synergy mimics quite well the interaction of dorsal and ventral sides of the embryo. In a series of experiments, Slack, Forman, Dale and Smith (see Dale and Slack, 1987) showed that dorsal marginal zones (DMZ) explanted at blastula stages make abundant muscle but ventral marginal zone (VMZ) explants make ventrolateral tissues such as blood and mesothelium and little or no muscle. This was surprising because the normal fate of VMZ is to contribute substantially to the muscles of the somites. Conjugation of lineage-labelled DMZ with VMZ explants showed that the DMZ can cause VMZ to make muscle and much less blood. This interaction was termed "dorsalisation" of the VMZ. Since DMZ-VMZ conjugates are, in effect, reconstructions of the intact marginal zone, it may be inferred that dorsalisation goes on in vivo. The famous "Three Signal Model" (Smith, Dale et al., 1985) summarises the results of these experiments. Signals one and two elicit (distinct) dorsal and ventral mesoderm (DMZ and VMZ) while signal three is the mediator of dorsalisation. If one were to analyse dorsalisation by measuring the amount of muscle actin produced it would be obvious that DMZ and VMZ are "synergistic" in the formation of muscle. A parallel with the Kimelman and Kirschner (1988) is now very clear. The next step is to test whether activin behaves like TGF-beta in "synergy" with FGF. Since activin on its own is capable of inducing muscle, unlike TGF-beta, the analysis and interpretation are more complex. However, preliminary experiments in our laboratory suggests that activin and FGF do potentiate one another (unpublished data).

## PROSPECTS: MARKERS AND MAPS, BEFORE AND AFTER

The application of factors to single cells clearly provides a powerful tool for the analysis of morphogenetic effects in amphibian mesoderm formation. The quantitative resolution of this tool is great and analysis of the results is now limited only by the number of available markers. As well as tissue-specific markers there is an increasing number of region-specific markers. Xhox3 and XlHbox genes have already been mentioned, while Xenopus twist, snail and brachyury (Hopwood, Pluck et al., 1989; Sargent and Bennett, 1990; Smith, Price et al., 1991) all have specific and distinct patterns of expression in early mesoderm. Responses of these genes to MIFs should provide a good test of the relevance of any gradient model for the blastula. Part of such a test will also be the creation and correlation of accurate maps of expression and cell fate.

Even if we are beginning to understand the possible roles of MIFs in embryogenesis, many questions are still open. It is certainly not clear whether MIFs are sufficient for specification of tissues such as pronephros or head mesoderm for which there are, as yet, no molecular markers. It is also not clear how mesoderm induction is initiated. There is evidence that there are at least two distinct steps in the establishment and localisation of the Organiser (Gimlich, 1986). Another inportant problem is how the correct distributions of factors might be set up in the first place. How is the accuracy and reliability of such distributions and the body pattern maintained within and among individuals? Likewise, there are open questions about events after mesoderm induction. What is the connection between mesoderm induction, the movements of gastrulation and the establishment of anteroposterior pattern? There is a body of evidence that suggests that much axial pattern is established after gastrulation is largely complete (Gerhart, Danilchik et al., 1989; see also Green, 1990) and activin certainly induces gastrulation-like movements (Symes and Smith, 1987), so perhaps mesoderm induction specifies some elements of pattern via the specification of gastrulation movements.

Since the discovery of mesoderm inducing factors five years ago there has been new excitement and acceleration in studies of mesoderm formation in amphibia. Further work on Xenopus and studies of MIFs in other vertebrate species hold the promise of great advances in our qualitative and quantitative understanding of this process.

Acknowledgement: I would like to thank Jim Smith, in whose lab this work was done, for help with the experiments and for his critical reading of the manuscript.

REFERENCES

Asashima, M., H. Nakano, H. Uchiyama, H. Sugino, T. Nakamura, Y. Eto, D. Ejima, S.-I. Nishimatsu, N. Ueno and K. Konishita, 1991, Presence of activin (erythroid differentiation factor) in unfertilised eggs and blastulae of Xenopus laevis, Proc Natl Acad Sci U S A, 88:6511-6514.

Cho, K. W. Y. and E. M. DeRobertis, 1990, Differential activation of *Xenopus* homeobox genes by mesoderm-inducing growth factors and retinoic acid, Genes & Dev., 4:1910-1916.

Cooke, J., J. C. Smith, E. J. Smith and M. Yaqoob, 1987, The organization of mesodermal pattern in Xenopus laevis: experiments using a Xenopus mesoderm-inducing factor, Development, 101:893-908.

Dalcq, A. M., 1938, "Form and Causality in Early Development", Cambridge University Press, Cambridge.

Dale, L. and J. M. Slack, 1987, Regional specification within the mesoderm of early embryos of Xenopus laevis, Development, 100:279-95.

Gerhart, J., M. Danilchik, T. Doniach, S. Roberts, B. Rowning and R. Stewart, 1989, Cortical rotation of the Xenopus egg: consequences for the anteroposterior pattern of embryonic dorsal development, Development, 107 Supplement:37-52.

Gimlich, R. L., 1986, Acquisition of developmental autonomy in the equatorial region of the Xenopus embryo, Dev Biol, 115:340-52.

Green, J. B. A., 1990, Retinoic acid: morphogen of the main body axis?, BioEssays, 12:437-9.

Green, J. B. A., G. Howes, K. Symes, J. Cooke and J. C. Smith, 1990, The biological effects of XTC-MIF: quantitative comparison with Xenopus bFGF, Development, 108:229-38.

Green, J. B. A. and J. C. Smith, 1990, Graded changes in dose of a Xenopus activin A homologue elicit stepwise transitions in embryonic cell fate [see comments], Nature, 347:391-4.
Green, J. B. A. and J. C. Smith, 1991, Are gradients and thresholds sufficient for body pattern formation in vertebrates?, Trends Genet., 7:245-250.
Hamburger, V., 1988, "The Heritage of Experimental Embryology", Oxford University Press, New York.
Hopwood, N. D., A. Pluck and J. B. Gurdon, 1989, A Xenopus mRNA related to Drosophila twist is expressed in response to induction in the mesoderm and the neural crest, Cell, 59:893-903.
Kimelman, D.,J. A. Abraham, T. Haaparanta,T. M. Palisi and M.W.Kirschner, 1988, The presence of fibroblast growth factor in the frog egg: its role as a natural mesoderm inducer, Science, 242:1053-6.
Kimelman, D. and M. Kirschner, 1987, Synergistic induction of mesoderm by FGF and TGF-beta and the identification of an mRNA coding for FGF in the early Xenopus embryo, Cell, 51:869-77.
Needham, J., 1942, "Biochemistry and Morphogenesis", Cambridge University Press, Cambridge.
Nieuwkoop, P. D., A. G. Johnen and B. Albers, 1985, "The Epigenetic Nature of Early Chordate Development", Cambridge University Press, Cambridge.
Ruiz i Altaba, A. and D. A. Melton, 1989, Interaction between peptide growth factors and homoeobox genes in the establishment of antero-posterior polarity in frog embryos, Nature, 341:33-8.
Saha, M. S., C. L. Spann and R. M. Grainger, 1989, Embryonic lens induction: more than meets the optic vesicle, Cell Differ Dev, 28:211-7.
Sargent, M. G. and M. F. Bennett, 1990, Identification in *Xenopus* of a structural homologue of the *Drosophila* gene *Snail*, Development, 109:967-973.
Slack, J. M., B. G. Darlington, J. K. Heath and S. F. Godsave, 1987, Mesoderm induction in early Xenopus embryos by heparin-binding growth factors, Nature, 326:197-200.
Slack, J. M. W. and H. V. Isaacs, 1989, Presence of basic fibroblast growth factor in the early *Xenopus* embryo, Development, 105:147-53.
Smith, J. C., 1987, A mesoderm-inducing factor is p;roduced by a Xenopus cell line, Development, 99:3-14.
Smith, J. C., L. Dale and J. M. Slack, 1985, Cell lineage labels and region-specific markers in the analysis of inductive interactions, J Embryol Exp Morphol, 89 supplement:317-331.
Smith, J. C., B. M. Price, N. K. Van and D. Huylebroeck, 1990, Identification of a potent Xenopus mesoderm-inducing factor as a homologue of activin A, Nature, 345:732-4.
Smith, J. C., B. M. J. Price, J. B. A. Green, D. Weigel and B. G. Herrmann, 1991, Expression of a Xenopus homolog of *Brachyury* (*T*) is an immediate-early response to mesoderm induction, Cell, 67:79-87.
Symes, K. and J. C. Smith, 1987, Gastrulation movements provide an early marker of mesoderm induction in Xenopus, Development, 101:339-349.
Wolpert, L., 1989, Positional information revisited, Development, 107 Supplement:3-12.

# ROLES FOR TGFß1 IN CHICK EMBRYO CELL TRANSFORMATION

Esmond J. Sanders

Department of Physiology
University of Alberta
Edmonton, Alberta, T6G 2H7
Canada.

## INTRODUCTION

Early embryonic development is characterized by the occurrence of sequential epithelial-to-mesenchymal and mesenchymal-to-epithelial cell transformations, each of which results in the appearance of a novel cell population. The factors that trigger and influence these early cell transformations are unknown, but may rely on changes in the complement of receptors at the cell surface, changes in the surrounding extracellular matrix, or changes in the influence of local soluble factors. Such events are well-illustrated by the differentiation of the various divisions of the early mesoderm in the avian embryo (Sanders, 1989; 1991). In this sequence, cells of the epithelial epiblast are first transformed into mesenchyme by passage through the primitive streak -- a region in which as yet undisclosed cell events result in localized phenotypic transformation (Bellairs, 1986; Sanders, 1986). After emergence from the primitive streak, the mesodermal cells align paraxially to form the transient segmental plate from which, by mesenchymal-to-epithelial transformation, somites form (Bellairs, 1979). Further differentiation of this tissue necessitates dispersal of its ventromedial portion into sclerotome, as a result of a transformation back into the mesenchymal phenotype. Nothing is known of the factors that precipitate this dispersal.

Also important for this discussion, is that at the same time that the sclerotome is dispersing medially from the somites, cells of the mesonephros are also in the process of differentiation.

Although a number of different influences must be operating simultaneously in the control of these transformations, it is becoming apparent that growth factors play a critical role even in the earliest phases of development (Whitman and Melton, 1989; Cross and Dexter, 1991; Melton, 1991). Of these factors, the transforming growth factor ß (TGFß) family of molecules has become of

*Formation and Differentiation of Early Embryonic Mesoderm*
Edited by R. Bellairs *et al.*, Plenum Press, New York, 1992

particular interest as possible regulators of early developmental events, because of their multiple effects on cell differentiation (Massague, 1987; Roberts et al., 1990a) and on the deposition of extracellular matrix (Rizzino, 1988; Heine et al., 1990; Roberts et al., 1990b). At the very earliest stages of development, the TGFß group has been implicated in inductive processes, most clearly in amphibians (Rosa et al., 1988), but also possibly in the chick (Mitrani and Shimoni, 1990; Mitrani et al., 1990; Cooke and Wong 1991), in such a way as to specify the overall embryonic axis. It is also becoming clear, however, that TGFß has influences over various aspects of morphogenesis and the differentiation of cells, for example in the heart (Potts and Runyan, 1989; Akhurst et al., 1990; Choy et al., 1991) and limb buds (Lucas and Caplan, 1988; Kulyk et al., 1989; Carrington and Reddi, 1990; Hayamizu et al., 1991) in both mammalian and chick embryos.

Among such studies are reports that TGFß is capable of influencing epithelial-mesenchymal transformations (Potts and Runyan, 1989), invasiveness (Welch et al., 1990), matrix deposition (Ignotz and Massague, 1986), and cell-substratum adhesiveness (Ignotz and Massague, 1987). Because all of these are relevant to the phenotypic transformations that occur in the mesodermal cell populations of early avian development, it was considered appropriate to investigate the possibility that TGFß might be involved in these events. In this work it is shown that TGFß is appropriately located to play such a role in several tissues, and that treatment of cells *in vitro* and embryos *in vivo* with TGFß or a neutralizing antibody result in perturbations supportive of such a role.

## EXPRESSION OF TGFß IN EARLY AVIAN DEVELOPMENT

It has previously been shown, by means of immunocytochemistry using an antibody to an amino-terminal fragment of human TGFß1 (Sanders and Prasad, 1991), that this factor is present in, or between, cells of Hensen's node and the primitive streak in embryos at stage 5 of Hamburger and Hamilton (1951). The appearance of the punctate fluorescent labelling in these locations gave the clear impression of an intracellular location. Labelling for TGFß1 was particularly strong in the endodermal component of the node and primitive streak and, more laterally, in the cells of the mesoderm (Fig. 1). The mesodermal cells appeared to gain staining intensity, in comparison with the overlying epiblast, after their departure from the immediate region of the primitive streak.

Embryos examined after three days of incubation (Hamilton and Hamburger stage 18) showed strong immunoreactivity in several clearly defined regions. Cells of the mesenchymal sclerotome tissue were particularly strongly labelled (Fig. 2), while those of the dermomyotome and overlying ectoderm were not. Also, cells of the mesonephric mesenchyme were positive in comparison with the cells of the adjacent mesonephric duct (Fig. 3).

Previous results (Sanders and Prasad, 1991) have indicated that at stage 5 the immunoreactivity is not

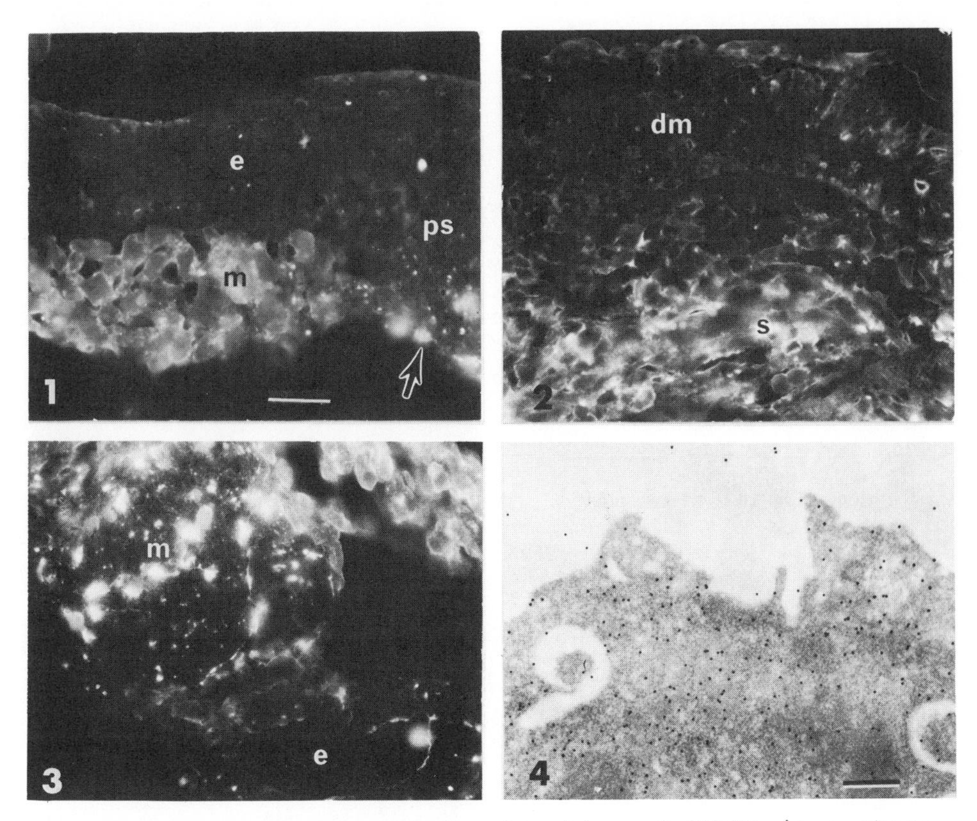


Fig. 1. Immunofluorescent localization of TGFß1 in a stage 5 embryo. Labelling is strong in the mesoderm (m) and in the endodermal component of the primitive streak (arrow). The primitive streak itself (ps) shows punctate labelling, while the epiblast (e) is not labelled. Bar = 20$\mu$m.

Fig. 2. TGFß1 localization in a somite at stage 18. The sclerotome (s) is immunoreactive, but the dermamyotome (dm) is not.

Fig. 3. TGFß1 localization in mesonephric tissue at stage 18. Mesenchymal tissue (m) is labelled, but the epithelial tissue (e) is not.

Fig. 4. Ultrastructural immunogold localization of TGFß1 in a sclerotome cell indicates that the localization of the factor is probably primarily intracellular. Bar = 2$\mu$m.

associated with hyaluronic acid, a major component of the extracellular matrix at this time. Examination of reactive tissues at stage 18 using immunogold cytochemistry (Fig. 4) confirms the probable intracellular localization. The possibility of an extracellular location is, however, not entirely ruled out, especially in view of previous reports that TGFß1 may bind to fibronectin (Mooradian et al., 1989) or to the type IV collagen of basement membranes (Paralkar et al., 1991).

## EFFECTS OF TGFβ AND A NEUTRALIZING ANTIBODY ON MESENCHYMAL CELL PHENOTYPE *IN VITRO*

### Addition of exogenous TGFβ

When mesoderm cells from a stage 5 embryo are explanted onto a substratum composed of fibronectin, they spread as a coherent sheet and with an epithelial morphology (Sanders, 1980). In the presence of TGFβ1 (50ng/ml), however, these cells on fibronectin disperse as individual cells with fibroblastic morphology (Figs. 5 and 6; Table 1).

On the other hand, when these cells are explanted on laminin in normal culture medium they tend to spread with a fibroblastic morphology (Brown and Sanders, 1991). In the presence of TGFβ1 such cells tend to transform into an epithelial phenotype (Sanders and Prasad, 1991; Table 1).

### Addition of a neutralizing antibody to TGFβ

Monoclonal antibody 1D11.16 is able to neutralize the biological activities of TGFβ1 and TGFβ2 (Dasch et al., 1989). When added exogenously to cultures of mesoderm cells from stage 5 embryos, or to cultures of sclerotome cells, both grown on a fibronectin substratum, this antibody produced a dose-dependent response (10 - 100μg/ml) in which cells spread progressively more poorly (Figs. 7 and 8). This was accompanied by decreased deposition of fibronectin (not illustrated). Controls incubated with similar concentrations of normal mouse IgG spread in the same way as the controls in normal medium.

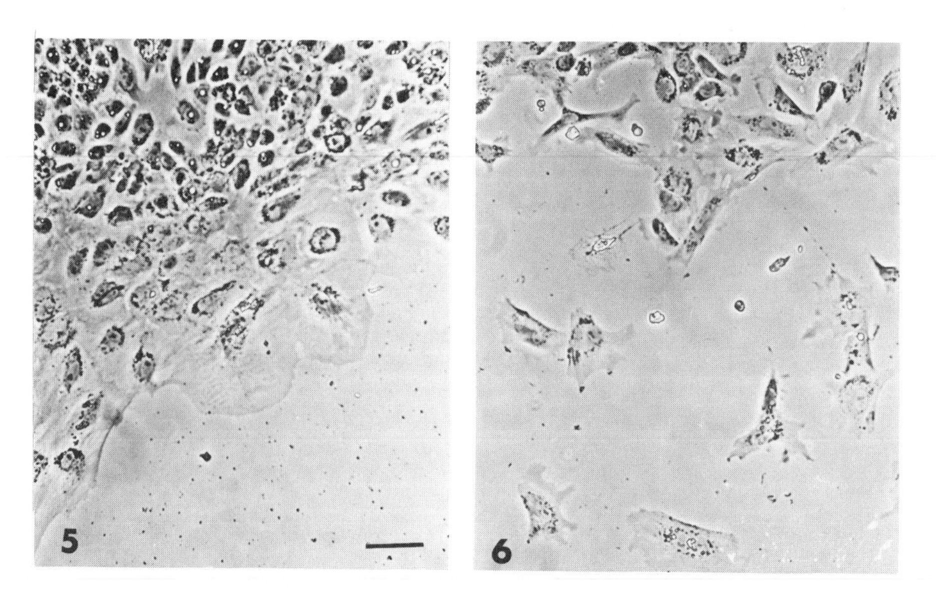


Fig. 5. Mesoderm cells cultured on a fibronectin substratum spread with an epithelial morphology. Bar = 100μm.
Fig. 6. In the presence of TGFβ1, mesoderm cells on fibronectin spread as individual fibroblast-like cells.

TABLE 1. Effects of TGFβ1 on early mesoderm cells in culture

| Parameter | Substratum | Effect |
|---|---|---|
| Phenotype<br>24 hrs | Fibronectin<br>Laminin | epithelial→fibroblastic<br>fibroblastic→epithelial |
| FN deposition<br>3 days | Fibronectin<br>Laminin | FN ↓<br>FN ↑ |
| LN deposition<br>3 days | Fibronectin<br>Laminin | No effect<br>LN ↑ |
| FN receptor<br>3 days | Fibronectin<br>Laminin | FNr ↓<br>No effect, FNr undetectable |

FN = fibronectin; FNr = fibronectin receptor; LN = laminin
Modified from Sanders and Prasad (1991).

EFFECTS OF TGFβ ON EXTRACELLULAR MATRIX DEPOSITION *IN VITRO*

Immunofluorescent examination of the deposition of fibronectin and laminin by stage 5 mesoderm cells *in vitro* showed a substratum-dependent response to the addition of TGFβ (Table 1). When cultured on a fibronectin substratum for a period of three days, mesoderm cells normally deposit a network of fibronectin over the entire explant (Fig. 9). In the presence of TGFβ, however, this network was markedly reduced in extent (Fig. 10). By contrast, on a laminin substratum these cells would normally produce little fibronectin, while in the presence of TGFβ the network appeared to be increased (Table 1). Similarly, laminin deposition was also enhanced by TGFβ in these cells when grown on a laminin substratum.

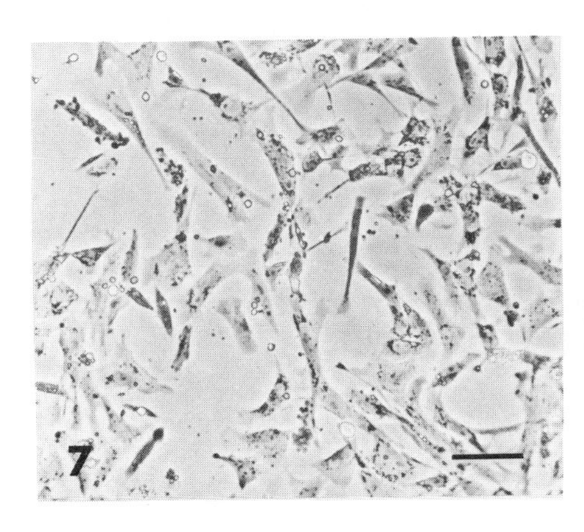


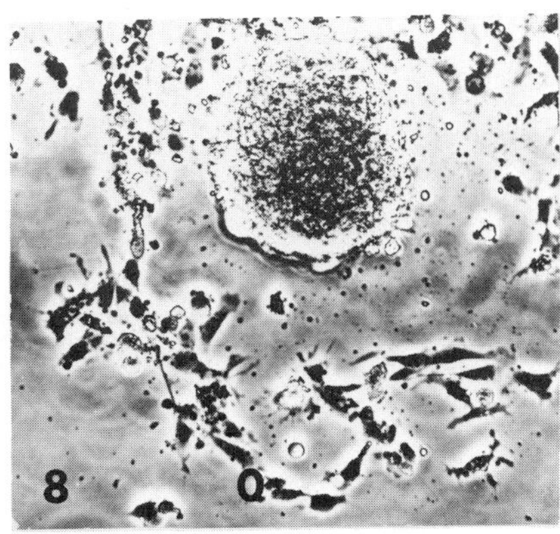


Fig. 7. Sclerotome cells in culture grow extensively and spread rapidly with a fibroblast-like morphology.
Fig. 8. Monoclonal antibody 1D11.16 (50μg/ml for 24 hrs) inhibits the spreading of sclerotome cells. Bar = 100μm.

Further, using an antibody that recognises all ß1 integrins (Sanders and Prasad, 1991), it was possible to demonstrate a down-regulation in the levels, or redistribution, of the fibronectin receptor in the presence of TGFß (Figs. 11 and 12).

EFFECTS OF TGFß AND A NEUTRALIZING ANTIBODY ON GASTRULATING EMBRYOS *IN VIVO*

In order to determine if the perturbations demonstrated above are of significance *in vivo*, TGFß1 and the 1D11.16 monoclonal antibody were introduced into living embryos maintained in culture according to the method of New (1955). Direct microinjections of approximately 25 nl of fluid were made into the mesodermal tissue space adjacent to the primitive streak using a "Picoinjector" (Medical Systems Corporation, USA). Injections were made on embryos at stage 5 close to the primitive streak and about half way along its

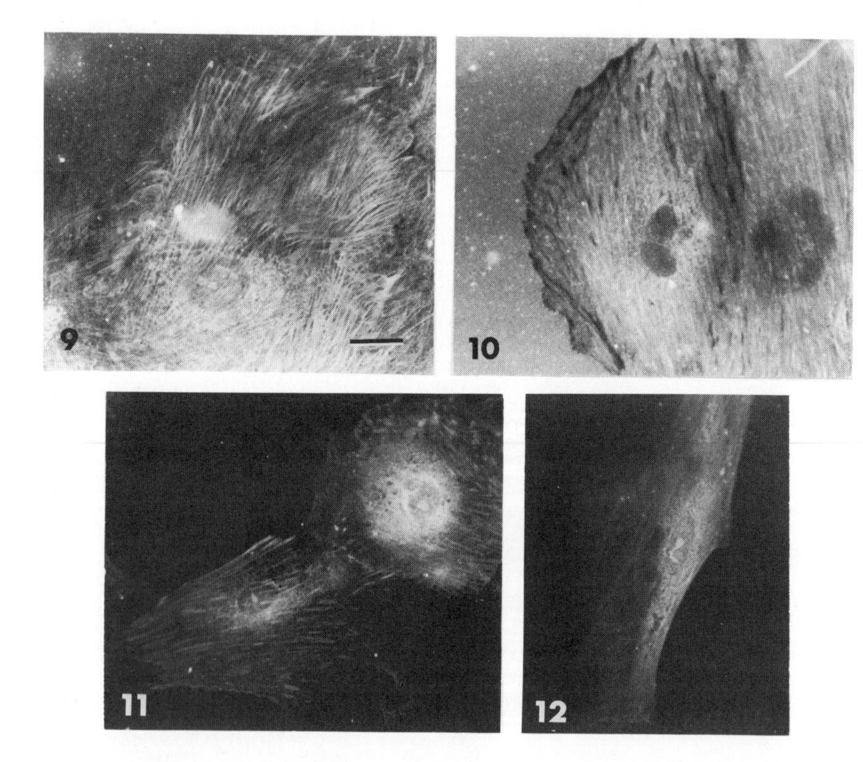


Fig. 9. Stage 5 mesoderm cultured for 3 days on fibronectin deposits a network of fibronectin over the entire explant. Bar = 20$\mu$m.
Fig. 10. Fibronectin deposition is down-regulated in the presence of TGFß1.
Fig. 11. Fibronectin receptor is present in streaks resembling focal contacts in mesoderm cells after 3 days in culture.
Fig. 12. In the presence of TGFß1, the fibronectin receptor is no longer detectable.

length. Embryos were incubated for a further 6 hrs before fixation for histological examination. With this method, perturbations were obtained at low frequency using the antibody (Fig. 13), but not with TGFß1. Control embryos, injected with similar concentrations of normal mouse IgG were not affected. The clearest effect with the antibody appeared to be the induction of sites of ectopic ingression (Fig. 13).

The poor reproducibility of this approach lead to the use of implants of carrier microbeads which had been previously soaked in TGFß1 or 1D11.16 antibody. The beads employed were ion-exchange resins obtained from Bio-Rad Laboratories of the type AG1-X2 in the chloride or formate form (Stern et al., 1990; Wedden et al., 1990). Beads of

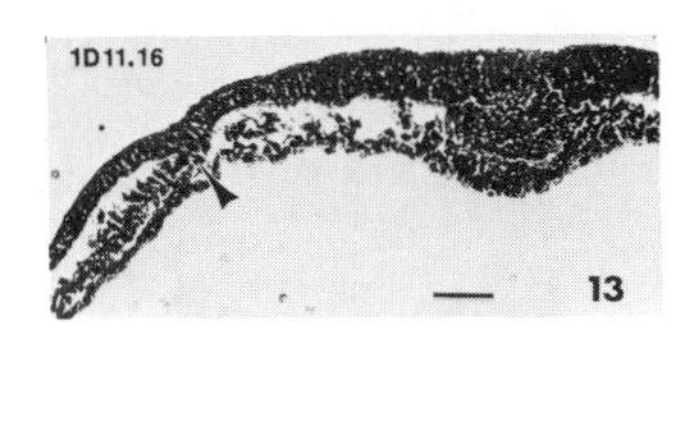


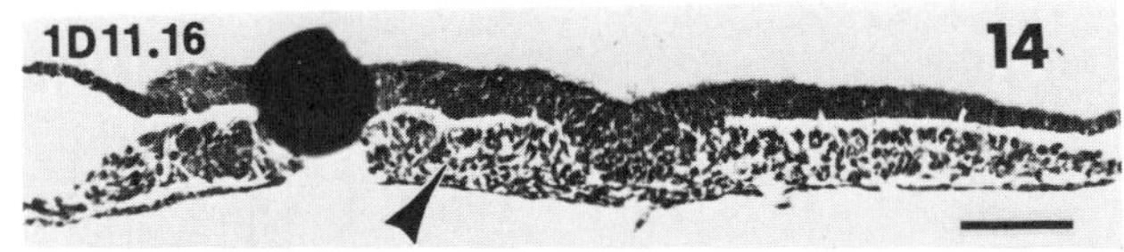

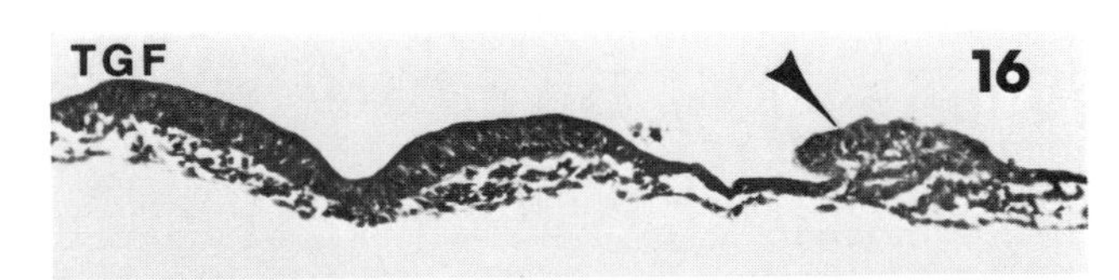


Fig. 13. Microinjection of antibody 1D11.16 adjacent to the primitive streak of a stage 5 embryo can cause the appearance of apparently secondary sites of ingression (arrow). Bar = 50$\mu$m.

Fig. 14. Ion-exchange bead soaked in antibody 1D11.16 and implanted for 6 hrs. The mesoderm appears to have proliferated on the implant side (arrow) in comparison with the control side. The bead has moved during sectioning. Bar = 100$\mu$m.

Fig. 15. As Fig. 14, but with TGF-treatment of the bead. Again, the mesoderm seems to have proliferated on the implant side (arrow).

Fig. 16. Occasionally, TGF-treated beads cause proliferation of the overlying epiblast (arrow). In this example the beads are in adjacent sections.

the Affi-Gel Blue type (Hayamizu et al., 1991) were found to be unsuitable.

Examination of specimens in which beads had been implanted indicated that the most common perturbation induced by the 1D11.16 antibody was an accumulation of mesenchymal cells at the site of the implant (Fig. 14). Beads soaked in TGFß1 also induced proliferation of mesodermal cells (Fig. 15), occasional compaction of the mesoderm, and overgrowth of the overlying epiblast (Fig. 16). Untreated beads had no such perturbing effects.

## DISCUSSION AND CONCLUSIONS

The TGFß1 expression patterns during gastrulation shown here and earlier (Sanders and Prasad, 1991) indicate that this factor is temporally and spatially placed appropriately for it to play a role in the promotion and/or maintenance of the mesenchymal phenotype following passage of cells through the primitive streak. The distributions, and the *in vitro* effects on cell morphology, are consistent with the suggestion that the presence of this factor among the mesoderm cells may reinforce the phenotypic change. Whether or not the presence of TGFß1 in the cells of Hensen's node and the primitive streak itself is significant in actually precipitating the change is not clear. The demonstrated modulations in extracellular matrix deposition caused by exogenous TGFß1 *in vitro*, indicate that the maintenance of the mesenchymal phenotype may be related to the capacity of the cells to synthesize a specific program of matrix molecules. Further, the regulation of the fibronectin receptor by TGF suggests that synthesis, deposition and cell binding of the matrix molecules are co-ordinated. A similar developmental role for TGFß has been proposed in the epithelial-mesenchymal cell transformations that occur in the embryonic heart (Potts and Runyan, 1989).

The experiments in which TGFß or its neutralizing antibody were directly introduced into gastrulating embryos *in vivo*, provided further support for an influence of the factor in modulating the epithelial or mesenchymal phenotype. Presence of the 1D11.16 antibody, either directly microinjected or on microbead carriers, tended to promote the appearance and possibly proliferation of the mesoderm. Beads soaked in TGFß also tended to promote the appearance of mesenchyme, but also showed some tendency towards epithelialization effects. By reference to the *in vitro* studies described above (Table 1), such an epithelialization effect of TGFß might be expected if laminin were a significant component of the extracellular matrix. Evidence has been found to suggest that laminin is present in the region of the mesoderm cells at this stage of development (Zagris and Chung, 1990).

The data presented here do not directly address the question of whether TGFß is able to act as an inducer of the embryonic axis (Mitrani and Shimoni, 1990; Mitrani et al., 1990; Cooke and Wong, 1991), however, it may be significant that the endoderm below Hensen's node and the primitive streak is strongly immunoreactive for TGFß1. This tissue appears to have the ability to influence cell differentiation at gastrulation (Veini and Hara, 1975).

Examination of embryos at three days of incubation indicated that two mesenchymal tissues were particularly immunoreactive for TGFß1. The presence of TGF in the sclerotome portion of the somite, but not the dermamyotome provides further evidence for a link with epithelial-mesenchymal transformation, since the sclerotome differentiates from the epithelial somite. The stimulus for the breakdown of the ventro-medial wall of the epithelial somite, giving rise to the sclerotome, is unknown. The precise localization of TGFß in the somite before sclerotome differentiation is currently being studied.

The mesonephric mesenchyme was also found to be intensely immunoreactive for TGFß1, while the adjacent epithelial duct was not. The possibility that this distribution of TGF is significant for kidney development is supported by other observations in which growth factors (including TGF) have been implicated in the induction of tubulogenesis (Taub et al., 1990; Weller et al., 1991).

## REFERENCES


Akhurst, R.J., Lehnert, S.A., Faissner, A., and Duffie, E., 1990, TGFß in murine morphogenetic processes: the early embryo and cardiogenesis, Development, 108:645-656.

Bellairs, R., 1979, The mechanism of somite segmentation in the chick embryo, J. Embryol. exp. Morph., 51:227-243.

Bellairs, R., 1986, The primitive streak, Anat. Embryol., 174:1-14.

Brown, A.J., and Sanders, E.J., 1991, Interactions between mesoderm cells and the extracellular matrix following gastrulation in the chick embryo, J. Cell Sci., 99:431-441.

Carrington, J.L., and Reddi, A.H., 1990, Temporal changes in the response of chick limb bud mesodermal cells to transforming growth factor ß-type 1, Expl. Cell Res., 186:368-373.

Choy, M., Armstrong, M.T., and Armstrong, P.B., 1991, Transforming growth factor-ß1 localized within the heart of the chick embryo, Anat. Embryol., 183:345-352.

Cooke, J., and Wong, A., 1991, Growth-factor-related proteins that are inducers in early amphibian development may mediate similar steps in amniote (bird) embryogenesis, Development, 111:197-212.

Cross, M., and Dexter, T.M., 1991, Growth factors in development, transformation, and tumorigenesis, Cell, 64:271-280.

Dasch, J.R., Pace, D.R., Waegell, W., Inenaga, D., and Ellingsworth, L., 1989, Monoclonal antibodies recognizing transforming growth factor-ß, J. Immunol., 142:1536-1541.

Hamburger, V., and Hamilton, H.L., 1951, A series of normal stages in the development of the chick embryo, J. Morph., 88:49-92.

Hayamizu, T.F., Sessions, S.K., Wanek, N., and Bryant, S.V., 1991, Effects of localized application of transforming growth factor ß1 on developing chick limbs, Dev. Biol., 145:164-173.

Heine, U.I., Munoz, E.F., Flanders, K.C., Roberts, A.B., and Sporn, M.B., 1990, Colocalization of TGF-ß1 and collagen I and III, fibronectin and glycosaminoglycans

during lung branching morphogenesis, Development, 109:29-36.
Ignotz, R.A., and Massague, J., 1986, Transforming growth factor-ß stimulates the expression of fibronectin and collagen and their incorporation into the extracellular matrix, J. Biol. Chem., 261:4337-4345.
Ignotz, R.A., and Massague, J., 1987, Cell adhesion protein receptors as targets for transforming growth factor-ß action, Cell, 51:189-197.
Kulyk, W.M., Rodgers, B.J., Greer, K., and Kosher, R.A., 1989, Promotion of embryonic chick limb cartilage differentiation by transforming growth factor-ß, Dev. Biol., 135:424-430.
Lucas, P.A., and Caplan, A.I., 1988, Chemotactic response of embryonic limb bud mesenchymal cells and muscle-derived fibroblasts to transforming growth factor-ß, Conn. Tiss. Res., 18:1-17.
Massague, J., 1990, The transforming growth factor-ß family, Ann. Rev. Cell Biol., 6:597-641.
Melton, D.A., 1991, Pattern formation during animal development, Science, 252:234-241.
Mitrani, E., and Shimoni, Y., 1990, Induction by soluble factors of organized axial structures in chick epiblasts, Science, 247:1092-1094.
Mitrani, E., Ziv, T., Thomsen, G., Shimoni, Y., Melton, D.A., and Bril, A., 1990, Cell, 63:495-501.
Mooradian, D.L., Lucas, R.C., Weatherbee, J.A., and Furcht, L.T., 1989, Transforming growth factor-ß1 binds to immobilized fibronectin, J. Cell. Biochem., 41:189-200.
New, D.A.T., 1955, A new technique for the cultivation of the chick embryo *in vitro*, J. Embryol. exp. Morph., 3:320-331.
Paralkar, V.M., Vukicevic, S., and Reddi, A.H., 1991, Transforming growth factor ß type 1 binds to collagen IV of basement membrane matrix: implications for development, Dev. Biol., 143:303-308.
Potts, J.D., and Runyan, R.B., 1989, Epithelial-mesenchymal cell transformation in the embryonic heart can be mediated, in part, by transforming growth factor ß, Dev. Biol., 134:392-401.
Rizzino, A., 1988, Transforming growth factor-ß: multiple effects on cell differentiation and extracellular matrices, Dev. Biol., 130:411-422.
Roberts, A.B., Flanders, K.C., Heine, U.I., Jakowlew, S., Kondaiah, P., Kim, S.J., and Sporn, M.B., 1990a, Transforming growth factor-ß: multifunctional regulator of differentiation and development, Phil. Trans. R. Soc. Lond., 327:145-154.
Roberts, A.B., Heine, U.I., Flanders, K.C., and Sporn, M.B., 1990b, Transforming growth factor-ß. Major role in regulation of extracellular matrix, Ann. N.Y. Acad. Sci., 580:225-232.
Rosa, F., Roberts, A.B., Danielpour, D., Dart, L.L., Sporn, M.B., and Dawid, I.B., 1988, Mesoderm induction in amphibians: the role of TGFß2-like factors, Nature, 239:783-785.
Sanders, E.J., 1980, The effect of fibronectin and substratum-attached material on the spreading of chick embryo mesoderm cells *in vitro*, J. Cell Sci., 44:225-242.
Sanders, E.J., 1986, Mesoderm migration in the early chick embryo, in: "Developmental Biology. A Comprehensive

Synthesis," volume 2, L. Browder, ed., Plenum, New York, pp. 449-480.
Sanders, E.J., 1989, "The Cell Surface in Embryogenesis and Carcinogenesis," Telford Press, Caldwell, New Jersey.
Sanders, E.J., 1991, Morphogenesis of the mesoderm in early avian development: sequential phenotypic transformations, in: "Growth Regulation and Carcinogenesis," volume 2, W.R. Paukovitz, ed., CRC Press, Boca Raton, pp. 233-245.
Sanders, E.J., and Prasad, S., 1991, Possible roles for TGFß1 in the gastrulating chick embryo, J. Cell Sci., 99:617-626.
Stern, C.D., Ireland, G.W., Herrick, S.E., Gherardi, E., Gray, J., Perryman, M., and Stoker, M., 1990, Epithelial scatter factor and the development of the embryonic axis, Development, 110:1271-1284.
Taub, M., Wang, Y., Szczesny, T.M., and Kleinman, H.K., 1990, Epidermal growth factor or transforming growth factor $\alpha$ is required for kidney tubulogenesis in matrigel cultures in serum-free medium, Proc. natn. Acad. Sci. U.S.A., 87:4002-4006.
Veini, M., and Hara, K., 1975, Changes in the differentiation tendencies of the hypoblast-free Hensen's node during 'gastrulation' in the chick embryo, Wilhelm Roux Arch., 177:89-100.
Wedden, S., Thaller, C., and Eichele, G., 1990, Targeted slow-release of retinoids into chick embryos, Methods Enzymol., 190:201-209.
Welch, D.R., Fabra, A., and Nakajima, M., 1990, Transforming growth factor ß stimulates mammary adenocarcinoma cell invasion and metastatic potential, Proc. natn. Acad. Sci. U.S.A., 87:7678-7682.
Weller, A., Sorokin, L., Illgen, E-M., Ekblom, P., 1991, Development and growth of mouse embryonic kidney in organ culture and modulation of development by soluble growth factors, Dev. Biol., 144:248-261.
Whitman, M., and Melton, D.A., 1989, Growth factors in early embryogenesis, Ann. Rev. Cell Biol., 5:93-117.
Zagris, N., and Chung, A.E., 1990, Distribution and functional role of laminin during induction of the embryonic axis in the chick embryo, Differentiation, 43:81-86.

CHARACTERISTICS OF THE PRECARDIAC MESODERM IN THE CHICK

Ruth Bellairs*, Heather Easton*, P. Govewalla*,
G.D. Harrison*, M.Y. Palmer*, R.C. Pollock and
M. Veini+

* Department of Anatomy and Developmental Biology
University College London, Gower Street
London, WC1E 6BT, U.K.
+ Department of Zoology, University of Athens
Athens, Greece

The precardiac mesoderm, the mesoderm which gives rise to the heart, is identifiable by stage 5 of Hamburger and Hamilton (1951) in the chick embryo (fig.1). It is a crescent-shaped region whose arms lie one on either side of the head process and are joined by a narrow bridge immediately anterior to the head process (Rosenquist and DeHaan, 1966). These precardiac cells migrate in an antero-medial direction over the underlying endoderm and fuse together to form a simple tube. Linask and Lash (1986, 1988) showed that fibronectin secreted by the endoderm played a critical role in this migration. Furthermore, they demonstrated that there was a gradient in concentration of the fibronectin with the highest level at the most anterior end. They concluded that the cells migrated up a gradient of increasing adhesiveness, ie. up a "haptotactic" (Carter, 1965) gradient. Further support came from the demonstration of Linask and Lash (1990) that the precardiac mesoderm cells possess a fibronectin receptor which is probably integrin.

These important results left certain questions unanswered. In particular, they provided no direct information about the reactivity of the cells to different concentrations of fibronectin, nor as to whether the cells in the different parts of the precardiac mesoderm possess different characteristics which may affect their response to fibronectin. We have therefore posed two questions:

## 1. DO THE CELLS RESPOND DIFFERENTLY TO DIFFERENT CONCENTRATIONS OF FIBRONECTIN?

Precardiac mesoderm (fig. 1) was dissected free from the ectoderm using tungsten needles, and explanted onto glass coverslips which had been coated with plasma fibronectin at concentrations of 50,

*Formation and Differentiation of Early Embryonic Mesoderm*
Edited by R. Bellairs *et al.*, Plenum Press, New York, 1992

100 or 250 µg/ml. The ability of the cells to settle and spread in sitting drops was compared for the three concentrations. The culture medium consisted of Medium 199 (GIBCO), fetal calf serum (Northumbria Biochemicals Ltd.) and penicillin/streptomycin, in the ratio of 9:1:0.5.

Three groups of control experiments were performed. The first was designed to test whether the explanted tissue was indeed precardiac mesoderm. Twenty seven donors from which the precardiac mesoderm had been extirpated were explanted and grown in culture using the New technique (New, 1955), and all exhibited deficiencies in the hearts which developed. It was concluded therefore that the correct tissue had been explanted. The second group of control experiments was designed to test whether the small piece of endoderm which often adhered to the mesoderm had any significant effect on the results. Fifteen embryos were immersed in a 1% trypsin solution for 15 minutes at 37°C to facilitate removal of the endoderm. No differences in results were apparent between explants taken from these embryos and from embryos not treated with trypsin. This step was therefore ommitted in the main experimental series. The third group of control experiments was designed to test whether the results obtained in the experimental series described below could be affected by fibronectin secreted by the explanted cells or by the small amount of fibronectin in the culture medium. 24 specimens were explanted onto glass coverslips which had not been coated with a substratum of fibronectin. Three of these explants

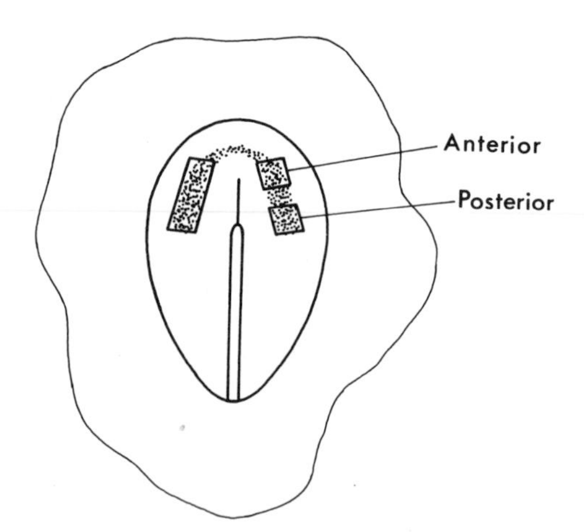


Figure 1. Diagram of a chick embryo at stage 5 to show the location of the precardiac mesoderm (stipple). The rectangle outlined on the left of the drawing indicates the location of the tissue removed and explanted in the experiments in Section 1. The squares outlined on the right indicate the tissue removed and explanted in the experiments in Section 2.

appeared to attach briefly but subsequently rolled up and failed to spread. The remainder failed to attach or spread, though some formed pulsating vesicles over the next 24 hours.

Before comparing the ability of the cells to spread on the different concentrations of fibronectin, we must first consider whether there were differences in the speed of settling. In other words, if the explants spread further on a higher concentration than on a lower one in the same period of time, was this because they attached more quickly on the higher concentration and so were able to spread for an appreciably longer period?

Speed of settling

The differences in the time taken to settle successfully on the three concentrations of fibronectin are shown in figs. 2 and 3, which are based on 101 explants, (44 on 50µg/ml, 31 on 100µg/ml and 26 on 250µg/ml). The settling time for the lowest concentration peaked at about 18 minutes whilst that for the highest concentration was about 5 minutes. All had settled by 35 minutes. It was concluded that the differences in settling time were insufficient to account for the results obtained (below) with the spreading of the explants.

Spreading

Two measurements were made on each explant at regular intervals: the maximum diameter and the diameter at right angles to it. The

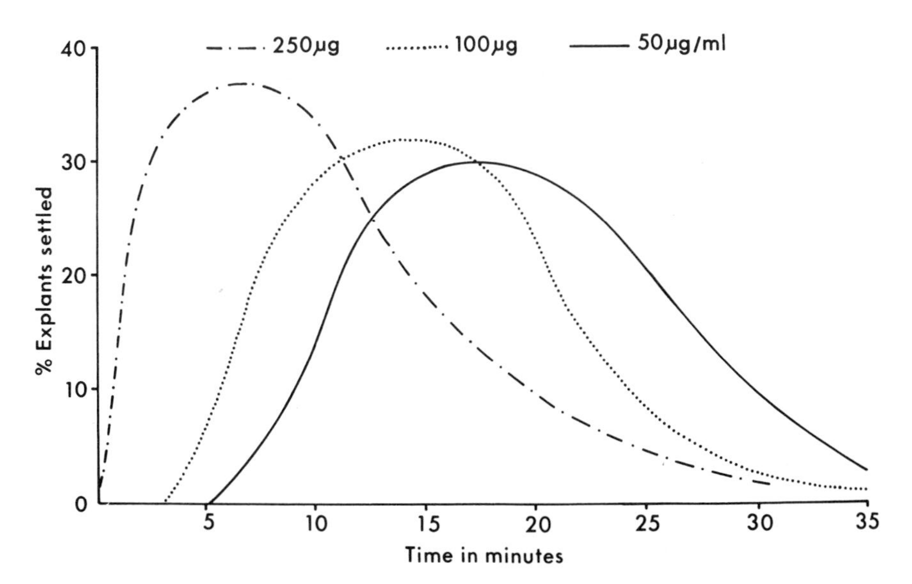


Figure 2. Graph showing the percentage of explants which had just settled at each time interval. Individual explants settled more quickly the higher the concentration of fibronectin.

mean diameter and the surface area were then calculated (fig. 4). Attempts were made to explant pieces of tissue of standard size, the mean diameter at the start of the experiment being about 0.28mm. The spread of each explant was calculated as a percentage increase on the initial size and/or on its size after 24 hours of incubation. Fig. 5 shows the percentage increases in surface area based on 67 explants (24 at 50 µg/ml; 27 at 100 µg/ml; and 16 at 250µg/ml). At all time points, the mean spread of the explants was fastest on the highest concentration of fibronectin and, with the exception of 48h, slowest on the lowest. When the data was analysed by t-test (taking into account that the sample numbers are not identical in the groups being compared), these differences were found to be significant at 72 hours ($p<0.01$).

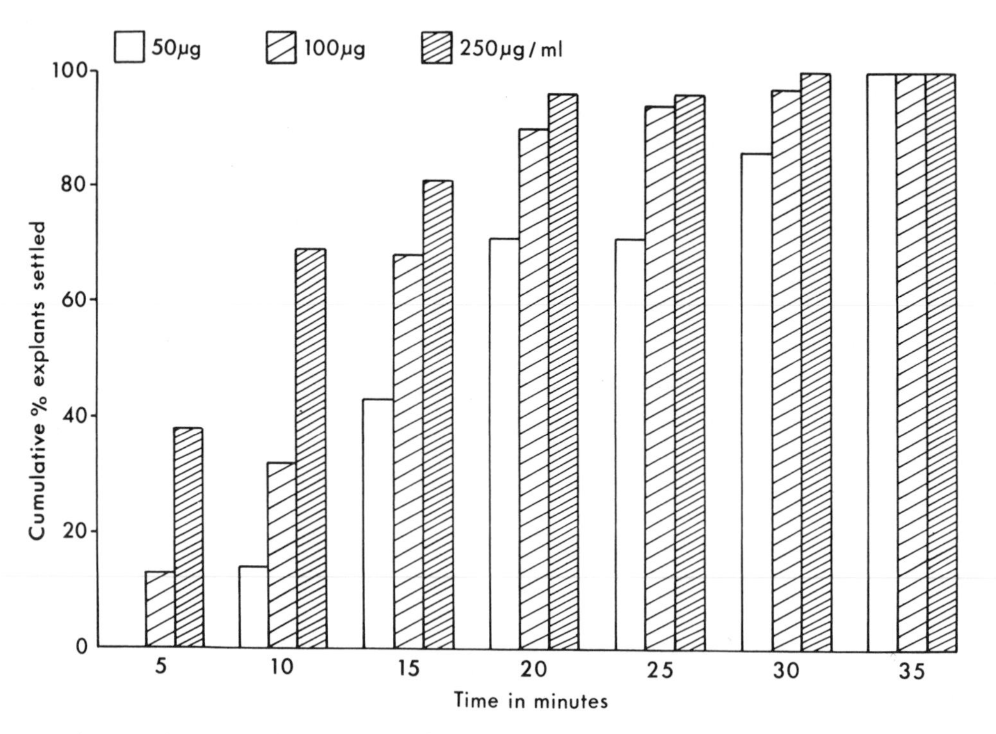


Figure 3. Bar chart showing the accumulative percentage of explants which had settled by each time point. During the first 30 minutes the proportion of settled explants was greater the higher the concentration of fibronectin.

In an additional series of experiments, 183 explants were placed at the junction of two patches of fibronectin, one being 50 µg/ml and the other 250 µg/ml (fig. 6). The junction was scored with a fine line on the under-surface of the cover slip and the spread was measured as the distance from each side of this line (fig.7). The differences in the measurements were calculated and found to be significant at 24, 48, 72 and 96 h, using the t-test and $p<0.01$.

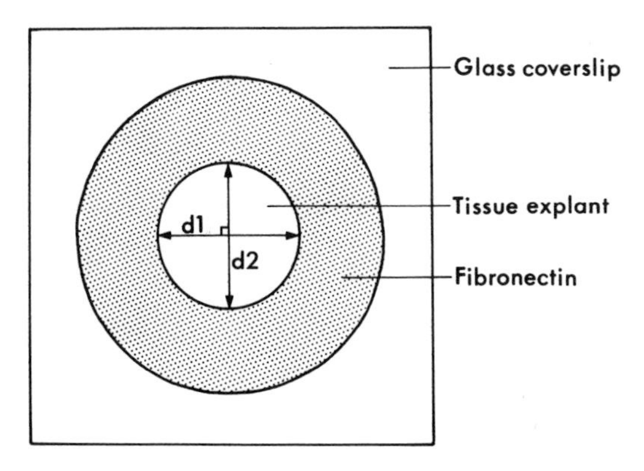


Figure 4. Diagram of an explant which has undergone some spreading on a fibronectin substratum on a glass coverslip. Measurements were made of the maximum diameter (d1) and the diameter at right angles to it (d2). The mean diameter and surface area were then calculated.

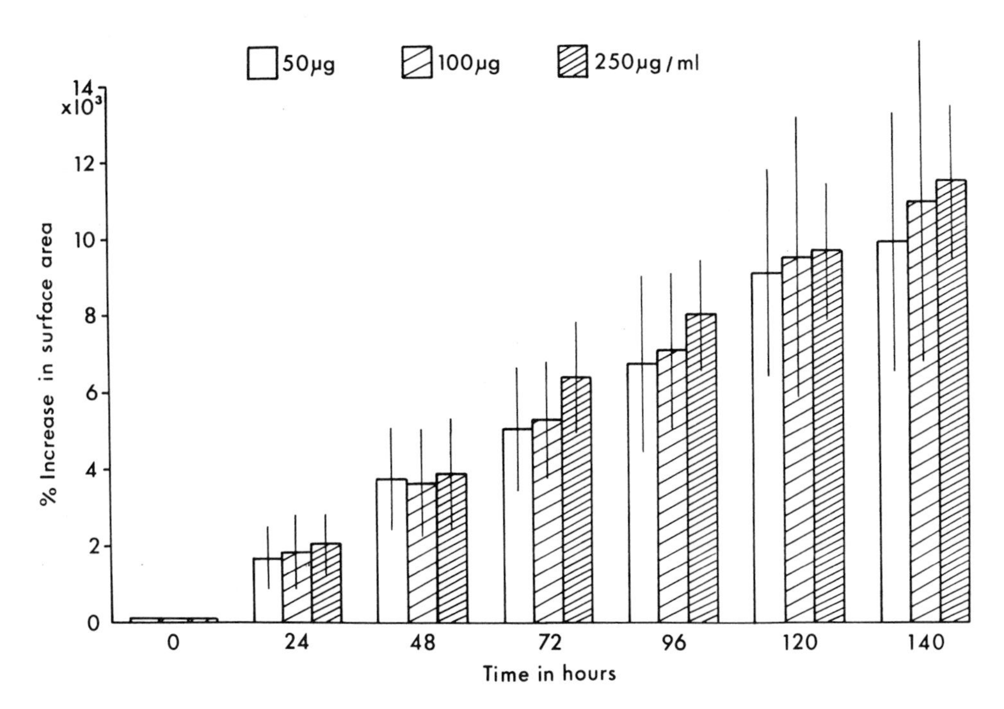


Figure 5. Bar chart showing the percentage increases in surface area of explants at each time point. The mean spread of the explants was fastest on the highest concentration of fibronectin.

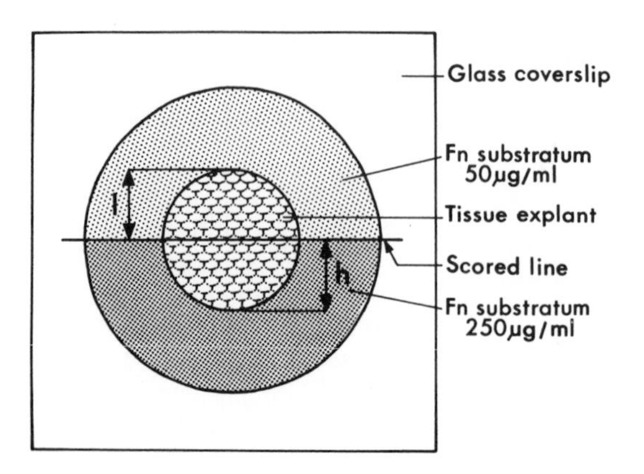


Figure 6. Diagram of an explant on a substratum consisting of two abutting patches of fibronectin, the junction between them marked by a fine line applied to the undersurface of the coverslip. The furthest point that the explant had spread from the line was measured for the lower (l) and the higher (h) concentrations of fibronectin at intervals of time (see fig. 7).

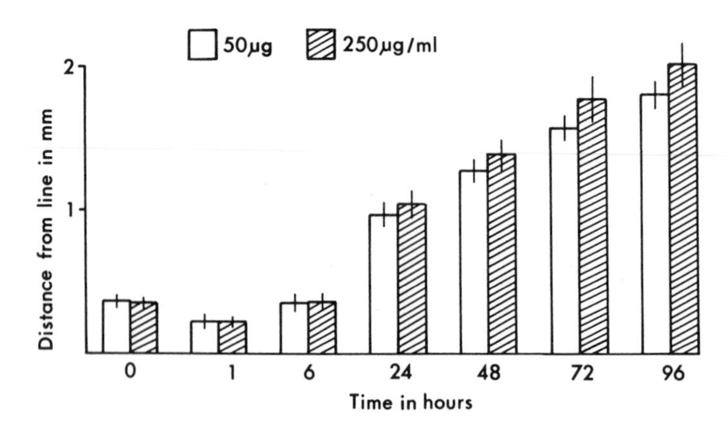


Figure 7. Bar chart showing the maximum distance from the scored line which explants have spread on a substratum consisting of two abutting concentrations of fibronectin (see fig. 6).

Conclusion 1

The precardiac mesoderm cells not only settled more quickly, but migrated significantly faster the higher the concentration of fibronectin in the substratum.

2. DO CELLS AT THE ANTERIOR END OF THE PRECARDIAC REGION DIFFER FROM THOSE AT THE POSTERIOR END?

We present two sets of evidence.

a. Tissue culture. Experiments were carried out using the sitting drop method described above, except that two small explants (0.17 x 0.17 mm) were used, one taken from the anterior end (41 explants) and one from the posterior end (57 explants) of the precardiac mesoderm of embryos at stage 5/6.5 (fig.1). They were explanted on separate coverslips. Similar explants (20 anterior and 24 posterior) were taken from older embryos (stage 7/8). In each case the concentration of fibronectin used was 50μg/ml.

The percentage increases in surface area of the explants are shown in fig. 8. The posterior pieces taken from stage 5/6 donors had expanded at a significantly faster rate than the anterior pieces from the same stage after both 48 and 72h ($p<0,05$). This difference was not seen in explants taken from the older embryos ($p<0.05$). It is concluded that there are significant differences at the time of migration between the anterior and posterior regions of precardiac mesoderm.

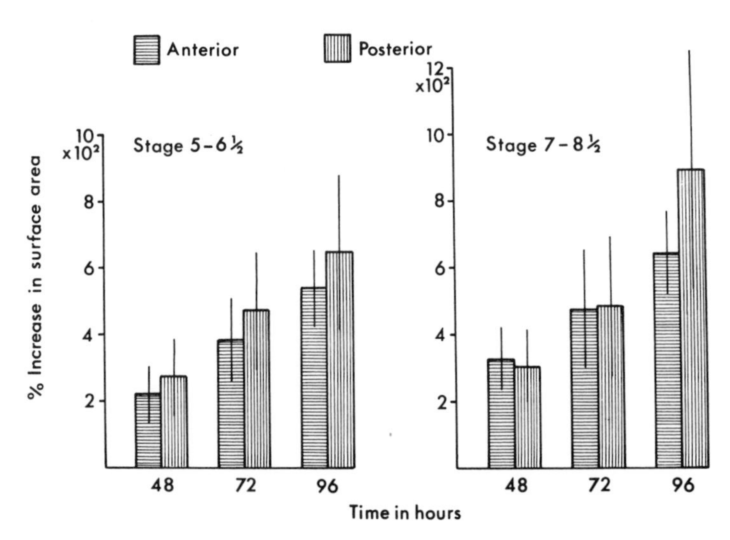


Figure 8. Bar chart to show the relative percentage increase in surface area of explants taken from anterior and posterior regions of precardiac mesoderm (see fig. 1). The posterior pieces spread at a significantly faster rate than the anterior pieces when taken from the younger embryos, but this difference was less apparent with tissue from the older donors during the early stages of the experiment.

b. Microsurgery. In a recent paper, Easton et al. (1991), we have shown differences in the anterior and posterior mesoderm of the precardiac region which have an effect on the subsequent looping of the heart. After the simple tube has formed it undergoes looping to the right. It is not known how this looping is controlled though it appears that the process has been initiated by the time a simple heart tube has formed, since if this is extirpated it nevertheless undergoes some degree of looping (Castro-Quezada et al., 1972). There is evidence that there are differences in the distribution of the microfilaments in different regions of the the tubular heart (Itasaki et al.,1989). In our experiments (Easton et al., 1991) we took explants from either the anterior or the posterior region of quail precardiac mesoderm and grafted them into either the anterior or the posterior regions of chick precardiac mesoderm, (fig. 9). Thus four different experiments were performed: QACP (quail anterior into chick posterior), QPCA (quail posterior into chick anterior), QACA (quail anterior into chick anterior) and QPCP (quail posterior into chick posterior).

Although several heart abnormalities formed, only one occurred with sufficient frequency to be significant. We called this "compact heart" and it presented in a range of severity (fig. 10). Essentially, the head appeared not to have extended forward and the heart not to have looped. When the proportion of compact hearts to the number of experiments performed was compared for each of the four groups of experiment, the compact hearts were

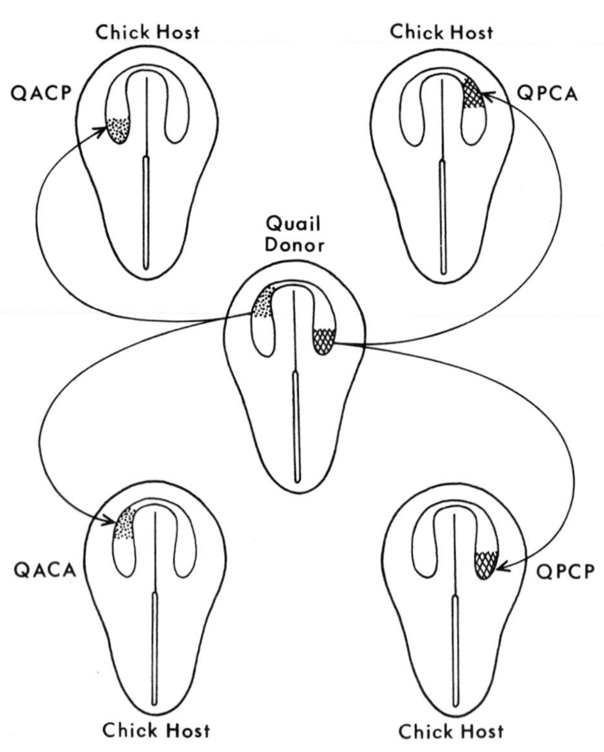


Figure 9. Diagram to show the microsurgical operations. Tissue was taken from the anterior (A) or the posterior (B) precardiac region of a quail (Q) embryo and inserted into either the anterior or the posterior precardiac region of a chick (C) embryo.

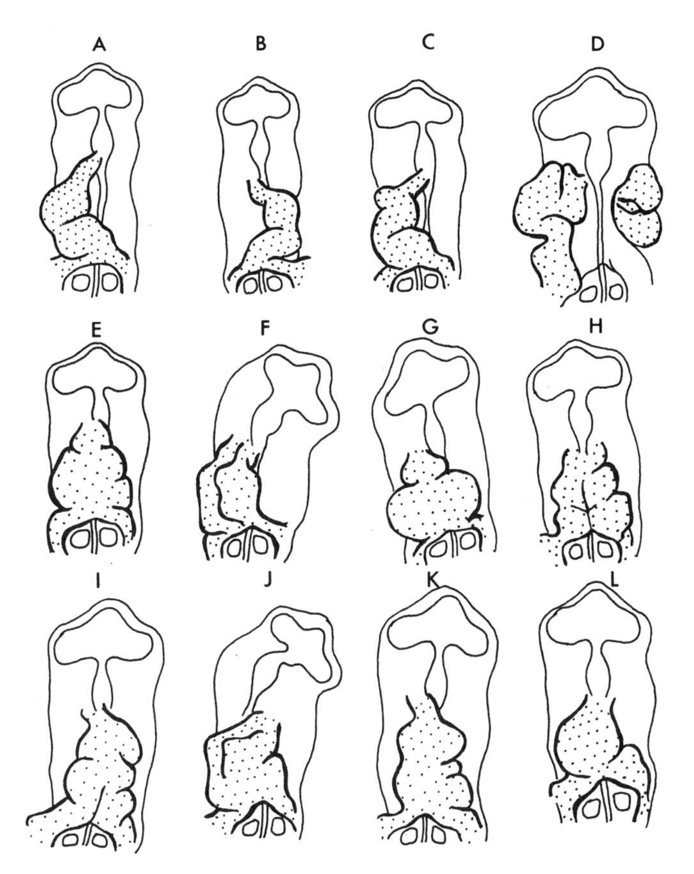


Fig. 10. Drawings to show the heart anomalies produced by chimaeras following grafting as shown in fig. 9. All embryos are viewed from the ventral side. A; near normal; B: inverted heart; C: extra tissue; D: double heart; E-L: show a range of compact (non-looped) hearts.

found to occur significantly more frequently when the posterior material was grafted into the anterior region ($p < 0.01$) (fig. 11). It was concluded that this could not be attributed to a mere mechanical disturbance caused by the operation, since no compact hearts were obtained when a graft taken from the anterior end of the precardiac region was inserted into the same region (QACA). This conclusion was supported by the results of mock operations in which surgical interference was carried out but no graft was inserted and no compact hearts formed. The compact hearts were obtained irrespective of whether the graft was inserted into the left or right side of the host. Moreover, the effect could not be attributed to the fact that the grafts were of quail tissue, since the same result was obtained when chick grafts were used. When the sections were examined, no correlation could be found between the final position of the graft and the occurrence of compact hearts. We concluded therefore that there are differences between the grafts taken from the anterior and posterior regions of the precardiac mesoderm, and that these differences are responsible for the results. We have tentatively suggested that a morphogen is produced by the tissue at the caudal end of the precardiac region which inhibits looping.

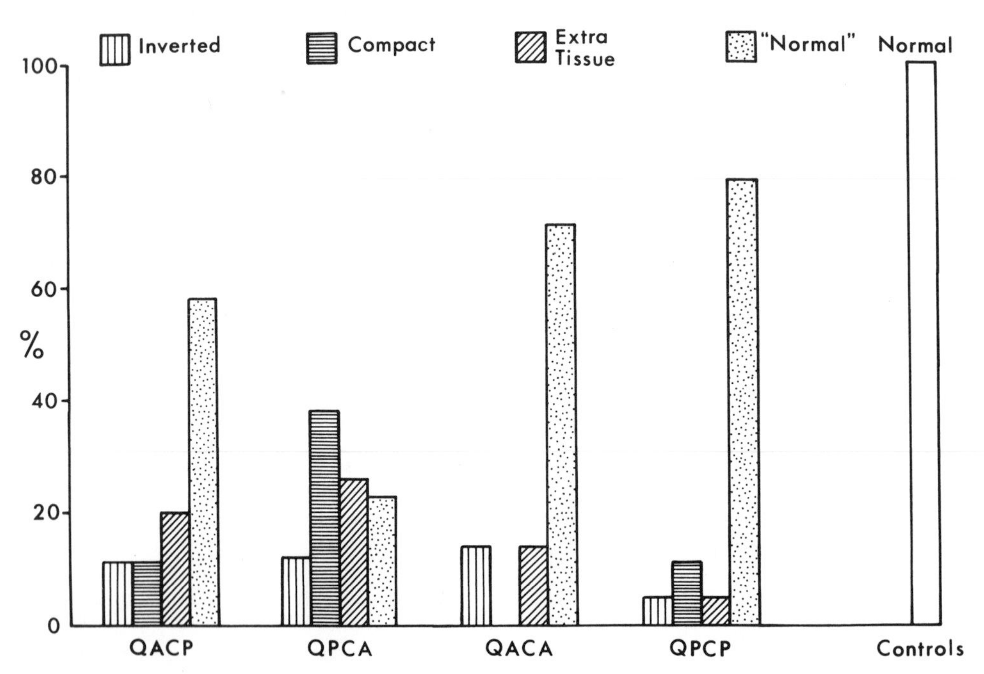


Figure 11. Bar chart to show the incidence of the different heart anomalies according to the type of experiment. Compact hearts formed significantly more frequently when grafts of posterior precardiac mesoderm were implanted into the anterior precardiac mesoderm (QPCA) than in any other combination.

Summary The answers to the two questions we have posed are:

1. The cells responded to an increased concentration of fibronectin in the substratum by settling slightly more quickly, but also, and more importantly, by migrating faster.

2. There are behavioural differences between the anterior and posterior regions of the precardiac mesoderm at stage 5/6.

These results, whilst supporting the concept that migration of the precardiac mesoderm is controlled by the concentration of fibronectin in the substratum, imply in addition that regional differences exist between cells within the precardiac region.

Acknowledgements: We are most grateful to Mrs R. Cleevely for her excellent technical support, to the British Heart Foundation for a grant to R.B. and to the Royal Society and EMBO for travel assistance for M.V.

## References

Carter, S.B. (1965). Principles of cell motility; the direction of cell movement and cancer invasion. Nature, Lond. 208: 1183-1187.

Castro-Quezada,A., Nadal-Ginard, B.,De La Cruz, M.(1972). Experimental study of the bulbo-ventricular loop in the chick. J. Emb. Exp. Morphol. 27: 623-637.

Easton, H., Veini, M., Bellairs, R. (1991) Cardiac looping in the chick embryo: the role of the posterior precardiac mesoderm. Anat. Emb. 185: 249-258.

Hamburger, V. Hamilton, H.L. (1951) A series of normal stages in the development of the chick embryo. J.Morphol. 88: 49-92.

Itasaki, N., Nakamura,H., Yasuda, M. (1989) Changes in the arrangement of actin bundles during heart looping in the chick embryo. Acta Emb. 180:413-420.

Linask, K.K., Lash, J.W. (1986) Precardiac cell migration: fibronectin localization at mesoderm:endoderm interface during directional migration. Develop. Biol. 114: 87-101.

Linask, K.K., Lash, J.W.(1988) A role for fibronectin in the migration of avian precardiac cells. Rotation of the heart-forming region during different stages and its effects. Develop. Biol. 129: 324-329.

Linask, K.K., Lash, J.W. (1990) Fibronectin and integrin distribution on migrating precardiac mesoderm cells. Annals New York Acad. Sci. 588:417-420.

New, D.A.T.(1955) A new technique for the cultivation of chick embryos in vitro. J. Embryol. Exp.Morph. 3: 326-331.

Rosenquist, G.C., DeHaan, R.L. (1966) Migration of precardiac cells in the chick embryo: a radioautographic study. Carnegie Inst. Wash. Contr. Embryol. 263: 71-110.

# THE EFFECTS OF RETINOIC ACID ON EARLY HEART FORMATION AND SEGMENTATION IN THE CHICK EMBRYO

Mark Osmond

Department of Anatomy
Downing Street
Cambridge, U.K. CB2 3DY

## Introduction

The teratogenic effects of the vitamin A derivative retinoic acid have been recognised for many years, particularly since the experiments of Kalter and Warkany (1961) which showed that the retinoid causes craniofacial abnormalities in mouse embryos. Shenefelt (1972) has shown, in the hamster embryo, that almost every organ, or tissue system can be affected by retinoic acid if exposed at its "critical period" of development. A wide variety of retinoid-induced abnormalities have been demonstrated in mammalian embryos of several species (see Sporn *et al.*, 1984), including humans (Lammer *et al.*, 1985). However, the mechanisms involved in retinoid action are not well understood.

One of the mechanisms by which retinoids may have their teratogenic effects is through the disruption of cell migration. There are numerous examples in which retinoic acid acts to inhibit the migration of mesenchymal cells, particularly neural crest cells, both *in vitro* (Morriss, 1975; Thorogood *et al.*, 1982; Smith-Thomas *et al.*, 1987) and *in vivo* (Poswillo, 1975; Hassell *et al.*, 1977). In several such studies, it has been suggested that this inhibited migration is related to an alteration in the interactions between the cells and components of the extracellular matrix.

More recently, retinoic acid has been shown to act not only as a teratogen, but also as a morphogen, respecifying the positional values of cells and altering the pattern formation of developing tissues. This is particularly apparent in the limb bud of chick embryos (Tickle *et al.*, 1982, 1985) and in the regenerating limb of amphibians (Maden, 1985). The discovery of retinoid binding proteins (see Chytil and Ong, 1984) and nuclear retinoic acid receptors (Giguere *et al.*, 1987; Petkovich *et al.*, 1987) indicates the molecular basis for such modes of retinoid action. There is evidence that retinoic acid has a similar effect on the mouse embryonic axis, altering the positional information of cells at a given axial level so that they give rise to structures characteristic of another level (Kessel and Gruss, 1991).

The aim of this largely morphological study was to determine how retinoic acid affects the development of mesodermal tissues in the chick embryo looking, first, at very early heart formation and, second, at various stages of somitogenesis. Heart development begins, at stage 5 of Hamburger and Hamilton (1951), with the formation of two bilateral heart-forming areas (Rawles, 1943) made up of precardiac mesoderm cells that have migrated into the lateral plate during gastrulation (Rosenquist and DeHaan, 1966; Rosenquist, 1970). Soon after this, the

*Formation and Differentiation of Early Embryonic Mesoderm*
Edited by R. Bellairs *et al.*, Plenum Press, New York, 1992

cells of these precardiac regions undergo a directed craniomedial migration until the two sides meet in the mid-line over the head fold (DeHaan, 1963). The two branches of this cardiogenic crescent then condense and form bilateral heart tubes which come together and start to fuse in a craniocaudal direction, forming a single mid-line heart (Manasek, 1968; Stalsberg and DeHaan, 1969). During this process, the underlying endoderm invaginates to form the anterior intestinal portal, the entrance to the foregut. As the bilateral heart tubes fuse and form the sinus venosus caudally, the anterior intestinal portal is drawn caudalwards, thus lengthening the foregut (see Osmond *et al.*, 1991 for a summary of early heart development).

The major force responsible for the directed migration of the precardiac mesoderm has been shown, by Linask and Lash (1986), to be a gradient of fibronectin in the extracellular matrix between the precardiac mesoderm and underlying endoderm. This gradient, which appears after stage 5, draws the precardiac cells craniomedially via a haptotactic mechanism involving differential adhesiveness. Disruption of the interaction between the cells and extracellular fibronectin inhibits migration of the precardiac mesoderm and formation of the cardiogenic crescent, resulting in hearts that are stunted cranially, or show various degrees of bifurcation (Linask and Lash, 1988a,b).

Segmentation of the axial mesoderm commences during the same early stages described above. The somites, of which the first pair appears at about stage 7, are the first segmented structures in the developing embryo. They form in a craniocaudal sequence and lay down the segmentation pattern for other subsequent segmented systems such as the vertebrae, spinal nerves and blood vessels. The presumptive somitic mesoderm comes to lie on either side of Hensen's node which, initially, is at the anterior end of the primitive streak (Bellairs, 1982). The node then regresses down the streak, laying down a trail of notochord (Nicolet, 1970), while the presumptive somitic mesoderm regresses on either side and becomes arranged into two strips of tissue, the segmental plates. Cells leaving the primitive streak, as the node regresses, are continually added to the caudal end of the segmental plates (Bellairs and Veini, 1984; Ooi *et al.*, 1986). The somites bud off from the cranial end of the segmental plates as the cells elongate and form a single layer around a central lumen (Bellairs, 1979). Soon after forming, the somites grow in size and undergo a secondary differentiation into dermatome, which contributes to the dermis; myotome which forms much of the musculature; and sclerotome, forming the axial skeleton. Eventually, the segmentation process reaches the tail bud, which is mainly derived from the remains of Hensen's node and the primitive streak (Holmdahl, 1925; Sanders *et al.*, 1986) and contributes to the formation of the caudal somites (Schoenwolf, 1978; Tam, 1981; Ooi *et al.*, 1986).

The factors ultimately controlling segmentation are still not well understood but, at the cellular level, somite formation involves an increase in cell-cell adhesiveness (Bellairs *et al.*, 1978; Cheney and Lash, 1984) and a concurrent increase in extracellular material, particularly of collagen (Hay, 1973; Bellairs, 1979) and of fibronectin (Ostrovsky *et al.*, 1983; Lash and Ostrovsky, 1986), although other cell adhesion molecules have also been implicated (Duband *et al.*, 1987, 1988). As in heart development, disruption of cell-cell and cell-matrix interactions in the axial mesoderm - for example, with antibodies to fibronectin - can have a detrimental effect on somitogenesis (Lash *et al.*, 1984).

The main objective of the experiments described below was to determine whether the teratogenic effects of retinoic acid were related to a similar disruption of cell interactions and subsequent behaviour, but also to see if there was an indication that the retinoid acts at any other level (eg. by altering pattern formation) in these tissues. Three types of experiment were done: (1) early embryos of different stages were exposed to various concentrations of all-*trans*-retinoic acid (RA) in New (1955) culture; (2) embryos in either New culture or *in ovo* were exposed to a local application of RA via an anion exchange bead (Eichele *et al.*, 1984); and (3) explants of mesoderm were cultured in the presence of RA. For all three experiments, the RA was dissolved in dimethyl sulphoxide (DMSO) and then, for (1) and (3), added to tissue culture medium. Controls were exposed to the same concentration of DMSO only. All developmental stages refer to those of Hamburger and Hamilton (1951). See Osmond *et al.* (1991) for a more detailed description of the methods.

## Cardiac anomalies resulting from whole embryo exposure to retinoic acid

In this series of experiments, embryos of stages 3 to 8, in New culture, were exposed to five different concentrations of RA ($10^{-6}$ M to $10^{-4}$ M) in tissue culture medium for 24 hours at 37.5° C. A variety of heart malformations resulted from this treatment, ranging from the complete absence of any heart development to retarded growth of the heart tube, or structural abnormalities, such as *situs inversus* (heart tube bent to the left instead of to the right). The most frequently occurring anomaly, however, was a heart tube that was stunted cranially, but enlarged caudally. In these, the bulbus and ventricle could be moderately stunted, or completely absent, as in Fig. 1B and D, while the atrium and sinus venosus were large and swollen. Usually associated with this heart morphology were an abnormally shaped and sized anterior intestinal portal and "winged" somites, both of which can be seen in Fig. 1D. The winged somites, so named because of their wing-like craniolateral extensions, appear to be pericardial folds extending from the somites to the heart region (see Fig. 1C).

Both concentration and stage effects were seen in the range of heart abnormalities produced by this retinoid treatment. The incidence and the severity of malformations increased as the concentration of the retinoid increased and the developmental stage of the embryo, at time of treatment, decreased. Thus, all of the specimens that showed no heart tube formation had been treated with the higher concentrations at stages 3 to 5, while those with cranially deficient hearts had been treated with lower concentrations, at these stages, or higher concentrations at the later stages (see Osmond *et al.*, 1991). The controls, which were exposed to DMSO in medium, showed none of the anomalies described above.

## Local application of retinoic acid to the precardiac mesoderm

In order to determine the effect of a controlled dose of RA on the precardiac mesoderm, AG1-X2 anion exchange beads, of approximately 100 μm in diameter, were soaked in one of three different concentrations of RA (1.0 mg/ml, 0.1 mg/ml, or 0.01 mg/ml) and then implanted into the left heart-forming area of stage 5 to 7 embryos, in New culture (see Fig. 2A). Controls were treated in the same way except that beads were soaked in DMSO alone before implantation. In both cases, embryos were incubated at 37.5° C for 24 hours before observation. As in the experiments described in the previous section, a local application of RA produced a range of heart malformations. The incidence and severity of these effects depended on the concentration of the retinoid used in the bead implants, while the developmental stage at which the implant was made appeared to affect only the severity of the defect produced. Embryos treated at later stages and with lower concentrations of RA usually had single mid-line heart tubes that were stunted, or of an anomalous shape and occasionally displayed *situs inversus*. At the earlier stages and higher concentrations, however, various degrees of cardia bifida were seen, ranging from partially bifurcated hearts (Fig. 2C and D) to two completely separate hearts, one on either side of the embryo (Fig. 2B).

These bifurcated hearts were similar to those seen in embryos in which a piece of precardiac mesoderm had been removed from across the left heart-forming area (Osmond *et al.*, 1991). In these experiments, there was an irreparable discontinuity in the precardiac mesoderm which always gave rise to full cardia bifida, usually with a larger heart on the right. In most of these embryos, the anterior intestinal portal and foregut were reduced to a small flap of endoderm in the mid-line. In others, these structures appeared only to be associated with the right heart which had sometimes developed a rudimentary sinus venosus. In retinoid-induced cardia bifida, on the other hand, the hearts were usually of about the same size and it was possible for either one, or both, to form a sinus venosus-like structure with an associated anterior intestinal portal and foregut (Fig. 2B). One similarity between the two experiments was that the two hearts always beat independently, with the left one usually beating faster. In the partial bifurcations resulting from RA implants, usually the left side beat first, followed closely by the right side. These partially bifurcated hearts, which were associated with later stages and lower retinoid concentrations, consisted of two distinct areas of cardiac tissue joined in the mid-line, allowing the development of a single prominent sinus venosus and anterior

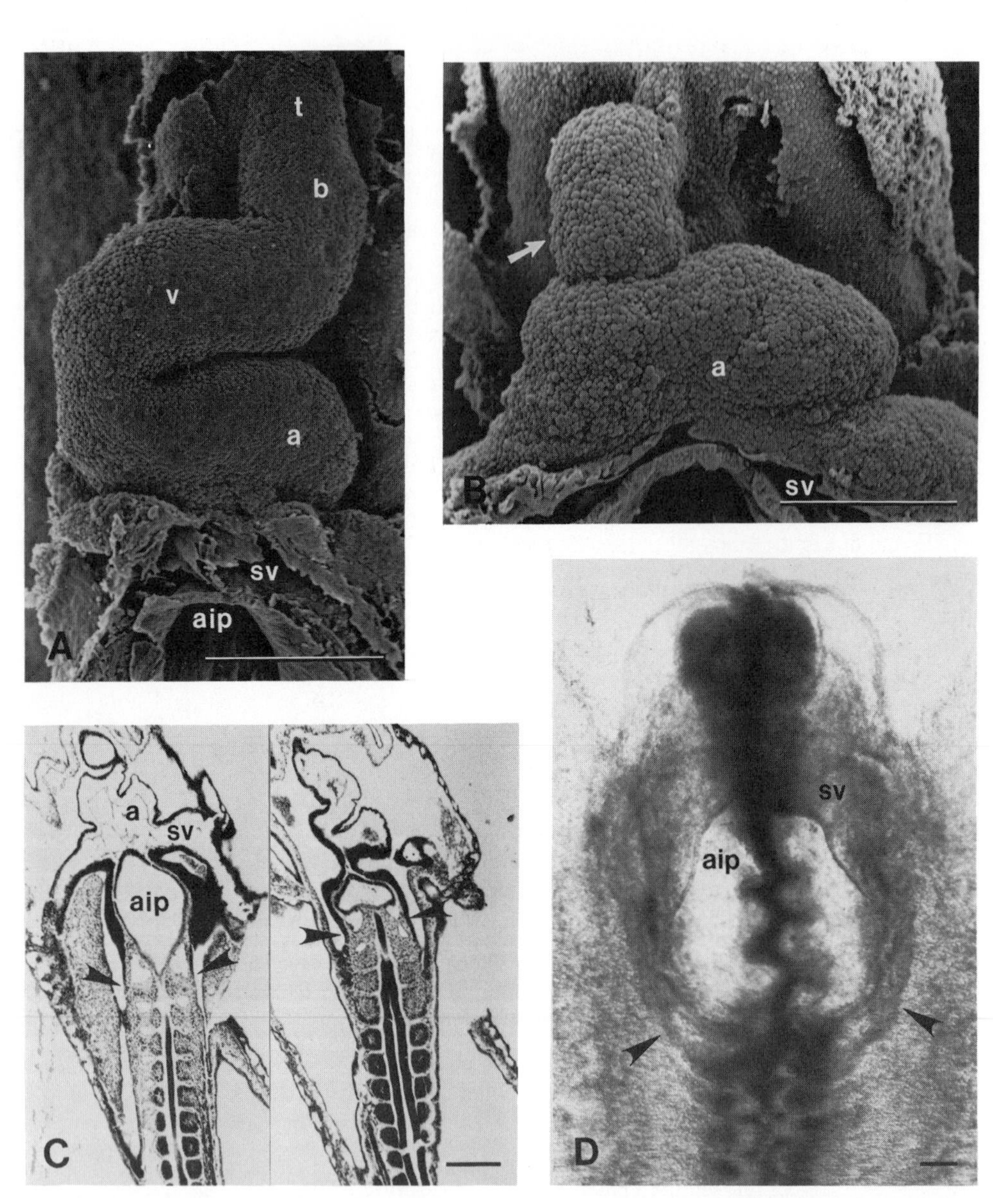


**Fig. 1.** Whole embryo exposure to either DMSO or RA. (A) Scanning electron micrograph (SEM) of a stage 13 embryo which had been treated with DMSO at stage 8, displaying a normal heart tube that has lengthened and started to bend to its right. The chambers of the primitive heart are labelled: t, truncus; b, bulbus cordis; v, ventricle; a, atrium; sv, sinus venosus; aip, anterior intestinal portal. (B) SEM of a stage 12 embryo which had been treated at stage 7 with $10^{-4}$ M RA. Caudally, the heart (atrium and sinus venosus) is well developed, but cranially is very stunted (arrow). (C) Two coronal sections of a stage 11 embryo which had been exposed to $10^{-5}$ M RA at stage 7, resulting in a heart similar to those in B and D, with winged somites extending craniolaterally. Here, the somites appear to be continuous with folds of pericardium (arrows). (D) An embryo which had been treated with $10^{-4}$ M RA at stage 5 and should now be stage 12-13, with a very stunted head, no heart development except for a large sinus venosus and its associated wide, square anterior intestinal portal and prominently winged somites, which appear to be continuous with the sinus venosus. Scale bar = 200 μm

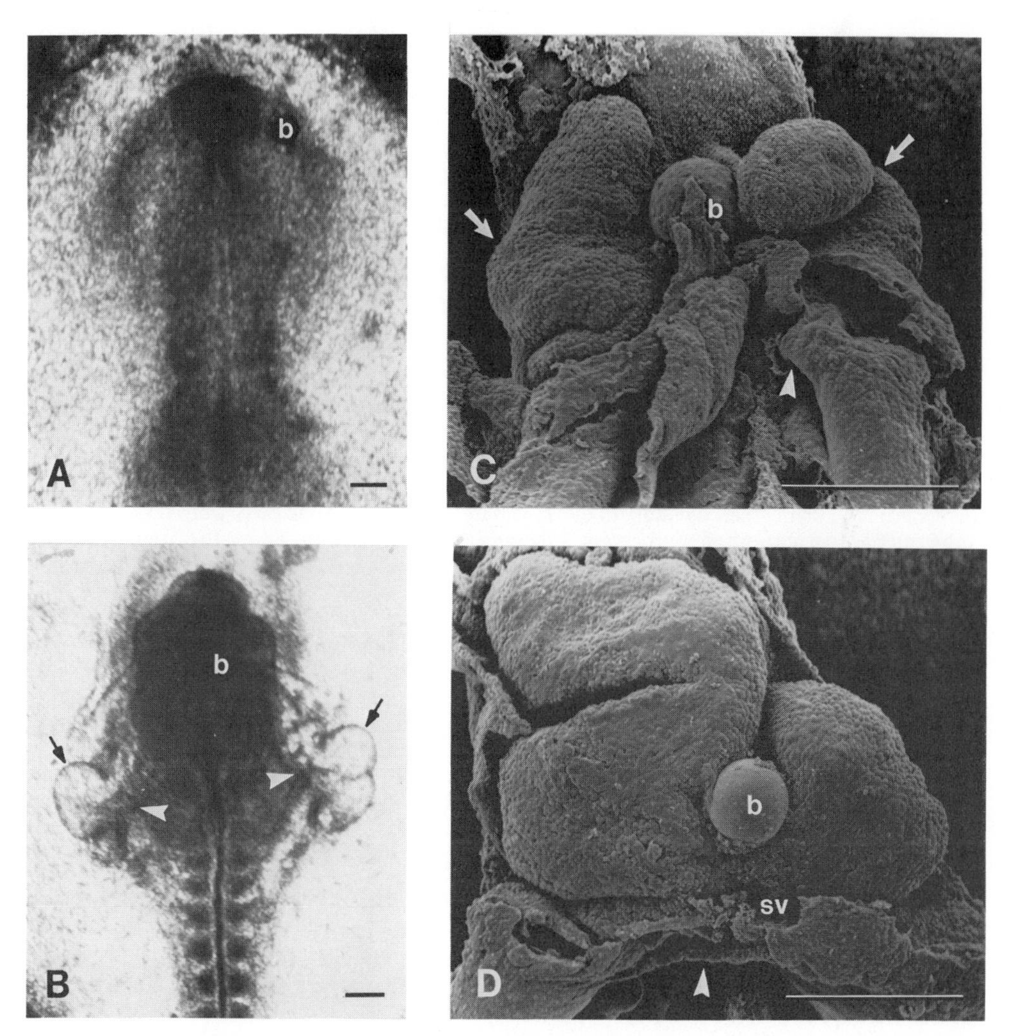


**Fig. 2.** Embryos after a bead (b) soaked in RA has been implanted into the left heart-forming area (on the right, in this ventral view). (A) Stage 7 embryo immediately after the bead soaked in 1.0 mg/ml has been implanted. (B) 24 h later, the same embryo has two hearts (arrows) which were beating independently. There is no mid-line anterior intestinal portal, or foregut, but both left and right hearts appear to have a rudimentary sinus venosus-like structure (arrowheads). (C) SEM of a stage 13 embryo which had been treated with 0.1 mg/ml RA at stage 7. There is a partial cardia bifida with the two hearts (arrows) connected over the portal (arrowhead) and the bead positioned between them. (D) As in C, but treated at stage 8. There is a less severe division of the heart than in C and there is a common mid-line sinus venosus (sv) and aip (arrowhead). Scale bar = 200 μm

intestinal portal (Fig. 2C and D). The anomalies described here are unlikely the result of the bead implant itself, as almost all of the DMSO-treated controls were normal. Rather, it appears to be the RA that impedes the formation of the cardiogenic crescent.

## Retinoid action on the precardiac mesoderm *in vitro*

The results of the whole embryo exposure to RA and locally applied RA experiments indicate that the retinoid inhibits the migration of the precardiac cells. In order to test this directly, explants of precardiac mesoderm were removed from the heart-forming areas of stage 5 to 7 embryos and exposed to RA in culture. Explants were placed in a drop of tissue culture medium on a plastic Falcon dish, or glass coverslip which had previously been coated with fibronectin or extracellular material from chick hypoblast (Sanders, 1980) to facilitate attachment and spreading of the cells. These were incubated for 24 hours before being treated with fresh medium containing either DMSO or one of three different RA concentrations ($10^{-6}$ M, $10^{-5}$ M, or $10^{-4}$ M) and incubated for a further 24 hours. A few explants continued to be incubated in medium only.

Within the first 24 hours, the precardiac mesoderm explant had usually attached to the substratum with the edge cells starting to migrate out from the centre. After a further 24 hours in untreated medium or medium containing DMSO, more cells had spread out, forming a monolayer around the explant with a diffuse outer edge. In some explants, the cells appear to have migrated out as a continuous sheet (Fig. 3A), while in others, they have migrated independently, or in small clusters. In either situation, the precardiac cells were flattened on the substratum and extended numerous substantial lamellipodia. After the total 48 hours incubation, the explant had usually undergone complete dispersal and formed a uniform sheet of well-spread cells.

When explants were treated with medium containing RA, the migration pattern of the precardiac cells was disrupted. In most cases, there was evidence of some outgrowth after treatment, suggesting a delay in the effects of the retinoid, while in others there was not. In both situations, the precardiac cells were not well-spread on the substratum, but were rounded and densely packed, extending few and very small lamellipodia, as in Fig. 3B. In many RA-treated explants, the outer edge had retracted to some degree and formed a distinct border, along which cells appear to have piled up. In some cases, the edge cells had retracted considerably around the entire perimeter, but inside this outer border of globular, densely packed cells were cells of apparently normal morphology (Fig. 3C).

A direct concentration effect was seen with the three RA concentrations used. Almost all of the explants treated with the highest concentration showed no further outgrowth from the time of treatment, while most of those treated with the intermediate concentration showed some subsequent spreading, but were all affected as described above. About half of the explants treated with the lowest concentration did not appear to be affected in any way and in those that were, the effects were usually less severe than those resulting from the higher concentrations. In several of these, there were some areas of affected cells, particularly edge cells that had retracted, but other areas in which the cells appeared to be of normal morphology. When cultures were replenished with fresh untreated medium, most of the explants which had been exposed to the highest concentration did not recover, while those treated previously with one of the two lower concentrations recuperated by 24 to 48 hours. There did not appear to be any correlation between the developmental stage of the embryo from which the explant was taken and the subsequent behaviour of the precardiac cells, both in those treated and not treated.

Many of the explants were fixed and stained with phalloidin, a phallotoxin which identifies actin microfilaments inside the cells, and anti-fibronectin, to show the fibronectin distribution. Phalloidin-stained untreated and DMSO-treated precardiac mesoderm cells displayed a well organised linear array of microfilaments spread throughout the cytoplasm, but with a higher concentration around the periphery, as in Fig. 3D. In many cells, there were several bright points close to the edge where many filaments converged. In explants treated with RA, however, most cells were indicated by small patches of diffuse staining, with little or

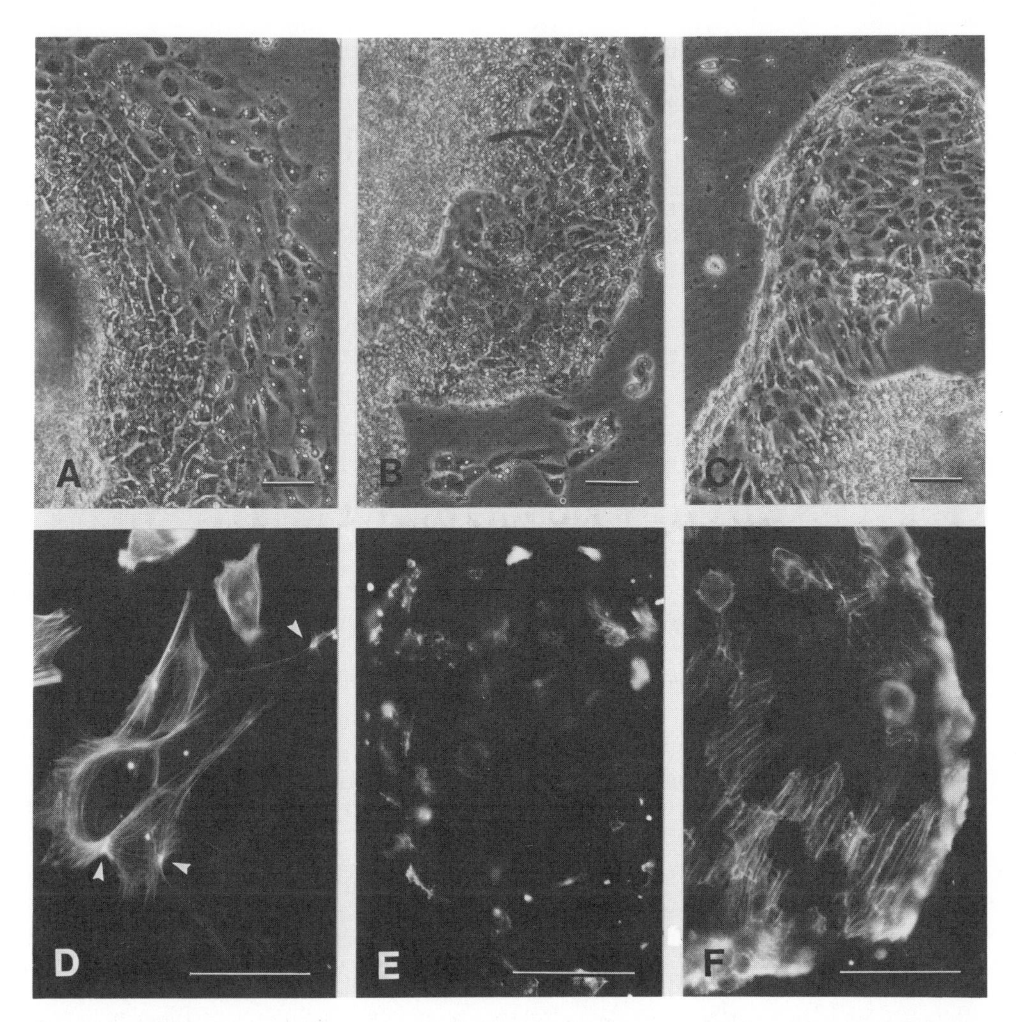


**Fig. 3.** Explants of precardiac mesoderm after 24 hours incubation in medium containing DMSO or RA, both living (A, B and C) and after staining with phalloidin-fluorescein (D, E and F) to show the actin microfilaments. (A) An explant treated with DMSO from which cells have spread out in a confluent sheet, with the flattened edge cells extending large lamellipodia. (B) After treatment with $10^{-5}$ M RA, there has been little or no further outgrowth from this explant and the cells are now rounded up, more densely packed and extending few lamellipodia. (C) Cells had migrated out from this explant as a sheet before treatment with $10^{-6}$ M RA, but soon after, the edge cells started to retract, forming a densely packed outer border. Within this border are cells with a normal, flattened morphology. (D) Phalloidin-stained edge cells of a DMSO control, in which there is a well-organised linear array of microfilaments concentrated mainly in the periphery of the cells. There are several bright points where numerous filaments converge (arrows). (E) The cells of this $10^{-5}$ M RA-treated explant show no microfilament organisation and staining is very diffuse. (F) Fluorescence of C shows the diffuse, but strong staining along the edge, where the cells have retracted, but a well-organised linear array of microfilaments throughout the cells within this border. Scale bar = 50 μm

no organisation in filament distribution (Fig. 3E). In some explants, particularly those exposed to the lowest RA concentration, the retracted edge cells showed strong, but diffuse staining along the outer border, while the cells within this border exhibited a well organised linear array of filaments (Fig. 3F). When stained with anti-fibronectin, both DMSO and RA-treated explants showed areas of strong, fibrillar fibronectin distribution and other areas where staining was weak and diffuse (data not shown). The RA-treated explants appeared to have slightly more of the latter, but this was difficult to quantify.

The results of these *in vitro* experiments provide further support to the suggestion that RA inhibits the migration of precardiac mesoderm cells. The morphology of the cells exposed to the retinoid and the associated disruption of cytoskeletal elements indicates that RA achieves this by interfering with cell-substratum interactions (see Discussion).

## Retinoid action on the axial mesoderm *in vitro*

In order to see if RA had similar effects on cells of the axial mesoderm, explants of segmental plate and tail bud mesoderm were removed from embryos of stages 13 to 14 and treated as described above for precardiac mesoderm explants. One difference, however, was that the explants of axial mesoderm usually had to be partially disaggregated before they would attach to the substratum and spread. A few somites were also explanted and these had to be almost completely disaggregated in order to overcome the strong cell-cell associations. Within 24 hours incubation, the pieces of segmental plate or tail bud mesoderm had usually attached to the substratum, with edge cells starting to spread out from the explant centres. After a further 24 hours incubation, in either untreated or DMSO-treated medium, there was substantial outgrowth. In the segmental plate explants, the cells tended to form a monolayer of confluent cells, while the cells from the tail bud appeared to migrate more independently of each other (data not shown). In both cases, the cells were well-flattened on the substratum and extended several large lamellipodia.

When treated with RA, both tissues were affected in largely the same way as the precardiac mesoderm explants (data not shown), with a similar concentration effect. Those treated with the highest concentration often showed a considerable retraction of the edge cells, so that there was a distinct border of "piled-up" cells containing densely packed, rounded cells within. The explants exposed to the intermediate concentration were less prone to lose their cell-substratum attachments, but still most became rounded, or sometimes spindle shaped. Most of those treated with the lowest concentration were less severely affected. When cultures were replenished with fresh untreated medium, only those treated with one of the two lower concentrations showed signs of recovery after 24 to 48 hours. Phalloidin staining of the microfilaments yielded results similar to those seen in the precardiac mesoderm. In the untreated and DMSO-treated explants, there was an organised array of filaments spread throughout most cells, but in the retinoid-treated cells, staining was weak and diffuse.

### *Whole segmental plate cultures*

In a further experiment, whole segmental plates (from last-formed somite to tail bud) with surrounding tissues (ie. lateral plate, notochord, endoderm and ectoderm) were cultured on agar-albumen-coated dishes in medium containing either DMSO or RA ($10^{-6}$ M, $10^{-5}$ M, or $10^{-4}$ M concentrations) and observed every three hours with respect to somite number and morphology. Differences in both parameters were seen between those exposed to RA and the controls within the first 9-10 hours, after which the tissues began to break up. The DMSO-treated specimens formed 1.0 to 1.5 (an incompletely formed somite counted as 0.5) somites every 3 hours, so that at the end of the 9 hour culture period, they had developed between 3 and 5 new somites, which were always of normal size and shape (see Fig. 4A). Most specimens treated with the highest RA concentration did not develop any more somites after treatment, but none had formed more than 1.5 during the culture period and these were always small and ill-defined. Those specimens treated with the intermediate concentration formed 0.5 to 1.0 somite every 3 hours, so that 1.5 to 3 new somites had formed after 9 hours. These somites were usually smaller than normal, flattened and with indistinct lateral borders, as in

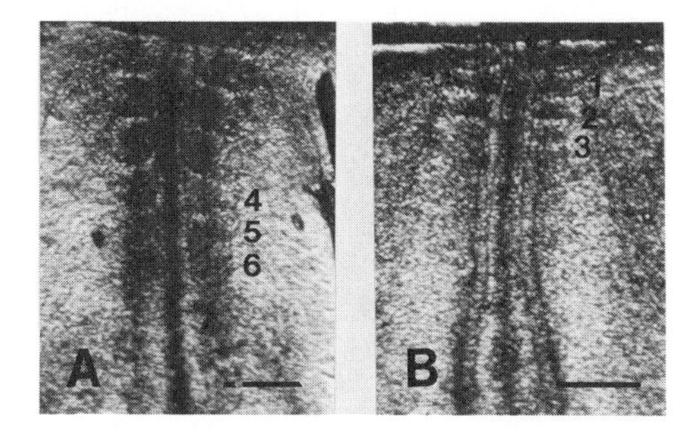


**Fig. 4.** Whole segmental plate cultures. Caudal tissues were removed from stage 11 embryos and cultured with DMSO or $10^{-5}$ M RA. (A) After 9 hours, this DMSO-treated specimen has 6 pairs of somites, all of which appear to be of normal morphology. (B) This RA-treated specimen has formed only 2 somites during the culture period, for a total of 3, and these last 2 appear to be flatter and have less distinct lateral borders than the first. Scale bar = 200 μm

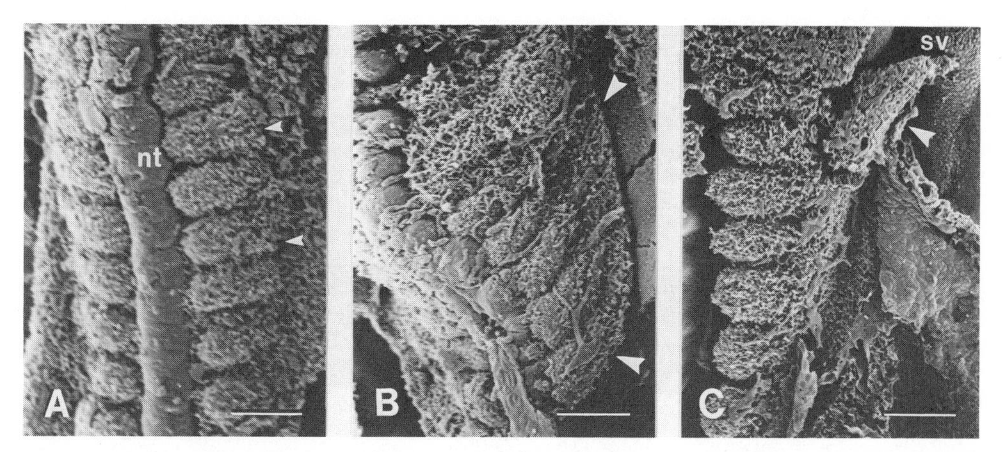


**Fig. 5.** Scanning electron micrographs of specimens after whole embryo exposure to either DMSO or RA. (A) Stage 13 embryo which had been treated with DMSO at stage 6, showing the neural tube (nt) and cranial somites. The somites are discrete blocks of tissue with distinct lateral borders (arrows). (B) Stage 12 embryo which had been treated with $10^{-4}$ M RA at stage 6, showing the same region as in A. The somites are individual blocks of tissue medially, but laterally they are joined together and are continuous with a large fold of tissue stretching cranially (arrow) to give the "winged" effect. (C) Stage 12 embryo given same treatment as B, showing the cranial somites which are continuous with a fold of mesoderm (arrow) that extends craniolaterally toward the sinus venosus (sv). Scale bar = 100 μm

Fig. 4B. About half the segmental plate cultures treated with the lowest RA concentration did not appear to be affected, while the other half responded very much like those treated with the intermediate concentration.

The results of these two experiments indicate that, in the axial mesoderm, RA can disrupt both cell-substratum and cell-cell interactions. The segmental plate and tail bud explants respond to the retinoid in a way that suggests a loss of cell-substratum attachments, similar to the situation in the precardiac mesoderm. In the RA-treated whole segmental plate cultures, the anomalous morphology and reduced number of somites reflect a deficiency in the cell-cell adhesiveness normally associated with somitogenesis (Cheney and Lash, 1984).

## Anomalies of the axial mesoderm after whole embryo exposure to retinoic acid

For this study, embryos were prepared and treated as described for the experiments resulting in cardiac anomalies after whole embryo exposure to RA. The effects of RA on somite morphology were largely dependent on concentration and developmental stage of the embryo at time of treatment and different types of defects were seen at different axial levels. Caudal somites were usually smaller and more rounded than normal (Fig. 6B) and were frequently widely and irregularly spaced. Further cranially, the somites were more often closer together, flattened and wide, with ill-defined lateral borders. A frequent variation of this defect was that of the winged somites, described earlier, where the first few (usually the first 5 or 6) somites extend craniolaterally and appear to be continuous with long, thin folds of tissue stretching down from the heart region (see Fig. 1C and D). The scanning electron micrographs in Fig. 5 show this anomaly more clearly. Embryos displaying this morphology were those which had been treated with intermediate to high RA concentrations at stages 3 to 6, before segmentation had commenced. In addition to abnormalities of somite morphology, retinoid-treated specimens appeared to produce fewer somites than the controls and this was again dependent on both concentration and developmental stage.

Another effect of RA on the axial mesoderm was a stunting of the tail region and, in this case, all stages of embryos treated with intermediate or high concentrations were affected in much the same way. The tail bud mesoderm appeared to be greatly reduced (Fig. 6D) and, when stained with Nile Blue sulphate to detect dead cells (see Jeffs and Osmond, 1992), premature cell death could sometimes be seen in this region, particularly after treatment with the highest concentration (Fig. 6B). Concurrent with this was a reduction in embryo length, which was due mainly to shortened segmental plates. The distance measured between the last-formed somite and the tip of the tail was signifcantly less in RA-treated specimens than in the controls and was the major factor in reducing the overall percentage length increase (see Fig. 7) of the embryo over the 24 hour treatment period. A striking exception to this concerned those embryos treated with the lowest RA concentration ($10^{-6}$ M) in that there was a substantial increase in this parameter. After 24 hours, these specimens showed an average percentage length increase of 131% (for all stages), as compared to 78% in the controls and 22-26% in those treated with the higher concentrations (Fig. 7). This stimulated length increase was due partially to an elongated area of unsegmented mesoderm (ie. segmental plates and tail bud), but was mainly in the segmented mesoderm, in which the somites appeared larger and spaced further apart than those in the controls.

## Local application of retinoic acid to the axial mesoderm

Similar to the RA implants in the precardiac mesoderm described earlier, beads soaked in one of four RA concentrations (10 mg/ml, 1.0 mg/ml, 0.1 mg/ml, or 0.01 mg/ml) were implanted in the axial mesoderm of stage 11-14 embryos *in ovo* at five levels: adjacent to the penultimate formed somite; at the cranial end, middle, and caudal end of the segmental plate; and in the tail bud. These experiments revealed two major trends. First, as in all the experiments described above, there was a distinct concentration effect, with higher

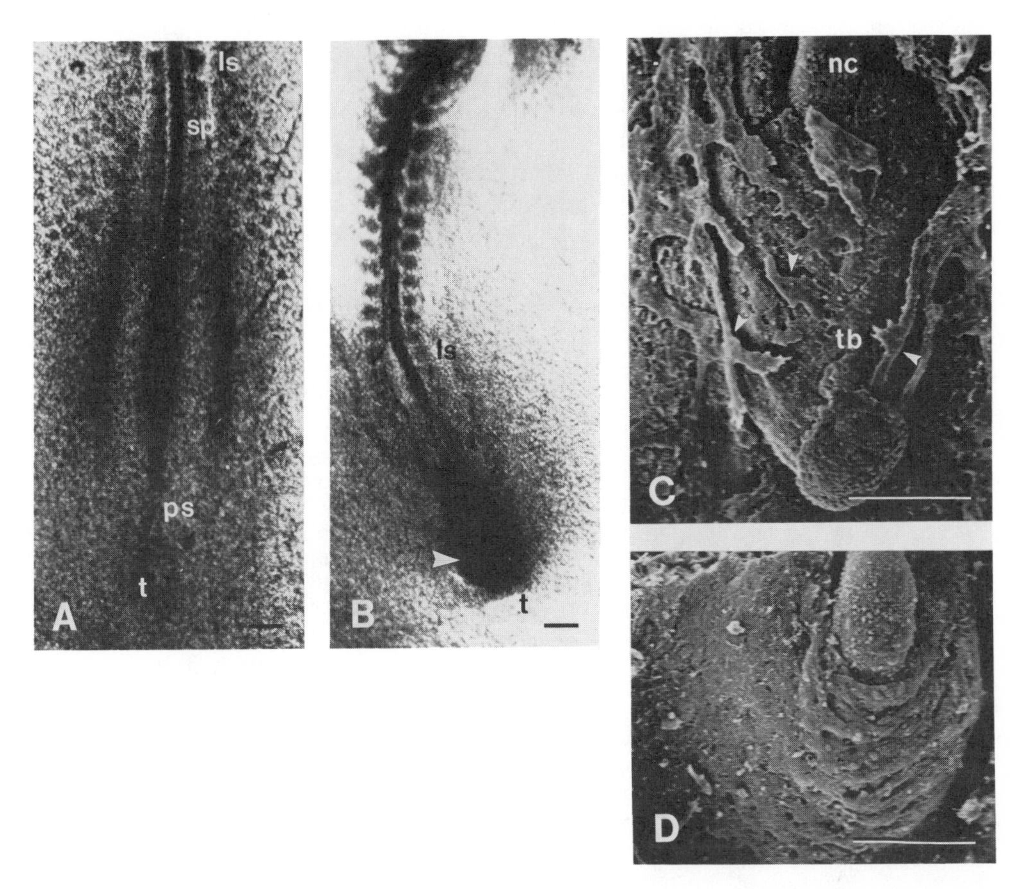


**Fig. 6.** Tail region of specimens after whole embryo exposure to either DMSO or RA. (A) Whole-mount of stage 11 embryo, which had been treated with DMSO at stage 6, stained with Nile Blue sulphate in order to detect dead cells. Note the long distinct segmental plates (sp), the remaining primitive streak (ps) in the tail region and the distance between the last pair of well-formed somites (ls) and the tail bud (t), which is free of dead cells. (B) Embryo of same stage as A, but treated with $10^{-4}$ M RA. The segmental plates are shortened and less distinct, and there is no visible primitive streak in the tail, which is characterised by a mass of dying cells (arrow). The somites are "winged" cranially, but small and rounded caudally and the distance between the last somite and the tail is less than half that in A. (C) Scanning electron micrograph of a stage 12 embryo, treated with DMSO at stage 8, showing the notochord (nc) extending into the tail bud (tb) which has an elongated tip and blood vessels (arrows) associated with it. (D) Embryo of same stage as C, but treated with $10^{-4}$ M RA, displaying a blunt tail bud with no growth at the tip, nor any associated blood vessels. Scale bar = 100 μm

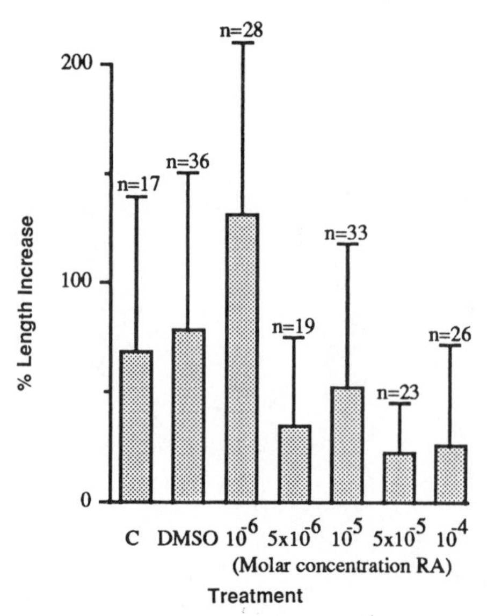


**Fig. 7.** Bar chart showing mean per cent increase in embryo length for each treatment (whole embryo exposure). Because there normally is so much inherent variation in embryo length, the most accurate way to make a quantitative comparison of this parameter, between different treatments, was to determine the difference in percentage length increase. This was calculated by subtracting the axial length of the embryo at time of treatment (using an eyepiece graticule) from the final length after 24 hours incubation. This length increase was expressed as a percentage of the initial length and the mean was calculated for each treatment (all stages grouped together). The large standard deviations, indicated by the error bars, reflect the wide range of values in each group. There is almost no difference in mean % length increase between the untreated and DMSO-treated controls, but the specimens treated with $10^{-6}$ M RA show a considerably higher value, which was found to be statistically significant using a t-test with probability $p=0.05$. Those specimens treated with $5x10^{-5}$ and $10^{-4}$ M RA display a mean % length increase that is significantly less than that of the DMSO controls. (n=no. of embryos)

concentrations having more influence on the axial mesoderm. Second, the further caudally the retinoid was applied, the greater the effect was on the development of local structures. However, for any given concentration, stage and implant position, there was still some variation in the types of anomalies produced and in the incidence at which they occurred.

*Implants in the segmented and unsegmented mesoderm*

When the bead was implanted adjacent to the penultimate formed somite, there was seldom any visible effect on the somites already formed, or on subsequent segmentation, even at the highest concentration. RA implants at the cranial end of the segmental plate resulted in local anomalies of usually the first one or two somites caudal to the bead and only on the treated side. In such cases, the somites were small and rounded, or had extra intersomitic clefts which made them appear to be out of alignment with the somites of the untreated side. Bead implants in the middle region of the segmental plate produced similar abnormalities to those described above, but these were further reaching and occurred more frequently. Usually, the two or three somites closest to the bead were very small and caudal to this, the somites were irregular in size, shape and spacing for several segments (Fig. 8B). Often, a few somites on the untreated side were also affected and these were sometimes several segments below the level of the bead. The defects were usually in the form of one or two smaller somites, or an extra intersomitic cleft and these could be seen more clearly in serial histological sections, particularly in the coronal plane (Fig. 8C). Occasionally, other structures in the region of the bead were affected, particularly blood vessels, such as the dorsal aorta and the omphalomesenteric artery, on the treated side (Fig. 8B).

Implants at the caudal end of the segmental plate produced a wider range and higher incidence of anomalies in both the somites and other caudal structures, particularly the leg buds and the tail. From one or two segments cranial to the bead, to several segments caudal to the bead, somites were often smaller, irregularly shaped and spaced apart, with extra intersomitic clefts. Somites on the untreated side were also affected in this way, in some cases. At the higher RA concentrations, the leg bud on the treated side was often stunted to varying degrees, as was the tail. Several of these specimens were stained with Nile Blue sulphate and the proportion and distribution of dead cells was no different from that seen in the control embryos. In a few embryos treated with the lower concentrations, the opposite effect was seen, in that the tail appeared to be elongated, as shown in Fig. 9B. This anomaly was sometimes accompanied by duplications of the leg bud(s), more often in the craniocaudal axis (Fig. 10A), but sometimes in the dorsoventral axis (Fig. 10C).

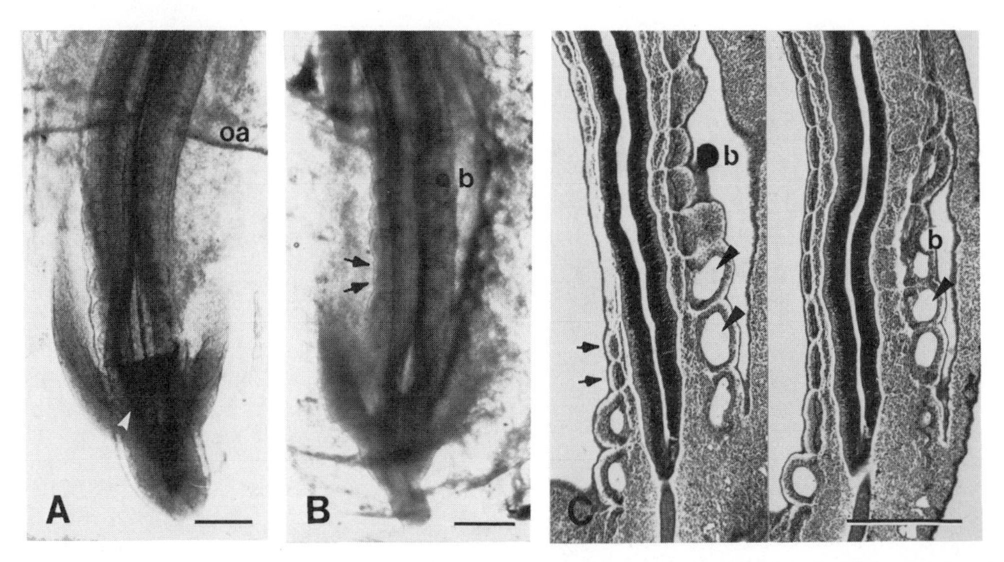


**Fig. 8.** Stage 18 embryos 24 hours after being treated *in ovo* at stage 13. (A) This untreated control has the appropriate number of normally shaped and spaced somites (34) which extend to the base of the tail (arrow). Note the bilateral omphalomesenteric arteries (oa). (B) This embryo had a bead soaked in 10 mg/ml RA implanted half way down the right segmental plate. The omphalomesenteric artery is missing on the treated side. The somites closest to the bead (b) are small and misshapen, while caudal to this they appear enlarged and widely spaced. Three segments below the bead on the untreated side, there is an extra intersomitic cleft, forming two small somites in the space of one (arrows). There is the appropriate number of somites on the left (untreated) side, extending to the base of the tail, but there appear to be fewer on the right. The leg buds and tail appear to be slightly deficient. (C) Coronal sections of embryo in B showing the position of the bead (b). The somites adjacent to the bead are smaller, while caudal to this, they are large and misshapen with bloated myoceles (arrowheads). The double somites on the untreated side are marked with arrows. The somites on the bead side appear not to extend as far caudally as on the untreated side. Scale bar = 500 μm

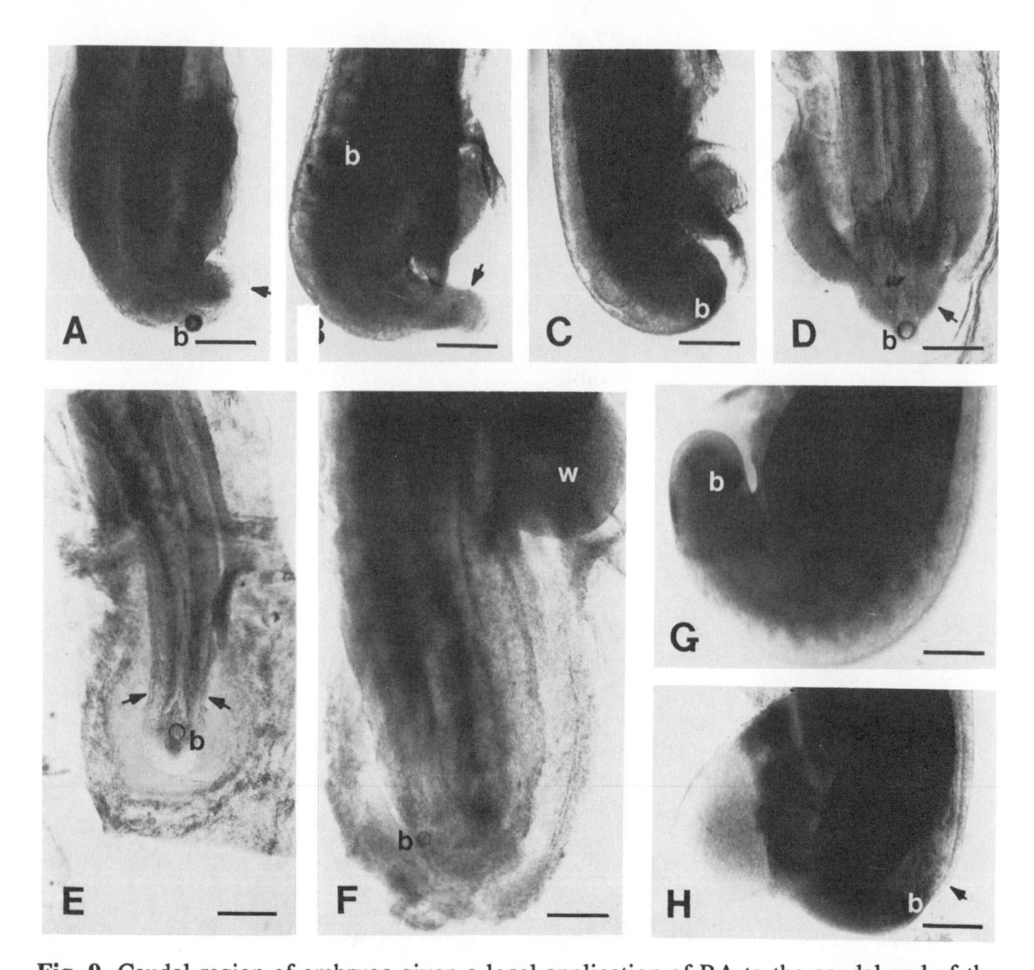


**Fig. 9.** Caudal region of embryos given a local application of RA to the caudal end of the segmental plate, or to the tail bud at stage 13-14. (A) Stage 18 embryo 24 hours after a bead (b) soaked in the lowest RA concentration (0.01 mg/ml) was implanted in the tail. There are 35 pairs of somites, which appear to be normal, extending almost to the bead. The tail is of normal size and shape for this stage (tip marked with arrow). (B) Embryo of the same stage as A, but had a 0.1 mg/ml RA implant at the caudal end of the segmental plate. The 34 pairs of somites appear to be normal, as are the leg buds, but there is a small extension of unsegmented mesoderm in the tail (arrow). (C) Stage 18 embryo 24 hours after a 1.0 mg/ml RA implant in the tail. There are 34 pairs of normal-looking somites extending almost to the bead, but the tail is considerably shortened. (D) Ventral view of stage 18 embryo treated as in C. The bead is at the tip of the stunted tail (compare to that in Fig. 8A) and staining with Nile Blue sulphate did not reveal any dead cells in this region. The leg buds are normal, as are the 33 pairs of somites of which the last are adjacent to the bead (arrow). (E) Stage 18 embryo 24 hours after a 10 mg/ml RA implant in the tail. Caudal structures are completely missing (compare to Fig. 8A) and there is a very small patch of darkly staining cells immediately caudal to the bead, which may indicate cell death. There are only 26 pairs of somites, but these all appear to be of normal shape and size and the last pair (arrows) are just above the bead. (F) Stage 22 embryo 48 hours after identical treatment to that in E. Cranial structures such as the wing bud (w) are normal, as are the somites which extend all the way to the bead, but caudal structures (eg. leg buds) are missing (Compare to G). There was no cell death detected in this region. There are only 23 pairs of somites, less than half the normal number. (G) Stage 21 embryo which had a 0.01 mg/ml RA implant in the tail. There are 44 pairs of somites, all of normal size and shape, but the tip of the tail is still unsegmented. (H) Stage 21 embryo treated with 1.0 mg/ml in the tail. The tail is very stunted and there are only 34 pairs of somites (10 less than G), the last pair (arrow) are just above the bead. Scale bar = 500 μm

*Implants in the tail*

The effects of RA when applied to the tail bud mesoderm differed from those described above in that the tail implants did not usually affect somite morphology. With higher concentrations, they inhibited further segmentation by reducing the amount of unsegmented mesoderm and stunting the tail. Lower concentrations, either had no effect, or stimulated tail length by increasing the amount of unsegmented mesoderm. In both cases, the somites which did form appeared to be normal. Almost all of the embryos treated with the highest concentration of RA in the tail displayed complete sacral agenesis, which was obvious within the first 24 hours (Fig. 9E). During this period, up to 10 pairs of somites had formed, but there was little or no remaining unsegmented mesoderm and the somites extended all the way to the caudal limit of the embryo, which is usually where the bead was still located. Axial length was thus considerably shortened in such specimens and there were no leg buds, or other caudal structures (Fig. 9F). Staining with Nile Blue sulphate revealed a very small patch of dying cells caudal to the bead in some of these embryos - less than is seen in controls of the same stage - but in most, none could be detected.

The results of the next highest RA concentration (1.0 mg/ml) applied to the tail were not as severe, but were almost as consistent as those of the higher concentration. After 24 hours, the normal number of somites (10 to 12) had formed for that period, in most cases, but there was little or no unsegmented mesoderm left, so that the somites extended to the caudal limit of the stunted tail (Fig. 9C and D). The somites were almost always of normal size and shape and development of other caudal structures, such as the leg buds, was also usually unaffected (Fig. 9D). When these embryos were allowed to develop further, the stunting in the tail became even more apparent. After 48 hours incubation, the embryos had the same number of somites as the similarly treated specimens which had been fixed 24 hours earlier (Fig. 9H).

The effects of the two lower RA concentrations on the tail were much less consistent. A few specimens treated with 0.1 mg/ml showed minor stunting of the tail, or small local defects such as *spina bifida*, while those treated with 0.01 mg/ml RA rarely displayed these anomalies. In a small number of cases, however, these lower concentrations appeared to have a stimulatory effect on the tail. Unlike the elongated tails resulting from implants in the caudal end of the segmental plate, mentioned above, a low dose of RA in the tail altered the normal pattern of development by producing a small supernumerary tail, as seen in Fig. 10D.

*Long term effects of retinoid implants in the axial mesoderm*

Many embryos from the above experiments were incubated for a total of 7 to 9 days and then stained with Alcian blue and Alizarin red (Dingerkus and Uhler, 1977) to show cartilaginous and ossified skeletal elements. Of particular interest, were the morphology and number of vertebrae after caudal RA implants. Skeletal anomalies were seldom seen in specimens which had the retinoid applied to the segmented mesoderm or the cranial end of the segmental plate. In a few cases, however, defects were apparent in the thoracic vertebrae and the ribs when the highest concentration was used (see Fig. 11B). The caudal vertebrae, in these embryos, appeared to be unaffected, as was vertebrae number. Implants in the middle of the segmental plate were more likely to cause a variety of skeletal anomalies. In some cases, the wings were affected, but more often there was a stunting of the legs (Fig. 11C). In many of these specimens, the ribs were deficient, resulting in an open ventral body wall and the thoracic vertebrae were often small and irregular. Usually, the lumbar and caudal vertebrae were more severely affected, with reduced and missing segments (Fig. 11C.) and the tail was often stunted with missing caudal and coccygeal vertebrae.

The implants into the caudal end of the segmental plate and the tail produced very similar skeletal defects, which often extended further cranially than was apparent when observed at the earlier stages. Most embryos treated with the highest RA concentration were greatly reduced, as in Fig. 9B, and did not usually survive for more than 2 or 3 days after treatment. Those which did, were all underdeveloped for their age and, in addition to being very deficient caudally, were often missing the ventral body wall, while the skeletal elements either did not stain, or stained very weakly. Similar caudal implants with the next concentration (1.0 mg/ml) allowed a higher survival rate, but also resulted in a greatly reduced vertebrae number, with the final number of segments comparable to the final number of somites seen in

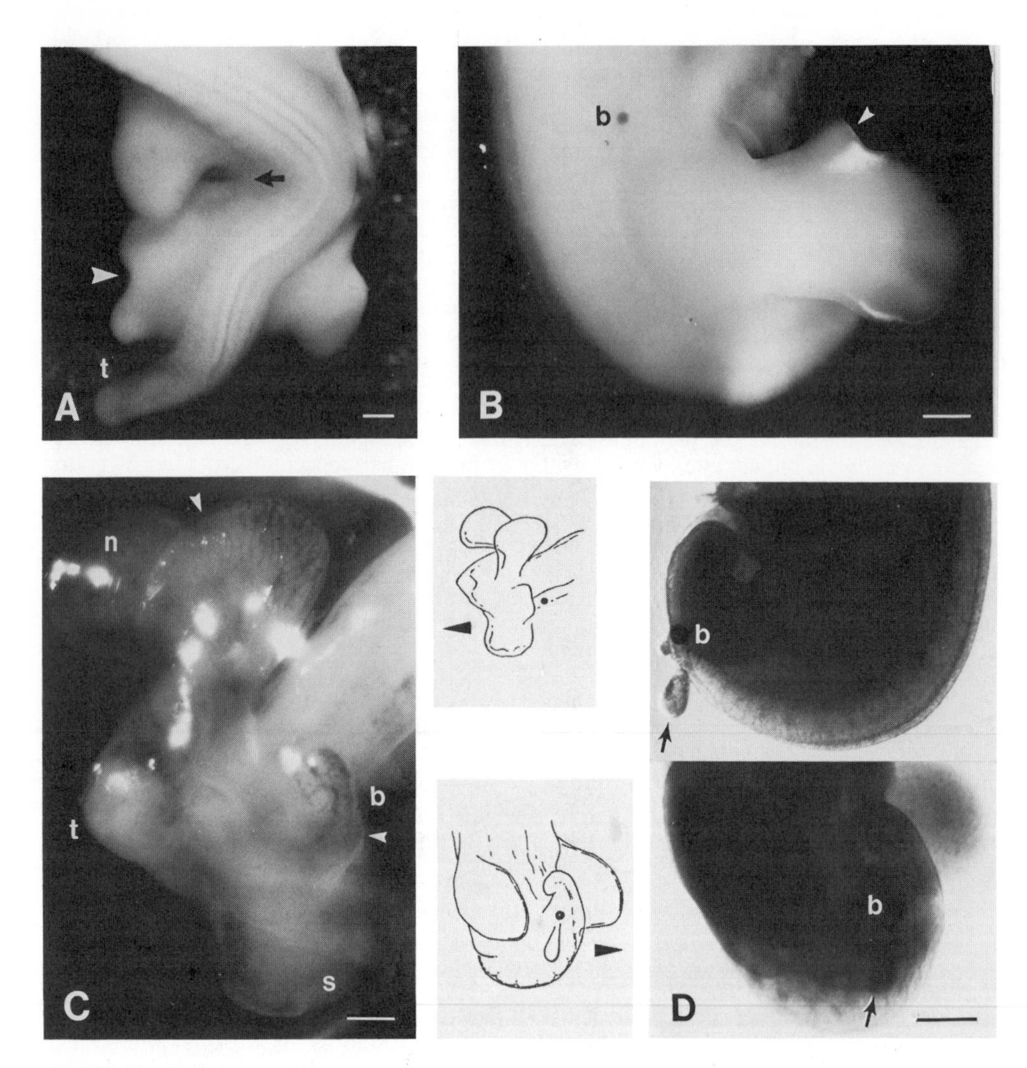


**Fig. 10.** Altered pattern formation in the tail region after RA implants. (A) Stage 22 embryo which had a 1.0 mg/ml RA implant at the caudal end of the segmental plate at stage 13. The bead (not visible) is embedded at the base of the left leg bud, at its cranial margin (arrow) and, at this level, the body has become severely twisted. There is a duplication of the left leg bud in the craniocaudal axis (arrowhead) and the tail is elongated, with an extension of unsegmented mesoderm at the tip (t). The somites, which are very small caudally, extend all the way to this and there are 55 pairs, 2-3 more than normal. (B) Stage 26 embryo which was incubated for 3 days after a 0.1 mg/ml RA implant at the caudal end of the segmental plate at stage 13. The bead (b) is embedded just cranial to the base of the right leg bud, which is partially duplicated with an extra digit (arrow). (C) Stage 25 embryo which had a 1.0 mg/ml RA implant at the caudal end of the segmental plate, at stage 13. The tail (t) is stunted, but there are four leg buds, a normal one (n) on the untreated side, a stunted one (s) on the bead (b) side and two supernumerary leg buds (arrows) growing out from the ventral aspect of the tail. (D) Two views of a stage 21 embryo which had a 0.01 mg/ml RA implant in the tail at stage 14. Everything is normal except that there is a tiny, unsegmented supernumerary tail (arrows) growing out from the base of the tail, very close to the bead. Scale bar = 500 μm

the earlier stages of such treated embryos. In some cases, the most caudal 12 to 15 vertebrae were missing, while the remaining lumbar and sacral vertebrae were all normal, but it was more usual for the last remaining vertebrae to be smaller, misshapen and partially fused, as in Fig. 11E and F. In some cases, the legs were stunted, while in a small number of specimens, leg and pelvic elements were completely missing. In many of these embryos, vertebral defects extended as far as the thoracic region with the ribs and wings also being affected, as in Fig. 11E. Caudal implants with the two lower concentrations sometimes resulted in small deficiencies and irregularities in the caudalmost vertebrae, but most specimens did not appear to be affected in any way. There were no cases of elongated tails, or extra segments observed at these stages.

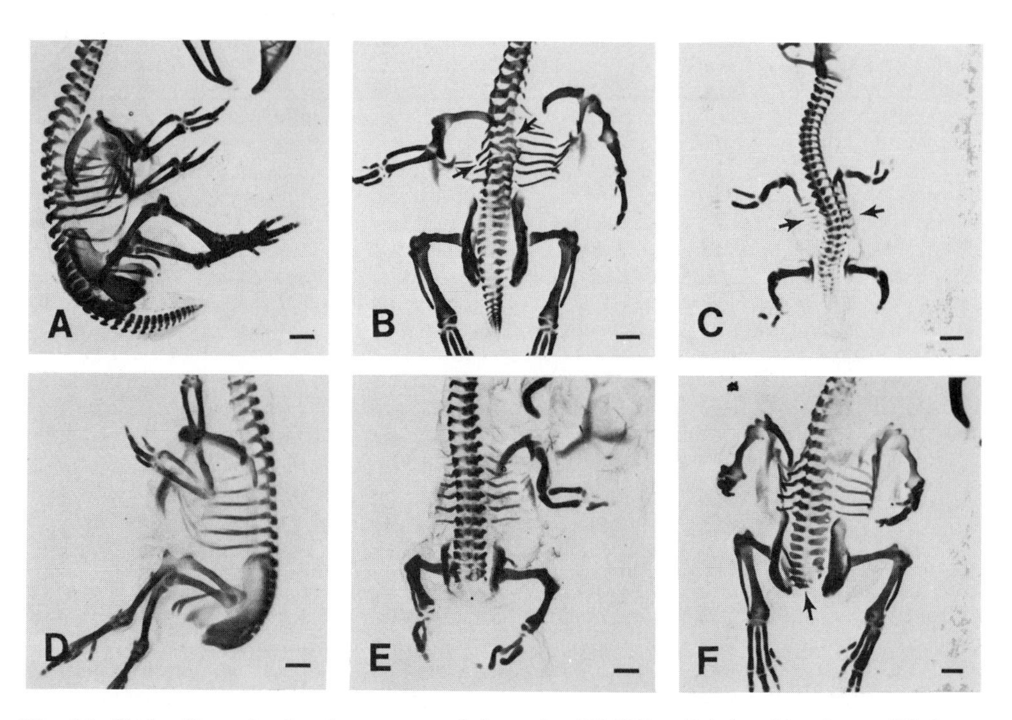


**Fig. 11.** Skeletally stained embryos several days after DMSO or RA bead implants. (A) Stage 34 (8 day) control which had a DMSO implant at the caudal end of the segmental plate at stage 14. There are 45 vertebrae, all of normal size and shape, with the full complement in the tail. The limbs and ribs also appear to be normal. (B) Embryo of same stage as A, which had a 10 mg/ml RA implant adjacent to the last formed somite at stage 14. The ribs are deficient on the left, which had led to a partially open ventral body wall. The cervical vertebrae are normal, but the thoracic vertebrae are jumbled and irregular with 2 hemi-vertebrae (arrows). The vertebrae caudal to this appear to be normal and there is a total of 44. (C) Stage 29 (7 day) embryo which had a 1.0 mg/ml implant half way down the segmental plate at stage 12. In whole-mount, it had no ventral body wall and it can be seen here that the ribs are very deficient (arrows), as are the limbs, pelvis and tail. There are only 31 vertebrae, of which the cervical are normal, but the thoracic are small and irregular, causing a lateral curvature in the spine. This anomaly becomes progressively worse caudally, in the lumbar and sacral vertebrae, while the caudal and coccygeal vertebrae are missing. (D) Embryo of same stage as in A, but given a 1.0 mg/ml RA implant in the tail bud at stage 15. Everything is normal, except that there are only 31 vertebrae. The lumbar and sacral vertebrae appeared to be of normal size and shape, but the 12 caudal and coccygeal vertebrae are missing (compare to A). (E) Embryo of same stage and given same treatment as D, but is more severely affected. The limb, rib and pelvic elements are all slightly shortened and there are only 27 vertebrae, of which the last two are partially fused. (F) Embryo of same stage and given same treatment as D and E. There are 30 vertebrae, the last three being hemi-vertebrae (arrow). Scale bar = 1mm

## Discussion

The effects of RA on development of the precardiac and axial mesoderm has been considered at three different levels. First, there is the gross morphological level at which structural changes in these tissues have been observed. Next, there is the cellular level, where retinoid effects on cell behaviour and morphology have been used to explain some of the gross anomalies. Lastly, there is the molecular level, where attempts have been made to relate changes in cell behaviour and morphology to retinoid action on cell metabolism, surface properties and on the genome. This study has directly addressed the first two of these levels, while speculating on the third.

*Retinoid action on the precardiac mesoderm*

A local application of RA to the heart forming area resulted in various degrees of cardia bifida, very similar to the effect of removing a piece of precardiac mesoderm from this region (Osmond *et al.*, 1991). It is suggested here that RA induces a zone of non-migrating cells in the precardiac mesoderm surrounding the bead, resulting in a functional, rather than a physical discontinuity in the cardiogenic crescent (see Fig. 12). Support for this hypothesis lies in the fact that almost all the DMSO controls were normal and that, in cases of full cardia bifida induced by RA, the two hearts were of approximately equal size. This indicates that there was no tissue eliminated (eg. by cell death) from the treated side. The amount of tissue affected by RA depends on the concentration. A high concentration will inhibit the migration of the entire right heart forming area, preventing formation of the cardiogenic crescent, so that two separate hearts develop. A lower concentration, however, may not affect the precardiac cells at the cranial end of the heart forming area, allowing them to join up with the left branch and form a partially bifurcated heart. The degree of bifurcation also depends on the stage at which the embryo is treated. Thus, at stage 5, when craniomedial migration of the precardiac mesoderm has not yet commenced, RA is more likely to produce two separate hearts, while at stage 7 the cardiogenic crescent has formed, so the retinoid is only able to cause a partial bifurcation.The degrees of cranially deficient hearts after whole embryo exposure to RA indicate a similar concentration and stage-dependent inhibition of precardiac cell migration, affecting both heart forming areas.

The anomalies produced in both types of experiments are very similar to those seen after disruption of the fibronectin gradient in the extracellular matrix between the precardiac mesoderm and underlying endoderm. Stunted or bifurcated hearts developed in embryos exposed to high concentrations of exogenous fibronectin, or antibodies to fibronectin (Linask and Lash, 1988a) suggesting that migration was prevented by blocking the interaction between the precardiac cells and extracellular fibronectin. Embryos exposed to other components of the extracellular matrix developed normally, indicating that it is predominantly fibronectin that is involved in the directional migration of the precardiac mesoderm. The possibility therefore exists that RA interferes with the interaction between the precardiac cells and fibronectin in the extracellular matrix.

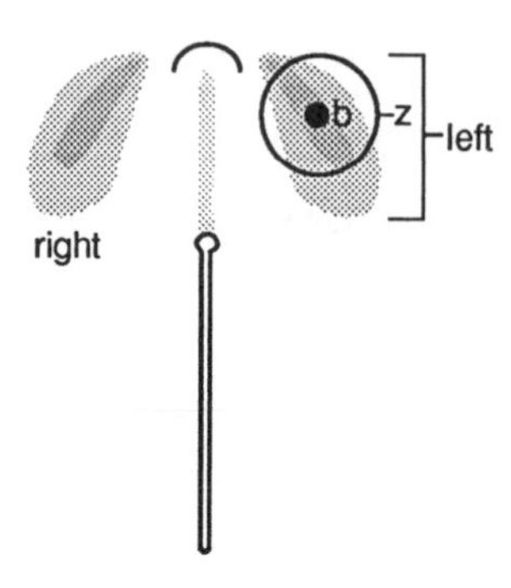


**Fig. 12.** Diagram illustrating the proposed action of RA on the precardiac mesoderm. The RA bead (b) implanted into the left heart forming area affects the precardiac mesoderm within a certain radius and thus sets up a zone (z) of non-migrating cells. These cells are prevented from joining with the rest of the cardiogenic crescent and form a separate heart on the left. The size of the non-migrating zone depends on the RA concentration. A high concentration inhibits the migration of the whole left heart forming area, but a lower concentration does not affect the more cranial precardiac cells, allowing them to contribute to the rest of the cardiogenic crescent.

Retinoids have been shown to inhibit migration in other mesodermal cell populations. Thorogood *et al.* (1982) found that in cranial neural crest cultured with retinol, cell morphology, fibronectin distribution, microfilament organisation and outgrowth from the explant were all severely affected. Smith-Thomas *et al.* (1987) have reported a similar response to RA. Likewise, in the RA-treated explants of precardiac mesoderm, cell outgrowth was inhibited and edge cells became rounded and often retracted. Associated with this was a disruption in microfilament organisation and a loss of adhesion plaques, where many filaments converge and which are known to be involved in cell-matrix contacts via transmembrane receptors (Burridge *et al.*, 1987). It was not shown conclusively that RA interfered with fibronectin distribution. However, the changes in cell morphology and migration are similar to those seen in mesoderm cells cultured in the presence of anti-fibronectin, by Sanders (1980), providing further evidence for the role of fibronectin in cell-substratum contacts. It may be more informative to look at the distribution of the transmembrane fibronectin receptor, integrin, in the cells. The fact that explants treated with the lower RA concentrations recovered after being replenished with untreated medium, indicates that these effects are reversible.

*Retinoid action on the axial mesoderm in culture*

RA was found to similarly affect cell migration, morphology and microfilament organisation in explants of segmental plate and tail bud mesoderm. In addition to these effects on cell-substratum interactions, the whole segmental plate cultures revealed that RA can also affect cell-cell adhesion. While almost all the controls developed several somites of normal size and shape, the RA-treated specimens produced significantly fewer somites and these were smaller, flattened and usually had diffuse lateral borders. This suggests that the retinoid may have a direct effect on the compaction and epithelialisation stages of segmentation. These processes involve an increase in cell-cell adhesion at the cranial end of the segmental plate and a concurrent increase in cellular fibronectin (Ostrovsky *et al.*, 1983). This might explain why, in the RA-treated cultures, the first one or two somites only partially succumbed to the effects of the retinoid, while further caudally epithelialisation was completely inhibited. The importance of fibronectin in somitogenesis is shown by the fact that exogenous cellular fibronectin stimulates compaction in segmental plate cultures and in the intact embryo, while anti-fibronectin inhibits this process (Lash *et al.*, 1984). However, the distribution of other extracellular matrix and cell adhesion molecules, such as collagen (Bellairs, 1979), laminin (Thiery *et al.*, 1982), cytotactin (Tan *et al.*, 1987), N-CAM, N-cadherin and A-CAM (Duband *et al.*, 1987, 1988) have also been shown to correlate temporally and spatially with somite formation.

*Whole embryo exposure to RA*

There were both similarities and differences in the effects of RA on the axial mesoderm, after whole embryo exposure, to those seen in the segmental plate cultures. The fact that the winged somites observed in many of these embryos were flattened, with no lateral borders and were always the most cranial somites, again implies that the retinoid acts directly on the compaction and epithelialisation stages of segmentation in this region. As RA has a relatively short active life, it will have its greatest effects in the first few hours of treatment, so it will only affect the first few pairs of somites to form, in this way. With incomplete epithelialisation, these somites appear to remain continuous with the lateral plate mesoderm so that, with body lengthening and heart formation, the lateral plate is drawn cranially and these connections are stretched with it (see Fig. 13).

The caudal somites were affected by RA in a different way. Unlike the winged cranial somites, these were distinct blocks of tissue which were small, rounded and sometimes irregularly shaped and spaced. Their small size might be explained in terms of an effect on cell migration and regression of Hensen's node. As the node regresses down the primitive streak and the segmental plates are laid down, cells in the post-nodal streak migrate laterally into the segmental plate and contribute to the somites (Bellairs and Veini, 1984; Ooi *et al.*, 1986). Somites can form without this contribution from the primitive streak, but these are smaller than normal (Bellairs and Veini, 1984). As discussed above, RA inhibits migration in cells of the axial mesoderm *in vitro* and may have a similar effect on the lateral migration of cells which would normally supplement the somitic mesoderm. The irregularities in size, shape and spacing, often seen in the caudal somites, may be due to a disruption of the somitic pre-pattern that is thought to exist in the segmental plate (Meier, 1979; Meier and Jacobson, 1982). Such a disruption of the segmental pattern (and evidence for its existence) has been seen after heat shock in amphibian embryos (Elsdale and Pearson, 1979) and in chick embryos (Veini and Bellairs, 1986). It has been suggested that this and other deleterious treatments act on a wave of cellular change which passes down the unsegmented mesoderm, setting up a periodic pre-pattern and programming the cells for eventual development into somites, as described in the "clock and wavefront" model of Cooke and Zeeman (1976). Evidence for this is given by the fact that after treatment, several pairs of normal somites form (8-10 in the chick), followed by one or more abnormal pairs. This could explain why, in RA-treated embryos, such irregularities in size and spacing were usually seen in only the most caudal somites.

The RA-induced anomalies in embryo length and somite number may be due to an effect on node regression and cell migration, similar to that discussed above. In addition to inhibiting the migration of cells from the post-nodal primitive streak into the somitic mesoderm, RA may also impede regression of the node itself and thus the laying down of the segmental plates. This could account for the shortened segmental plates (and thus axial length) seen in embryos treated with higher concentrations. It might also explain why only those embryos treated at the earlier stages, before the segmental plates were fully laid down, had a reduced somite number after 24 hours. These anomalies are similar to those seen in the *amputated* mutant mouse described by Flint *et al.* (1978). Mesoderm cells from these mutants were found to be less motile in culture, giving further support to the suggestion that such defects are due to a disruption in cell migration. The stimulatory effect of the lowest RA concentration on axial length is more difficult to explain, but is discussed below in the context of pattern formation.

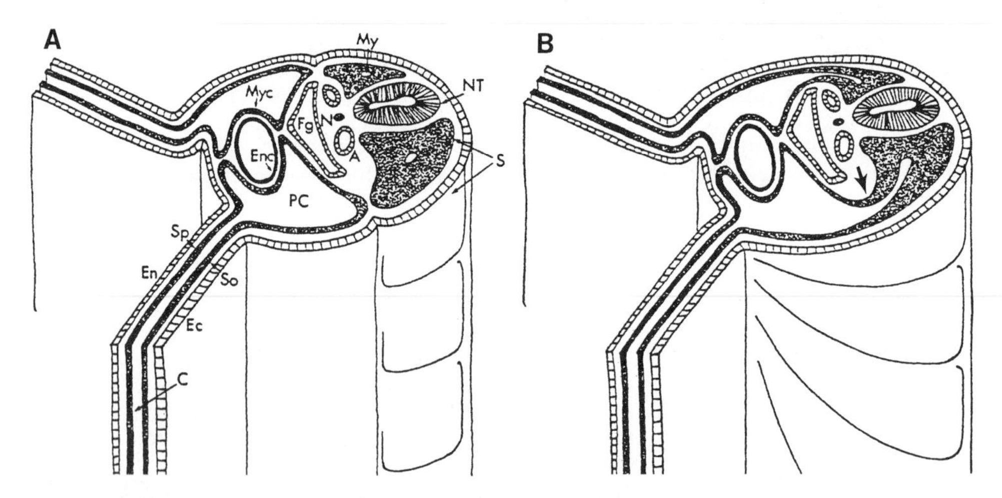


**Fig. 13.** Schematic transverse sections illustrating the "winged" morphology seen after retinoid treatment. (A) In a normal stage 11-12 embryo, the somites (S) are distinct blocks of mesoderm, with a central myocoele (My), and are separated from the lateral plate mesoderm. With body folding, the bilateral splanchnic branches (Sp) of the lateral plate, containing the heart forming areas, have moved toward the mid-line, foming the endocardium (Enc) and myocardium (Myc) of the heart. The remainder of the splanchnic mesoderm and the somatic branch (So) form the pericardium and enclose the associated coelom (C), creating the pericardial cavity (PC). Other labelled structures: NT, neural tube; N, notochord; A, dorsal aorta; Fg, foregut; En, endoderm; Ec, ectoderm. (B) In a retinoid-treated embryo displaying winged somites, a continuity persists between the somites and the lateral plate (arrow). The splanchnic and somatic branches have formed the pericardium around the heart, but both are still associated with the somitic mesoderm. In some cases, the myocoele may be continuous with the pericardial cavity. Externally, the somite does not have a lateral border and thus appears to extend craniolaterally.

Another factor which likely contributes to reduced axial length in embryos treated with high RA concentrations, is that of excessive cell death in the tail. There is evidence for both genetic and environmental factors involved in programmed cell death (Mills and Bellairs, 1989). There are mutants, such as *Rumplessness* in the chick (Zwilling, 1942) and *Brachyury* in the mouse (Wittman *et al.*, 1972) in which there is a deregulation of cell death in the tail, causing a shortened axis, while a similar anomaly has been induced in the chick with insulin (Mosely, 1947). In the chick limb bud, evidence for an environmental component to cellular death is shown by its inhibition by transplantion to another environment (Saunders *et al.*, 1962). The fact that RA has a cytotoxic effect only in the tail suggests that it is acting as a chemical signal which triggers premature death in genetically susceptible cells.

*Local retinoid action on the axial mesoderm*

Some differences in retinoid action on the axial mesoderm were seen, when applied locally and often these effects were not apparent until later in development. Implants in the segmented mesoderm, or at the cranial end of the segmental plate seldom had a visible effect on somite morphology in the first 2 days after treatement, even when high concentrations were used. However, when RA was applied further caudally down the segmental plate, there was a high incidence of anomalies in somite size, shape and spacing local to the bead. This supports the contention that there exists a somite pre-pattern in the segmental plate which is more difficult to disrupt at the cranial end, than further caudally. When specimens given cranial implants were incubated for a total of 7 to 9 days and skeletally stained, however, defects in further development of the somites became apparent. In many of these, defects in the thoracic vertebrae and the ribs, were observed and, in whole-mount, a deficiency in the ventral body wall was seen. It is possible that RA affects differentiation of the somite into sclerotome and dermomyotome, although this had not been detected in histological sections. Alternatively, the retinoid action may be manifest at a later stage, during resegmentation of the sclerotome, which may be involved in vertebral development (see Bagnall *et al.*, 1988), or during lateral migration of sclerotome and myotome to form the ribs and muscles of the body wall (see Jaffredo *et al.*, 1988). These processes are both associated with migration of somite-derived mesenchymal cells and it may be that RA is able to disrupt this later cell movement by some action at an earlier stage.

When implanted at the caudal end of the segmental plate and particularly in the tail, higher concentrations of RA appear to cause an immediate cessation in the laying down of the segmental plates and thus a severely truncated axis. Between 8 and 12 somites developed in the 24 hours after treatment, depending on the concentration used, but there was no unsegmented or tail bud mesoderm left at the end of this period and thus no further segmentation. The number of somites formed and the fact that, in most cases, these appeared to be of normal morphology suggests that RA does not disrupt the somitic pre-pattern thought to exist in the segmental plate. However, the 7 to 9 day skeletally stained specimens reveal that there are often defects in further differentiation of the somites, as shown by the anomalous caudal vertebrae.

Its effect on the tail is a prime example of how retinoid action appears to depend on the mode of application. In the whole embryo exposure experiments, high RA concentrations caused an increase in cell death in the tail region, but when applied locally, little or no cell death was detected. This may be reflecting a concentration effect, as the carrier beads release the RA slowly, instead of the tissue being exposed to a large concentration at once. Rather than eliminating the tail and other caudal structures with excessive cell death, RA applied locally appears to prevent their formation in the first place. Segmentation in the tail relies on the continued laying down of the segmental plate and this, in turn, depends on the contribution of the remaining primitive streak and Hensen's node to the tail bud mesoderm (Holmdahl, 1925; Schoenwolf, 1981) and on the subsequent proliferation and migration of this mesoderm (Tam, 1984; Ooi *et al.*, 1986). The effects of RA on migration of axial mesoderm cells have already been discussed. RA has been shown to reduce proliferation in various cell lines and Jetten (1984) has suggested that this is related to changes in cell adhesiveness, associated with alterations in cell-surface glycoproteins. It is thus possible that RA also has an anti-proliferative effect on the tail bud mesoderm.

In addition to a direct effect on cell growth and motility, there is an indication that RA may have a morphogenetic effect on the tail, similar to that seen in the chick limb bud. A local application of RA to the anterior margin of the limb bud mimics the action of the zone of polarizing activity (ZPA) to produce digit duplications (Tickle *et al.*, 1982). The degree of duplication depends on the RA concentration, but very high concentrations either reduce or prevent limb development. Tickle *et al.* (1985) proposed that RA is either a diffusible morphogen produced by the ZPA, which specifies digit pattern, or that it induces a ZPA in the adjacent tissue. Wanek *et al.* (1991) have provided evidence for this second suggestion, while Thaller and Eichele (1987) have demonstrated an endogenous gradient of RA in the chick limb, which supports the first. ZPA grafts in the tail result in truncations of varying degrees (C. Mills, personal communication), much like those produced by RA. This result is similar to that of larger ZPA grafts to the anterior margin of the limb bud, which also cause truncations (Tickle, 1981) and suggests that the ZPA has an analogous action on the tail. Additional support for RA having a morphogenetic effect on the tail is shown by the examples of tail extension. Low concentrations applied to the tail caused small, unsegmented supernumerary tails to grow from the region of the implant, while elongated tails resulted from implants to the caudal end of the segmental plate. The latter anomaly was sometimes accompanied by limb bud duplications, similar to those described by Tickle and co-workers. However, in the present study, the retinoid must be acting at a much earlier stage and is either setting up an apposing morphogenetic gradient, or inducing a new ZPA, well before the limb has started to develop.

*Mechanisms of retinoid action*

In both the precardiac and axial mesoderm, RA consistently alters cell-cell or cell-substratum interactions, with a subsequent effect on adhesiveness and migration of these cells. In the axial mesoderm, RA also appears to affect pattern formation, causing changes in the normal segmentation pattern and axial length. The molecular mechanism(s) by which retinoids have these effects is not clear. With the discovery of retinoid binding proteins (see Chytil and Ong, 1984) and retinoid nuclear receptors (Petkovich *et al.*, 1987; Giguere *et al.*, 1987), there is evidence that retinoids can effect changes in gene expression directly. With regard to RA effects on cell adhesion and migration, this may be the route by which retinoids alter the synthesis of cell-surface glycoproteins *in vitro*, as had previously been shown for fibronectin (Carlin *et al.*, 1983; Grover and Adamson, 1985; Maden, 1985), as well as laminin (Wang *et al.*, 1985) and collagen (Oikarinen *et al.*, 1985). However, retinoids can also act directly on the cell membrane and surface glycoproteins. Dingle and Lucy (1965) demonstrated that the lipid-soluble retinoids can enter the lipid phase of the membrane, causing changes in its viscosity and permeability, while both retinol (De Luca, 1977) and RA (Bernard *et al.*, 1984) have been shown to influence the glycosylation of membrane glycoproteins and, in particular, fibronectin. RA alters fibronectin levels in enucleated cells (Bolmer and Wolf, 1982) and in cells where protein synthesis has been inhibited (Zerlauth and Wolf, 1984), providing further evidence for direct retinoid action on a cell-surface target. The effects of RA are so many and varied, that it most likely acts at several different levels during development.

*A model for defining segmentation pattern and embryo length*

The stunted tails, smaller somites, reduced somite number and shortened embryo length seen after a local application, or whole embryo exposure to high RA concentrations have been discussed in terms of an effect on node regression and cell migration. The irregularly shaped, sized and spaced somites are more difficult to explain by this mechanism, as are the cases of elongated, or supernumerary tails, larger somites and increased embryo length seen after lower doses of RA. All of the above effects might be explained in terms of a morphogenetic gradient along the embryonic axis which determines the segmentation pattern. Flint *et al.* (1978) proposed the "wave form gradient" model in which somites and intersomitic boundaries are defined by peaks and troughs of morphogen concentration, respectively. The distance between peaks depends on the rate at which cells are added to the segmental plate and thus on the rate of node regression, so that the faster the node regresses, the larger the somites. Flint *et al.* (1978) presented the *amputated* mutant mouse, with its smaller somites and shorter body length, as support for this model when they found that mesoderm cells from these mutants were less motile than those from normal mice in culture. In the case of RA, it may be that a high concentration slows down node regression and cell migration, leading to smaller somites and a shortened axis, while a low concentration speeds up this process, having the opposite effect. Alternatively, the retinoid might achieve the same results by providing an extraneous source of morphogen, altering the concentration in the tissue, so that segmental plate cells are assigned new positional values and intersomitic boundaries are redefined. Similarly, irregularities seen in somite size and spacing could be due to RA disrupting the wave form pattern.

To modify and add to this model, it is suggested here that the wave form, by getting smaller further caudally, could define a second "steady-state" gradient which determines embryo length. So, in addition to a sinusoidal morphogen gradient defining somites and somite boundaries, cells are also assigned positional values according to their location in this second gradient and this determines their postition along the embryonic axis. If RA is the morphogen involved, or mimics its action, then the addition of exogenous RA could cause an upwards shift in this length gradient, so that cells become "proximalised" in their positional values. This would allow potential for further axial growth. The results of a low dose of RA implanted in the caudal segmental plate, or the tail support this idea. The elongated and supernumerary tails may be due to distal cells being "proximalised", so that extra distal cells are laid down. The irregularity with which these effects are produced suggests that exact timing and position of the retinoid implant is crucial in altering the normal pattern. It is more difficult to explain the stunted axes resulting from high RA concentrations, in these terms, but it might be that a high dose is antagonistic and negates the endogenous morphogen concentration gradient. This could effectively "distalise" the tail cells and prevent further axial growth.

The complexities of the wave form gradient model of Flint *et al.* (1978) make it more difficult to envisage than a single steady-state gradient along the embryonic axis. Thaller and Eichele (1987) have shown that there is a gradient of endogenous RA across the chick limb bud and have suggested that this gradient determines digit pattern. More recently, this has also been found to correlate with a gradient of homeobox gene (Hox) expression across the limb (Oliver, *et al.*, 1990; Nohno *et al.*, 1991). In the mouse embryo, RA has been found to cause shifts in Hox expression and corresponding vertebral transformations (Kessel and Gruss, 1991). It is thus possible that positional information along the axis is defined by the specific patterns of Hox expression and that these, in turn, are determined by the concentration gradient of a chemical morphogen, such as retinoic acid. The actual presence of such a diffusible morphogen along the embryonic axis, however, remains to be demonstrated.

## Acknowledgements

This work was supported by grants from the British Heart Foundation. I am grateful to Drs. A. Butler and F.C.T. Voon for their collaboration on some of the work presented here and to Mrs. R. Cleevely for her skilled technical assistance. I would also like to thank Prof. R. Bellairs, in whose laboratory this work was carried out and whose guidance made it possible.

## References

Bagnall, K. M., Higgins, S. J. and Sanders, E. J. (1988). The contribution made by a single somite to the vertebral column: experimental evidence in support of resegmentation using the chick-quail chimaera model. *Development* **103**, 69-85.

Bellairs, R. (1979). The mechanism of somite segmentation in the chick embryo. *J. Embryol. exp. Morph.* **51**, 227-243.

Bellairs, R. (1982). Gastrulation processes in the chick embryo. In *Cell Behaviour* (eds. R. Bellairs, A. Curtis and G. Dunn), pp. 395-427. Cambridge: Cambridge University Press.

Bellairs, R. and Veini, M. (1984). Experimental analysis of control mechanisms in somite segmentation in avian embryos. II. Reduction of material in the gastrula stages of the chick. *J. Embryol. exp. Morph.* **79**, 183-200.

Bellairs, R., Curtis, A. S. G. and Sanders, E. J. (1978). Cell adhesiveness and embryonic differentiation. *J. Embryol. exp. Morph.* **46**, 207-213.

Bernard, B. A., De Luca, L. M., Hassell, J. R., Yamada, K. M. and Olden, K. (1984). Retinoic acid alters the proportion of high mannose to complex type oligosaccharides on fibronectin secreted by cultured chondrocytes. *J. Biol. Chem.* **259**, 5310-5315.

Bolmer, S. D. and Wolf, G. (1982). Retinoids and phorbol esters alter release of fibronectin from enucleated cells. *Proc. Natl. Acad. Sci.* **79**, 6541-6545.

Burridge, K. Molony, L. and Kelly, T. (1987). Adhesion plaques: sites of transmembrane interation between the extracellular matrix and the actin cytoskeleton. *J. Cell Sci.* **8** (Suppl.), 211-229.
Carlin, B. E., Durkin, M. E., Bender, B., Jaffe, R. and Chung, A. E. (1983). Synthesis of laminin and entactin by F9 cells induced with retinoic and dibutyryl cyclic AMP. *J. Biol. Chem.* **258**, 7729-7737.
Cheney, C. M. and Lash, J. W. (1984). An increase in cell-cell adhesion in the segmental plate results in a meristic pattern. *J. Embryol. exp. Morph.* **79**, 1-10.
Chytil, F. and Ong, D. E. (1984). Cellular retinoid-binding proteins. In *The Retinoids*, vol. **2** (eds. M. B. Sporn, A. B. Roberts and D. S. Goodman), pp. 90-123. Orlando: Academic Press.
Cooke, J. and Zeeman, E. C. (1976). A clock and wavefront model for the control of the number of repeated structures during animal development. *J. Theor. Biol.* **58**, 455-476.
DeHaan, R. L. (1963). Organization of the cardiogenic plate in the early chick embryo. *Acta Embryol. Morph. exp.* **6**, 26-38.
De Luca, L. (1977). The direct involvement of vitamin A in glycosyl transfer reactions of mammalian membranes. In *Vitamins and Hormones,* vol. **35** (eds. P. L. Munson, E. Diczfalusy, J. Glover and R. E. Olson), pp. 1-57. New York: Academic Press.
Dingerkus, G. and Uhler, L. (1977). Enzyme clearing of alcian blue stained whole small vertebrates for demonstration of cartilage. *Stain Technol.* **52**, 229-232.
Dingle, J. T. and Lucy, J. A. (1965). Vitamin A, carotenoids and cell function. *Biol. Rev. Cambridge Philos. Soc.* **40**, 422-461.
Duband, J.-L., Dufour, S., Hatta, K., Takeichi, M., Edelman, G. M. and Thiery, J. P. (1987). Adhesion molecules during somitogenesis in the avian embryo. *J. Cell Biol.* **104**, 1361-1374.
Duband, J.-L., Volberg, T., Sabanay, I., Thiery, J. P. and Geiger, B. (1988). Spatial and temporal distribution of the adherens-junction-associated adhesion molecule A-CAM during avian embryogenesis. *Development* **103**, 325-344.
Eichele, G., Tickle, C. and Alberts, B. M. (1984). Micro-controlled release of biologically active compounds in chick embryos: Beads of 200 μm diameter for the local release of retinoids. *Anal. Biochem.* **142**, 542-555.
Elsdale, T. and Pearson, M. (1979). Somitogenesis in amphibian embryos. II. Origins in early embryogenesis of two factors involved in somite specification. *J. Embryol. exp. Morph.* **53**, 245-267.
Flint, O. P., Ede, D. A., Wilby, D. K. and Proctor, S. (1978). Control of somite number in normal and *Amputated* mutant mouse embryos: an experimental and theoretical analysis. *J. Embryol. exp. Morph.* **45**, 189-202.
Giguere, V., Ong, E. S., Segui, P. and Evans, R. M. (1987). Identification of a receptor for the morphogen retinoic acid. *Nature* **330**, 624-629.
Grover, A. and Adamson, E. D. (1985). Roles of extracellular matrix components in differentiating teratocarcinoma cells. *J. Biol. Chem.* **260**, 12252-12258.
Hamburger, V. and Hamilton, H. L. (1951). A series of normal stages in the development of the chick embryo. *J. Morph.* **88**, 49-92.
Hassell, J. R., Greenberg, J. H. and Johnston, M. C. (1977). Inhibition of cranial neural crest development by vitamin A in the cultured chick embyro. *J. Embryol. exp. Morph.* **39**, 267-271.
Hay, E. D. (1973). Origin and role of collagen in the embryo. *Amer. Zool.* 13, 1085-1107.
Holmdahl, D. E. (1925). Experimentelle Untersuchungen uber die Lage der Grenze Zwischen primarer und sekundarer korperntwicklung beim Huhn. *Anat. Anz. Bd.* **59**, 393-396.
Jaffredo, T., Horwitz, A. F., Buck, C. A., Rong, P. M. and Dieterlen-Lievre, F. (1988). Myoblast migration specifically inhibited in the chick embryo by grafted CSAT hybridoma cells secreting an anti-integrin antibody. *Development* **103**, 431-446.
Jeffs, P. S. and Osmond, M. (1992). A segmented pattern of cell death during development of the chick embryo. *Anat. Embryol.* In press.
Jetten, A. M. (1984). Modulation of cell growth by retinoids and their possible mechanism of action. *Fed. Proc.* **43**, 134-139.
Kalter, H. and Warkany, J. (1961). Experimental production of congenital malformations in strains of inbred mice by maternal treatment with hypervitaminosis A. *Am. J. Pathol.* **38**, 1-21.

Kessel, M. and Gruss, P. (1991). Homeotic transformations of murine vertebrae and concomitant alteration of *Hox* codes induced by retinoic acid. *Cell* **67**, 89-104.
Lammer, E. J., Chen, D. T., Hoar, R. M., Agnish, N. D., Benke, P. J., Braun, J. T., Curry, C. J., Fernhoff, P. M., Grix, A. W. Jr., Lott, I. T., Richard, J. M. and Sun, S. C. (1985). Retinoic acid embryopathy. *New Eng. J. Med.* **313**, 837-841.
Lash, J. W. and Ostrovsky, D. (1986). On the formation of somites. In *Developmental Biology: A Comprehensive Synthesis*, vol. **2** (ed. L. Browder), pp. 547-563. New York: Plenum Press.
Lash, J. W., Seitz, A. W., Cheney, C. M. and Ostrovsky, D. (1984). On the role of fibronectin during the compaction stage of somitogenesis in the chick embryo. *J. exp. Zool.* **232**, 197-206.
Linask, K. K. and Lash, J. W. (1986). Precardiac cell migration: Fibronectin localization at mesoderm-endoderm interface during directional movement. *Devl Biol.* **114**, 87-101.
Linask, K. K. and Lash, J. W. (1988a). A role for fibronectin in the migration of avian precardiac cells: I. Dose-dependent effects of fibronectin antibody. *Devl Biol.* **129**, 315-324.
Linask, K. K. and Lash, J. W. (1988b). A role for fibronectin in the migration of avian precardiac cells: II. Rotation of the heart-forming region during different stages and its effects. *Devl Biol.* **129**, 325-329.
Maden, M. (1985). Retinoids and the control of pattern in regenerating limbs. *Ciba Found. Symp.* **113**, 132-155.
Manasek, F. J. (1968). Embryonic development of the heart. I. A light and electron microscopic study of myocardial development in the early chick embryo. *J. Morph.* **125**, 329-366.
Meier, S. (1979). Development of the chick embryo mesoblast: Formation of the embryonic axis and establishment of the metameric pattern. *Devl Biol.* **73**, 25-45.
Meier, S. and Jacobson, A. G. (1982). Experimental studies of the origin and expression of metameric pattern in the chick embryo. *J. exp. Zool.* **219**, 217-232.
Mills, C. L. and Bellairs, R. (1989). Mitosis and cell death in the tail of the chick embryo. *Anat. Embryol.* **180**, 301-308.
Morriss, G. M. (1975). Abnormal cell migration as a possible factor in the genesis of vitamin A-induced craniofacial anomalies. In *New Approaches to the Evaluation of Abnormal Embryonic Development* (eds. D. Neubert and H. J. Merker), pp.678-687. Stuttgart: Thieme.
Mosely, H. R. (1947). Insulin-induced rumplessness of chickens IV. Early embryology. *J. exp. Zool.* **105**, 279-315.
New, D. A. T. (1955). A new technique for the cultivation of the chick embryo in vitro. *J. Embryol. exp. Morph.* **3**, 326-331.
Nicolet, G. (1970). Is the presumptive notochord responsible for somite genesis in the chick? *J. Embryol. exp. Morph.* **24**, 467-478.
Nohno, T., Noji, S., Koyama, E., Ohyama, K., Myokai, F., Kuroiwa, A., Saito, T. and Taniguchi, S. (1991). Involvement of the *Chox-4* chicken homeobox genes in determination of anteroposterior axial polarity during limb development. *Cell* **64**, 1197-1205.
Oikarinen, H., Oikarinen, A. I., Tan, E. M., Abergel, R. P., Meeker, C. A., Chu, M. L., Prockop, D. J. and Uitto, J. (1985). Modulation of procollagen gene expression by retinoids. Inhibition of collagen production by retinoic acid accompanied by reduced type I procollagen messenger ribonucleic acid levels in human skin fibroblast cultures. *J. Clin. Invest.* **75**, 1545-1553.
Oliver, G., De Robertis, E. M., Wolpert, L. and Tickle, C. (1990). Expression of a homeobox gene in the chick wing bud following application of retinoic acid and grafts of polarizing region tissue. *EMBO J.* **9**, 3093-3099.
Ooi, V. E. C., Sanders, E. J. and Bellairs, R. (1986). The contribution of the primitive streak to the somites in the avian embryo. *J. Embryol. exp. Morph.* **92**, 193-206.
Osmond, M. K., Butler, A. J., Voon, F. C. T. and Bellairs, R. (1991). The effects of retinoic acid on heart formation in the early chick embryo. *Development* **113**, 1405-1418.
Ostrovsky, D., Cheney, C. M., Seitz, A. W. and Lash, J. W. (1983). Fibronectin distribution during somitogenesis in the chick embryo. *Cell Diff.* **13**, 217-223.
Petkovich, M., Brand, N. J., Krust, A. and Chambon, P. (1987). A human retinoic acid receptor which belongs to the family of nuclear receptors. *Nature* **330**, 444-450.

Poswillo, D. (1975). The pathogenesis of the Treacher Collins syndrome (mandibulofacial dysosostosis). *Br. J. Oral. Surg.* **13**, 1-26.

Rawles, M. E. (1943). The heart-forming areas of the early chick blastoderm. *Physiol. Zool.* **6**, 22-44.

Rosenquist, G. C. (1970). Location and movements of cardiogenic cells in the chick embryo: The heart-forming portion of the primitive streak. *Devl Biol.* **22**, 461-475.

Rosenquist, G. C. and DeHaan, R. L. (1966). Migration of precardiac cells in the chick embryo: A radiographic study. *Carnegie Inst. Washington Contrib. Embryol.* **38**, 111-121.

Sanders, E. J. (1980). The effect of fibronectin and substratum-attached material on the spreading of chick embryo mesoderm cells *in vitro*. *J. Cell Sci.* **44**, 225-242.

Sanders, E. J., Khare, M. K., Ooi, V. E. C. and Bellairs, B. (1986). An experimental and morphological analysis of the tail bud mesenchyme of the chick embryo. *Anat. Embryol.* **174**, 179-185.

Saunders, J. W., Gasseling, M. T. and Saunders, L. C. (1962). Cellular death in morphogenesis of the avian wing. *Devl. Biol.* **5**, 147-178.

Schoenwolf, G. C. (1978). Effects of complete tail bud extirpation on early development of the posterior region of the chick embryo. *Anat. Rec.* **192**, 289-296.

Schoenwolf, G. C. (1981). Morphogenetic processes involved in the remodelling of the tail region of the chick embryo. *Anat. Embryol.* **162**, 183-197.

Shenefelt, R. E. (1972). Morphogenesis of malformation in hamsters caused by retinoic acid: Relation to dose and stage at treatment. *Teratology* **5**, 103-118.

Smith-Thomas, L., Lott, I. and Bronner-Fraser, M. (1987). Effects of isotretinoin on the behaviour of neural crest cells in vitro. *Devl Biol.* **123**, 276-281.

Sporn, M. B., Roberts, A. B. and Goodman, D. S., eds. (1984). *The Retinoids*, vol. 1 and 2. Orlando: Academic Press.

Stalsberg, H. and DeHaan, R. L. (1969). The precardiac areas and formation of the tubular heart in the chick embryo. *Devl Biol.* **19**, 128-159.

Tam, P. P. L. (1981). The control of somitogenesis in mouse embryos. *J. Embryol. exp. Morph.* **65** (Suppl.), 103-128.

Tam, P. P. L. (1984). The histogenetic capacity of tissues in the caudal end of the embryonic axis of the mouse. *J. Embryol. exp. Morph.* **82**, 253-266.

Tan, S.-S., Crossin, K. L., Hoffman, H. and Edelman, G. M. (1987). Asymmetric expression in somites of cytotactin and its proteoglycan ligand is correlated with neural crest cell distribution. *Proc. Natl. Acad. Sci.* **84**, 7977-7981.

Thiery, J. P., Duband, J.-L. and Delouvee, A. (1982). Pathways and mechanisms of avian trunk neural crest cell migration and localization. *Devl Biol.* **93**, 324-343.

Thaller, C. and Eichele, G. (1987). Identification and spatial distribution of retinoids in the developing chick limb bud. *Nature* **327**, 625-628.

Thorogood, P., Smith, L., Nicol, A., McGinty, R. and Garrod, D. (1982). Effects of vitamin A on the behaviour of migratory neural crest cells in vitro. *J. Cell Sci.* **57**, 331-350.

Tickle, C. (1981). The number of polarizing region cells required to specify additional digits in the developing chick wing. *Nature* **289**, 295-298.

Tickle, C., Alberts, B. M., Wolpert, L. and Lee, J. (1982). Local application of retinoic acid to the limb bud mimics the action of the polarizing region. *Nature* **296**, 564-565.

Tickle, C., Lee, J. and Eichele, G. (1985). A quantitative analysis of the effect of all-trans-retinoic acid on the pattern of chick wing development. *Devl Biol.* **109**, 82-95.

Veini, M. and Bellairs, R. (1986). Heat shock effects in chick embryos. In *Somites in Developing Embryos* (eds. R. Bellairs, D. A. Ede and J. W. Lash), pp. 135-145. New York: Plenum Press.

Wanek, N., Gardiner, D. M., Muneoka, K. and Bryant, S. V. (1991). Conversion by retinoic acid of anterior cells into ZPA cells in the chick wing bud. *Nature* **350**, 81-83.

Wang, S. Y., La Rosa, G. J. and Gudas, L. J. (1985). Molecular cloning of gene sequences transcriptionally regulated by retinoic acid and dibutyryl cyclic AMP in cultured mouse teratocarcinoma cells. *Devl Biol.* **107**, 75-86.

Wittman, K. S. Krupa, P. L., Pesetsky, I. and Hamburgh, M. (1972). Electron microscopy and histochemistry of tail regression in the *Brachyury* mouse. *Devl Biol.* **27**, 419-424.

Zerlauth, G. and Wolf, G. (1984). Kinetics of fibronectin release from fibroblasts in response to 12-0-tetradecanoylphorbol-13-acetate and retinoic acid. *Carcinogenesis* **5**, 863-868.

Zwilling, E. (1942). The embryology of a recessive rumpless condition of chickens. *J. exp. Zool.* **99**, 79-91.

REGULATORY ROLE OF CELL ADHESION MOLECULES IN EARLY HEART DEVELOPMENT

Kersti K. Linask

Division of Cardiology, Department of Pediatrics
The Children's Hospital of Philadelphia and
University of Pennsylvania School of Medicine
Philadelphia, Pennsylvania 19104-4399

**INTRODUCTION**

Pericardial coelom development is a critical step for early heart development in that the cardiac precursor cells are for the first time physically delineated from the somatic mesoderm cells. In the chick embryo this takes place between 20-24 hrs of development, i.e. between stages 5 and 7 (Hamburger and Hamilton, 1951). The pericardial coelom, also known as the amniocardiac vesicle, first appears as foci of smaller cavities that arise by the separation of mesoderm cells in discrete areas to eventually form two layers of mesoderm, the somatic (next to the ectoderm) and the splanchnic mesoderm (next to the endoderm). These cavities become larger and begin to coalesce with each other until two large amniocardiac vesicles are formed on either side of the embryo. These eventually fuse to form a single pericardial coelom. This process was first described in the chick embryo by Sabin (1920). Concomitant with the splitting of the mesoderm, the precursor cardiac cells localize to the splanchnic layer in the lateral regions. These cells also at this point in time undergo a mesenchymal to epithelial change in cell organization. These processes of cell sorting, as well as of cell shape changes, suggest that modulations of cell surface components and of cytoskeletal interactions may be involved.

The importance of cell adhesion molecules in developing systems has been historically an important area of study (cf. Townes and Holtfreter, 1955; Moscona, 1956; Steinberg, 1963). More recently the term *morphoregulatory* has been given to these adhesion molecules that by their spatial and temporal regulation in the embryo can guide changes in form and tissue patterns (Edelman et al., 1990). These morphoregulatory adhesion molecules essentially fall into three families: cell-cell adhesion molecules; cell-substratum adhesion molecules, and cell junctional molecules. In early heart development members of these families of adhesion molecules have been shown to be involved in the orchestration of several important processes. Fibronectin, a cell-substratum, has been shown to localize to the mesoderm-endoderm interface between stages 5-8 in the chick heart-forming region (Linask and Lash, 1986) and appears to be involved in the anterio-mesiad directional migration of precardiac mesoderm cells from the bilateral embryonic regions to the embryonic mid-line by a haptotactic mechanism (Linask and Lash, 1988a; Linask and Lash, 1988b). A similar localization of fibronectin has been observed in the amphibian and may be involved in the migration of precardiac cells by the same mechanism (Hirakow et al., 1987).

Cadherins are a family of homophilic calcium-dependent cell adhesion molecules (Takeichi, 1988). An underlying commonality for all of the cadherins is that they are protected from proteolysis in the presence of calcium (Takeichi, 1977). Three members of this family of proteins, E-cadherin, P-cadherin, and N-cadherin, have been well characterized and shown to be homologous at the protein level (Hatta et al., 1988) and to be involved in developing systems. The various cadherins appear to be involved in homophilic adhesive interactions through the extracellular portion of the molecule and interact with the cytoskeleton through the cytoplasmic domain (Ozawa et al., 1989). Furthermore, it appears that the cytoplasmic interaction is necessary for full functional activity of the molecule (Nagafuchi and Takeichi, 1989). In chicken cardiomyocytes grown *in vitro* it has been shown that N-cadherin associates with intracellular proteins similar to those associated with E-cadherin in epithelial cells (Wheelock and Knudsen, 1991). It is postulated that the coimmunoprecipitating proteins are involved in linking the cadherins to the cytoskeleton.

Recent work suggests that N-cadherin has a functional role in directing pericardial coelom formation and in its involvement in the mesenchymal-epithelial cell shape change seen to occur during the early steps of precardiac mesoderm delineation in the embryo. A detailed description of this research was presented elsewhere.

## METHODS

Embryos. White Leghorn chicken embryos were removed from the yolk between stages 5-8 (staged according to Hamburger and Hamilton, 1951). Embryos were fixed in 3% p-formaldehyde-PBS solution, permeabilized, immunostained, and embedded in araldite. The scanning electron microscopy was also carried out as previously described. The detailed procedures have been previously published (Linask and Lash, 1986). Immunohistochemical sections were cut at one micron and viewed with a Nikon Optiphot-2 epifluorescence microscope equipped with fluorescein and/or rhodamine filters. Photographs were taken with black and white Kodak Tmax ASA 400 film.

Antibodies. Antibodies were generously provided by the following laboratories: *N-cadherin antibodies*: Polyclonal and monoclonal N-cadherin (NCD) antibodies were produced and characterized by Dr. M. Takeichi (Kyoto University, Kyoto, Japan). Immunostaining was performed with the rabbit polyclonal antibody. *$Na^+$-$K^+$-ATPase monoclonal antibody-24*: This antibody was provided by Dr. Douglas M. Fambrough (Johns Hopkins University, Baltimore, MD) and specifically recognizes the ß subunit of a major form of the $Na^+$, $K^+$-ATPase in enriched preparations from chicken kidney (Fambrough and Bayne, 1983). *TGF-ß2 polyclonal antibody (B 1/29)*: This polyclonal antibody was generously provided by Collagen Corporation (Palo Alto, CA).

## RESULTS AND DISCUSSION

The description below will address immunolocalization of N-cadherin in the mesoderm within the anterior-lateral heart-forming regions only.

Immunostaining for N-cadherin in a stage 5 embryo shows a general localization on cell surfaces throughout the mesoderm and endoderm, but not ectoderm (Fig.1). As seen in this section, large, round cells are often observed within the lateral mesoderm that brightly stain for N-cadherin. Some of these round cells appear to form rosette type configurations in some sections that were observed. Endoderm expresses N-cadherin throughout the stages observed, i.e. between stages 5-8, and will not be described further.

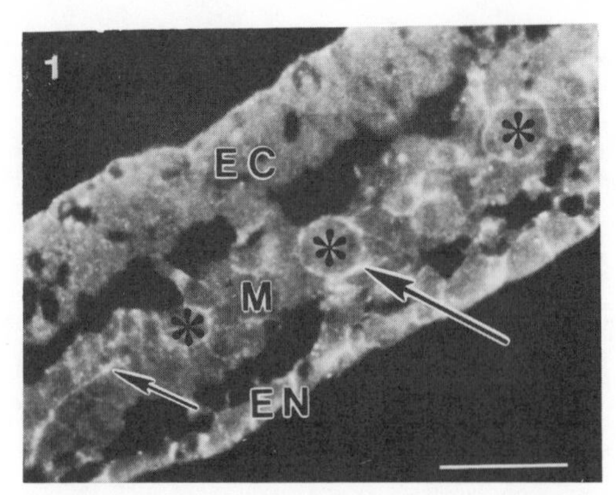


Fig. 1. Relatively even distribution of N-cadherin is seen at cell surfaces throughout the mesoderm (M) and endoderm (EN) in transverse sections through the heart forming region of an early stage 5 embryo. Asterisks mark cells within the mesoderm that have a very rounded appearance and stain brightly for N-cadherin. Ectoderm (EC) does not express N-cadherin. Small arrow points to almost a line of N-cadherin localization on the surfaces of cells in the central part of the mesoderm. Large arrow indicates the immunostaining for N-cadherin that surrounds a large, round cell in the mesoderm. Bar = 30 μm

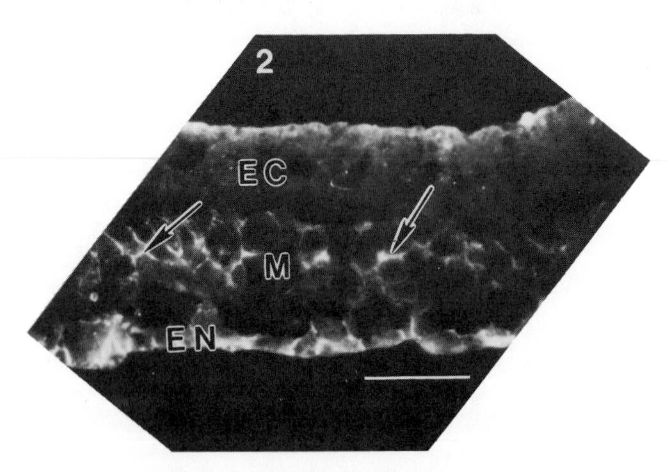


Fig. 2. At stage 6 the immunohistochemical localization of N-cadherin is now in the process of becoming more restricted to the central regions of the mesoderm (M) in the heart-forming region. Patch-like areas of N-cadherin staining (see large arrow) are beginning to appear. Endoderm (EN) continues to express N-cadherin uniformly. EC ectoderm. Bar=30 μm.

At stages 6-7, the N-cadherin immunostaining becomes more restricted in its localization to the central part of the mesoderm (Fig. 2). Often, the large round, brightly staining cells are associated with the bright N-cadherin patches seen in the mesoderm. Fig. 2 depicts a section where the restriction in localization is taking place and patches of staining are becoming apparent.

At stage 7 small foci of cavities begin to form within the mesoderm in the central regions. (These emerging cavities of the pericardial coelom can be seen in Fig. 5. Cavities are marked by asterisks). In a confocal study of these stages (submitted for publication) it appears that the cavities begin to form first in the regions where N-cadherin is localized in the central part of the mesoderm. The cavities appear to form adjacent to the relatively, brightly staining, round cells or within areas of high intensity of staining for N-cadherin. Computer-assisted image processing of sections with patch-like, bright immunostaining regions for N-cadherin allows measurement of the distance between the brightly fluorescing regions along a line drawn through the fluorescing areas. In this manner the distance between fluorescing regions is calculated to be approximately 35-48 microns. This correlates well with distances observed for the emerging foci of cavities.

The enlargement of the cavities to form the bilateral amniocardiac vesicles results in the precardiac cells sorting out to form the splanchnic mesoderm near the endoderm. Cells now situated on the other side of the emerging coelom form the somatic mesoderm. During this process the precardiac cells continue to express N-cadherin on their apical surfaces lining the coelom. The somatic cells initially express N-cadherin as well, but subsequently N-cadherin expression is down-regulated. As the smaller foci of cavities enlarge, they begin to coalesce with neighboring foci. Eventually by stage 8, the pericardial coelom is formed with N-cadherin evident on the apical surfaces of the precardiac cells, particularly at apical cell junctions (Fig. 3). Concomitant with the emerging apical localization of N-cadherin the cells take on an elongated appearance as a mesenchymal-epithelial transformation occurs to form a true epithelium (Manasek, 1976) of precardiac mesoderm.

In a scanning electron microscopy study of a stage 8 embryo (4 somites) fractured at the level of the anterior intestinal portal, the

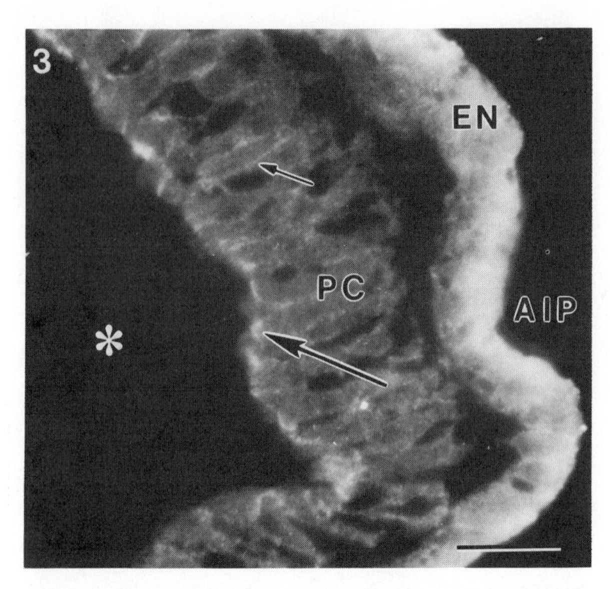


Fig. 3. Immunohistochemical localization of N-cadherin after coelom has formed. N-cadherin immunostains primarily at the apical cell-cell boundaries (large arrow) forming almost a line along the coelomic cavity (white asterisk). Small arrow points to N-cadherin expression that is also present on lateral cell surfaces. Somatic mesoderm and ectoderm are not included in this figure. Little or no expression is associated with the precardiac mesoderm (PC) basal cell surfaces facing the mesoderm-endoderm (EN) interface. AIP anterior intestinal portal. Bar = 20 μm.

closely associated precardiac apical cell surfaces suggest a cobblestone appearance showing smooth rounded surfaces facing the coelom (Fig. 4, see large arrow). The precardiac cell bodies away from the coelom appear not to be as tightly associated as are their apical surfaces. The cells have an elongated appearance with the long axis of the cell perpendicular to the lateral wall of the anterior intestinal portal (see also Linask and Lash, 1986). Finger-like projections on the basal aspect of the precardiac cells interact with the large fibrillar network covering the lateral developing foregut wall.

Fibronectin localizes in the fibrillar meshwork at the endoderm-mesoderm interface (Fig. 5). This localization is also apparent at the lateral developing foregut walls where precardiac cells stop migrating (see Linask and Lash, 1986). After it was observed that RGD-containing

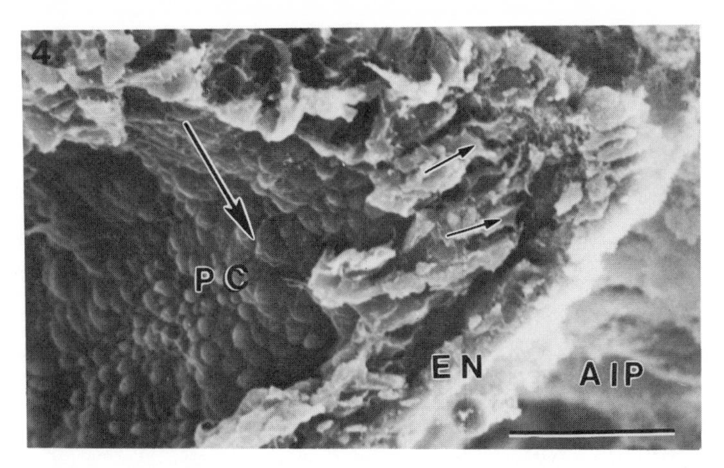


Fig. 4. SEM of a stage 8 embryo fractured through the anterior intestinal portal region (AIP). Stage of embryo and level of fracture within the embryo is similar to that seen in transverse section in Fig. 3. Precardiac mesoderm cells (PC) are seen oriented with their long axis perpendicular to the lateral endodermal wall (EN) of the developing foregut (see small arrow). Large arrow points to cobblestone appearance of apical surfaces of PC cells that are closely associated with neighboring cells as they line the pericardial coelom. Bar = 25 μm.

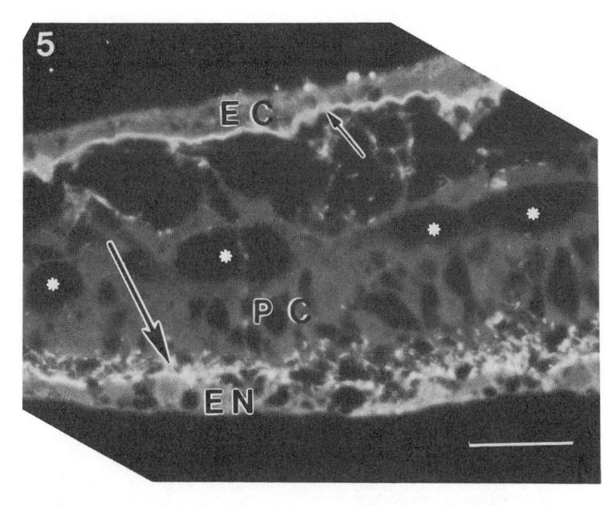


Fig. 5. Immunohistochemical localization of fibronectin of a stage $7^+$ embryo seen in transverse section within heart forming region. Intense fluorescence is associated with basal surface of the precardiac mesoderm cells (PC) at the mesoderm-endoderm (EN) interface (large arrow). Foci of developing coelomic cavities are apparent (asterisks). Fibronectin is also present in the basal lamina (small arrow) underlying the ectoderm (EC). Bar = 30 μm.

synthetic peptides perturb normal precardiac cell migration (Lash and Linask, 1987), it was shown that precardiac cells express integrin on their cell surfaces (Linask and Lash, 1990). *In vivo* an enrichment for the fibronectin receptor is seen on the basal aspect of the cells (submitted for publication). Hence, during epithelialization both cell-cell (apical N-cadherin) and cell-substratum (fibronectin-integrin complex) adhesion systems may regulate cell morphology through their interactions with the cytoskeleton from opposite poles of the cell. Both N-cadherin and integrin are known to interact with intracellular molecules that in turn bind to cytoskeletal components.

Members of the cadherin family apparently target proteins to specific plasma membrane domains. For example, it was suggested that E-cadherin functions in generating an asymmetrical distribution of $Na^+$, $K^+$-ATPAse to baso-lateral cell surfaces of transfected mouse fibroblast cell lines (McNeill et al., 1990). In early heart development N-cadherin may also affect a morphogenetic function in that $Na^+$, $K^+$-ATPase becomes localized primarily to lateral surfaces during the process of epithelialization of the precardiac mesoderm (Fig. 6, see arrows).

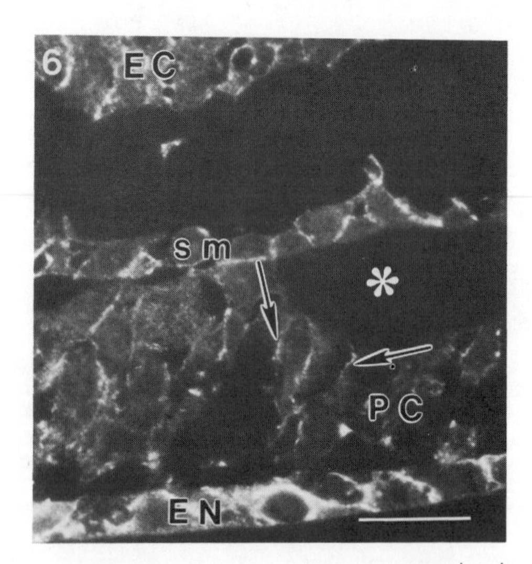


Fig. 6. Immunohistochemical localization of $Na^+$,$K^+$-ATPase during pericardial coelom formation (asterisk). Note localization of the sodium pump in lateral cell membranes of precardiac mesoderm (PC) (see arrows). sm somatic mesoderm; EC ectoderm; EN endoderm. Bar = 20 µm.

Hence, the emergence of an asymmetrical distribution of the sodium pump may be involved in pericardial coelom development, similar to the enzyme's functioning in mouse blastocoele development (Watson et al., 1990; Wiley and Obasaju, 1989).

The regulation of N-cadherin expression is not known. Growth factors have been shown to affect expression of cell adhesion molecules in a number of developing systems. Of significance are observations that transforming growth factor-ß1 (TGF-ß1) stimulates fibronectin and collagen synthesis (Ignotz and Massague, 1986). In the mouse TGF-ß1 mRNA is localized at high levels in the cardiac mesoderm of early embryos (Akhurst et al., 1990). In our recent immunohistochemical studies, the pattern of localization of TGF-ß2 suggests that this growth factor may be involved in developmental processes in the coelom region. TGF-ß2 localizes to cell surfaces of both somatic and splanchnic mesoderm facing the developing pericardial coelom (Fig. 7). The growth factor is not seen in the mesoderm posteriorly to the level of the embryo where the coelom is not yet forming. The significance of this localization and whether there may be effects on expression of cell adhesion molecules in this same region are areas of ongoing investigations.

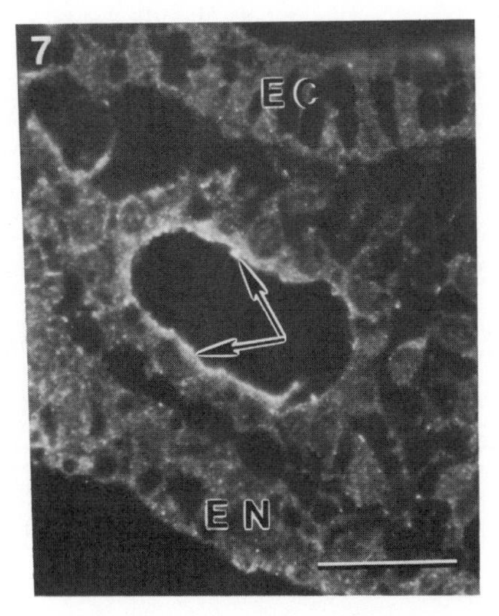


Fig. 7. Immunohistochemical localization of TGF-ß2 in the heart forming region of a stage 7 embryo. Note the immunostaining of TGF-ß2 at cell surfaces (arrows) lining the developing pericardial coelom. EC ectoderm; EN endoderm. Bar = 20 μm.

## SUMMARY

Pericardial cell epithelialization during coelom formation at around 24 hrs of development seems to be a critical step for early stages of heart organogenesis. Indeed it may be the first phenotypic sign of cardiomyocyte cellular differentiation in the early embryonic heart forming region. Soon after this step the first stages of overt cardiac myofibrillogenesis are apparent (Ruzicka and Schwartz, 1988; Tokuyasu and Maher, 1987) leading to sarcomere assembly. It is also after epithelialization that the onset of electrical activity is reported. A spontaneous action potential has been recorded in a six-somite embryonic chick heart, ie.at 28-29 hrs (Hiroto, et al., 1987).

Cadherins appear to be involved in establishing polarized epithelium by their induction of intercellular junctions (McNeill, et al., 1990). Subsequent to cell-cell junction formation, three distinct cell surface domains are set up, i.e. apical, lateral and basal. In developing cardiomyocytes N-cadherin becomes apically localized, $Na^+$, $K^+$- ATPase laterally , and integrin basally. These cell adhesion molecules are known to interact with the cytoskeleton. Indeed it is possible to speculate that these interactions may result in the elongation of the cells and ultimately the folding of the precardial epithelium into endocardial tubes. In addition the polarized distribution of $Na^+$, $K^+$-ATPase results in the vectorial transport of ions and solutes. Such transport is essential to coelom fluid accumulation by osmotic forces. Vectorial transport is also linked to contractility and electrical activity of cardiomyocytes. Experiments currently in progress should elucidate the underlying molecular and biochemical sequences in greater detail as they relate to early heart development.

## ACKNOWLEDGEMENTS

This work was supported by a grant from the American Heart Association, Southeastern Pennsylvania affiliate. The technical assistance of Rasheed A. Rasheed is greatly appreciated.

## REFERENCES

Akhurst, R. J., Lehnert, S., Faissner, A. and Duffie, E. (1990). TGF beta in murine morphogenetic processes: the early embryo and cardiogenesis. *Development*. **108**: 645-656.

Edelman, G. M., Cunningham, B. A. and Thiery, J. P. (1990). In"Morphoregulatory Molecules" (G.M. Edelman, B.A. Cunningham and J.P. Thiery, Eds.), A Neurosciences Institute Publication.

Fambrough, D. M. and Bayne, E. K. (1983). Multiple forms of (Na + K)-ATPase in the chicken. *J. Biol. Chem.* **258**: 3926-3935.

Hamburger, V. and Hamilton, H. L. (1951). A series of normal stages in the development of the chick embryo. *J. Morphol.* **88**: 49-92.

Hatta, K., Nose, A., Nagafuchi, A. and Takeichi, M. (1988). Cloning and expression of cDNA encoding a neural calcium-dependent cell adhesion molecule: Its identity in the cadherin gene family. *The Journal of cell biol.* **106**: 873-881.

Hirakow, R., Komazaki, S. and Hiruma, T. (1987). Early cardiogenesis in the newt embryo. *Scanning Micro.* **1**: 1367-1376.

Hirota, A., Kamino, K., Komuro, H. and Sakai, T. (1987). Mapping of early development of electrical activity in the embryonic chick heart using multiple-site optical recording. *J. Physiol. Lond.* **383**,711-728.

Ignotz, R. A. and Massague, J. (1986). Transforming growth factor-ß stimulates the expression of fibronectin and collagen and their incorporation into the extracellular matrix. *J. Biol. Chem.* **261**: 4337-4345.

Lash, J. W. and Linask, K. K. (1987). Synthetic peptides that mimic the adhesive recognition signal of fibronectin: Differential effects on cell-cell and cell-substratum adhesion in embryonic chick cells. *Dev. Biol.* **123**: 411-420.

Linask, K. K. and Lash, J. W. (1986). Precardiac cell migration:Fibronectin localization at mesoderm-endoderm interface during directional movement. *Dev. Biol.* **114**: 87-101.

Linask, K. K. and Lash, J. W. (1988a). A role for fibronectin in the migration of avian precardiac cells. I. Dose dependent effects of fibronectin antibody. *Dev. Biol.* **129**: 315-323.

Linask, K. K. and Lash, J. W. (1988b). A role for fibronectin in the migration of avian precardiac cells. II. Rotation of the heart-forming region during different stages and its effects. *Dev. Biol.* **129**: 324-329.

Linask, K. K. and Lash, J. W. (1990). Fibronectin and integrin distribution on migrating precardiac mesoderm cells. *Ann. N.Y. Acad. Sci.* **588**: 417-420.

Manasek, F. J. (1976). Heart development: interactions involved in cardiac morphogenesis. In "The Cell Surface in Animal

Embryogenesis and Development" (G. Post and G. L. Nicolson, Ed.), 1, Elsevier/North Holland Biomedical Press,

McNeill, H., Ozawa, M., Kemler, R. and Nelson, W. J. (1990). Novel function of the cell adhesion molecule uvomorulin as an inducer of cell surface polarity. *Cell*. **62**: 309-316.

Moscona, A. (1956). Heterotypic combinations of dissociated embryonic chick cells. *Proc. Soc. exp. Biol. Med.* **92**: 410-416.

Nagafuchi, A. and Takeichi, M. (1989). Transmembrane control of cadherin-mediated cell adhesion: A 94 kDa protein functionally associated with a specific region of the cytoplasmic domain of E-cadherin. *Cell Regulation*. **1**: 37-44.

Ozawa, M., Baribault, H. and Kemler, R. (1989). The cytoplasmic domain of the cell adhesion molecule uvomorulin associates with three independent proteins structurally related in different species. *EMBO J.* **8**: 1711-1717.

Ruzicka, D. L. and Schwartz, R. J. (1988). Sequential activation of alpha-actin genes during avian cardiogenesis: Vascular smooth muscle alpha-actin gene transcripts mark the onset of cardiomyocyte differentiation. *J. Cell Biol.* **107**: 2575-2586.

Sabin, F. R. (1920). Studies on the origin of blood-vessels and of red blood corpuscles as seen in the living blastoderm of chicks during the second day of incubation. *Carnegie Contrib. Embryol.* **9**: 213-262.

Steinberg, M. S. (1963). Reconstruction of tissues by dissociated cells. *Science*. **141**: 401-408.

Takeichi, M. (1977). Functional correlation between cell adhesive properties and some cell surface proteins. *J. Cell Biol.* **75**: 464-474.

Takeichi, M. (1988). The cadherins: Cell-cell adhesion molecules controlling animal morphogenesis. *Development*. **102**: 639-655.

Tokuyasu, K. T. and Maher, P. A. (1987). Immunocytochemical studies of cardiac myofibrillogenesis in early chick embryos. I. Presence of immunofluorescent titin spots in premyofibril stages. *J. Cell Biol.* **105**: 2781-2793.

Townes, P. L. and Holtfreter, J. (1955). Directed movements and selective adhesion of embryonic amphibian cells. *J. Exp. Zool.* **128**: 53-120.

Watson, A. J., Damsky, C. H. and Kidder, G. M. (1990). Differentiation of an epithelium: Factors affecting the polarized distribution of $Na^{+}$, $K^{+}$- ATPase in mouse trophectoderm. *Dev. Biol.* **141**: 104-114.

Wheelock, M. J. and Knudsen, K. A. (1991). N-cadherin-associated proteins in chicken muscle. *Differentiation*. **46**: 35-42.

Wiley, L. M. and Obasaju, F. (1989). Effects of Phlorizin and ouabain on the polarity of mouse 4-cell/16-cell stage blastomere heterokaryons. *Dev. Biol*.. **133**: 375-384.

# ANGIOGENIC CAPACITY OF EARLY AVIAN MESODERM

Jörg Wilting[1], Bodo Christ[1], Milos Grim[1+2] and Pascale Wilms[1]

Departments of Anatomy, Universities of Freiburg[1] and Prague[2]
Germany, Czechoslowakia

## INTRODUCTION

In the yolk sac, vascular development begins with the formation of blood islands - groups of splanchnic mesodermal cells - before blood vessels appear in the body of the embryo. Whereas the inner cells of the blood islands differentiate into blood-forming cells, the outer cells give rise to vascular endothelium (for review see Romanoff, 1960). There are indications that the differentiation of extraembryonic endothelial cells requires controlling influences from the underlying endoderm (Miura and Wilt, 1969, 1970; Flamme, 1989).

Later on endothelial cells can be found in the body of the embryo along a latero-medial differentiation gradient. According to Reagan (1915), Dieterlen-Lièvre (1984), Pardanaud et al. (1987) and Noden (1989), the intra-embryonic vessels develop in situ, independently of the yolk-sac-derived vessels. As shown in Fig 1, the early intra-embryonic vessels are formed in mesoderm closely adjacent to the endoderm.

With regard to the intra-embryonic vascular development, two processes can be distinguished by definition: angiogenesis and vasculogenesis (Risau and Lemmon, 1988; Pardanaud et al., 1989). Angiogenesis is the development of new blood vessels by sprouting from preexisting ones. The term vasculogenesis, on the other hand, is reserved for vessel formation in situ by segregation of angioblastic cells from the mesoderm, as it has been demonstrated by scanning and transmission electron microscopy (Hirakow and Hiruma, 1981, 1983). This mode of vessel formation is similar to that described for the yolk sac. Both involve the proximity of extra-embryonic and early intra-embryonic vessels to the endodermal layer (Fig 1). Pardanaud et al. (1989) have emphasized that mesodermal/endodermal organ rudiments undergo vasculogenesis whereas ectodermal/mesodermal organ rudiments are invaded by angioblasts, i.e. undergo angiogenesis. The latter process includes cell migration, cell orientation, and vessel morphogenesis. A further important aspect of vessel formation is the organ-specific differentiation of endothelial cells.

In this paper we address the following problems of embryonic neovascularization:

1. Which parts of the early avian mesoderm are able to contribute to vessel endothelium?
2. Does the development of endothelial cells depend upon controlling factors released by the underlying endoderm?
3. Are angioblastic cells able to migrate over long distances?
4. How does the organ-specific differentiation of angioblastic cells take place?

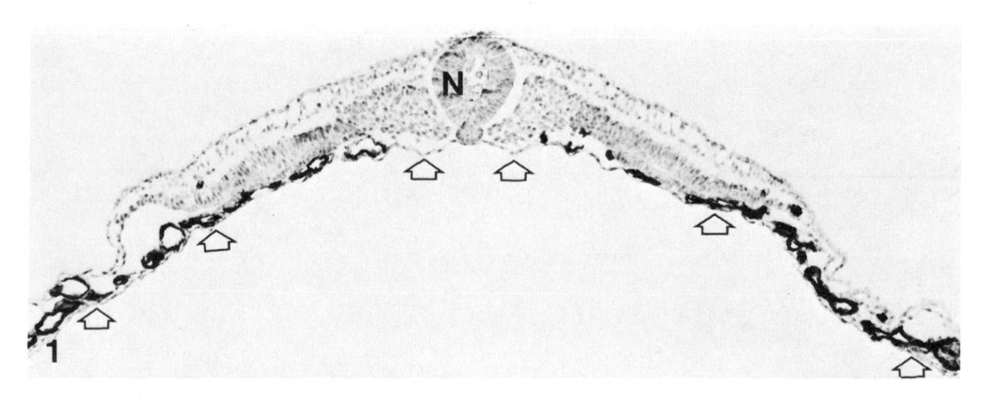


Fig. 1. Transverse section of a 2-day quail embryo stained with MB-1 antibody. Note the black-stained endothelial cells. Arrows: extra- and intraembryonic endoderm. N neural tube. X 150.

## RESULTS

### Angiogenic potency of initially avascular intra-embryonic mesoderm

Hensen's node, parts of the primitive streak or the prechordal mesoderm of quail embryos (HH-stages 3-13) were isolated and grafted into the wing buds of chick embryos (HH-stages 19-25). The grafts consisted of ectoderm, endoderm and non-segmented mesoderm, and were entirely devoid of blood vessels. After different periods of re-incubation, vascular endothelial cells were identified with the monoclonal antibodies MB-1 (Péault et al., 1983) and QH-1 (Pardanaud et al., 1987; Hybridoma Bank, Iowa), which specifically stain quail cells of the haemangiopoietic lineage. Endothelial cells have been found to differentiate from all grafted tissues. They participate in the formation of the endothelial lining of host vessels (Fig 2). These results provide evidence that, even long before overt vessel formation, the intra-embryonic mesoderm contains cells with angiogenic potency.

### Role of endoderm in vascular development

Fragments of blastodiscs from fresh, unincubated quail eggs were isolated and grafted into the wing buds of chick embryos (HH-stages 19-25). The grafts consisted of epiblast and hypoblast, but did not contain mesodermal or endodermal cells. As is shown in Fig 3, endothelial cells differentiated from these grafts. With the exception of leukopoietic cells, other cell types could not be observed. From these results it can be concluded that the first differentiation potency of blastodisc cells is to form endothelial cells.

In order to eliminate the possibility of a partial gastrulation of the grafts inside the host wing, unincubated quail eggs were treated with Cytochalasin B (4 to 6 µg / 1 ml of egg volume). It is known that Cytochalasin B inhibits morphogenetic cell movements and cytokinesis, but not karyokinesis (Tannenbaum, 1978; MacLean-Fletcher and Pollard, 1980). After an incubation period of 24 hours, fragments of the disordered blastodiscs were grafted into the wing buds of chick hosts. The adhesiveness of the cells was definitely reduced. Cytochalasin B - treated blastodiscs showed irregularly arranged cells which did not form separate germ layers. Some cells were multinucleated.

After re-incubation, the grafts consist mostly of mononucleated epithelially or mesenchymally arranged cells which do not participate in the formation of host connective tissue, muscle or cartilage. However, quail cells can be found within the endothelial lining of the vessels (Fig 4). Again, the results show that

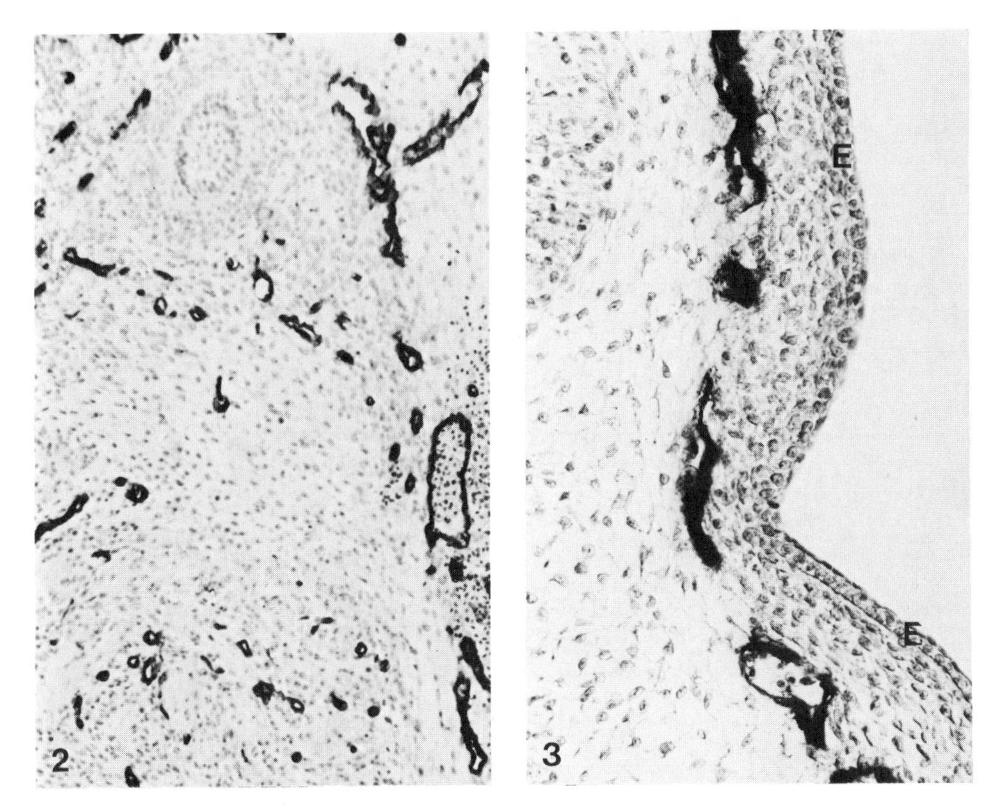


Fig. 2. Chick wing with grafted primitive streak from a quail. Endothelial cells of quail origin are QH-1-stained. X 220.

Fig. 3. Endothelial cells (QH-1-stained) which developed from a fragment of unincubated quail blastodisc inside the chick wing. E ectoderm of the wing. X 360.

endothelial precursors are very early committed to their lineage. Thus, the commitment of endothelial cells does not depend on signals from the endoderm.

Migration of angiogenic cells

In the experiments described above, quail endothelial cells were found in all kinds of blood vessels within the chick wing. They form a mosaic with host endothelial cells. Sections of limb areas proximal to the graft contain quail endothelial cells in the lining of the arteries. The accompanying veins do not possess any quail endothelium (Fig 5). On the other hand, in sections distal to the grafting site, quail endothelial cells are predominantly found in the lining of veins and capillary plexuses, but rarely in arteries (Fig 6). Graft-derived endothelial cells can be found up to several hundred µm from the graft site, depending on the time of re-incubation.

In another set of experiments, vascularized wing bud fragments were isolated from quail embryos of HH-stages 17-22, and grafted into the wing buds of chicks of corresponding stages. After a reincubation period of 2-5 days, the graft-derived quail endothelial cells were found to invade the chick host centripetally, favouring the arteries. They regularly reach the host aorta, where they contribute to the endothelium on the ipsilateral side (Fig 7). This also takes place when other vascularized mesodermal tissues are grafted into the wing

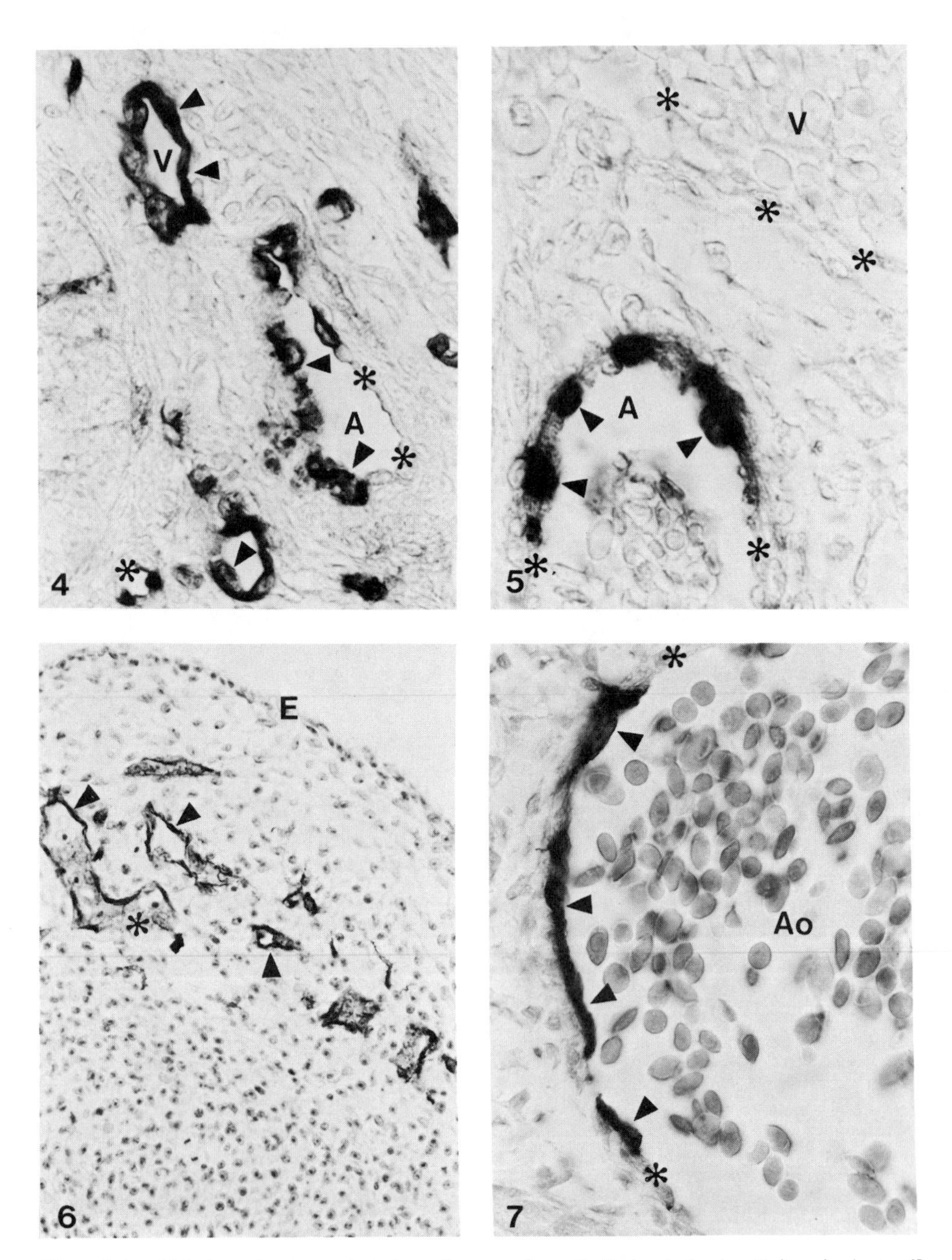


Figs. 4-6. Chick wings containing fragments of Cytochalasin B-treated quail blastodiscs. A artery; V vein. Arrowheads: QH-1-stained endothelial cells of quail origin; asterisks: endothelial cells of chick origin. Fig. 4. Section at the level of the graft. X 720. Fig. 5. Proximal to the graft. X 1200. Fig. 6. Distal to the graft. X 310.

Fig. 7. Aorta (Ao) of a chick host colonized by MB-1-stained quail endothelial cells (arrowheads) from grafted vascularized quail mesenchyme. Asterisks: chicken endothelial cells. X 790.

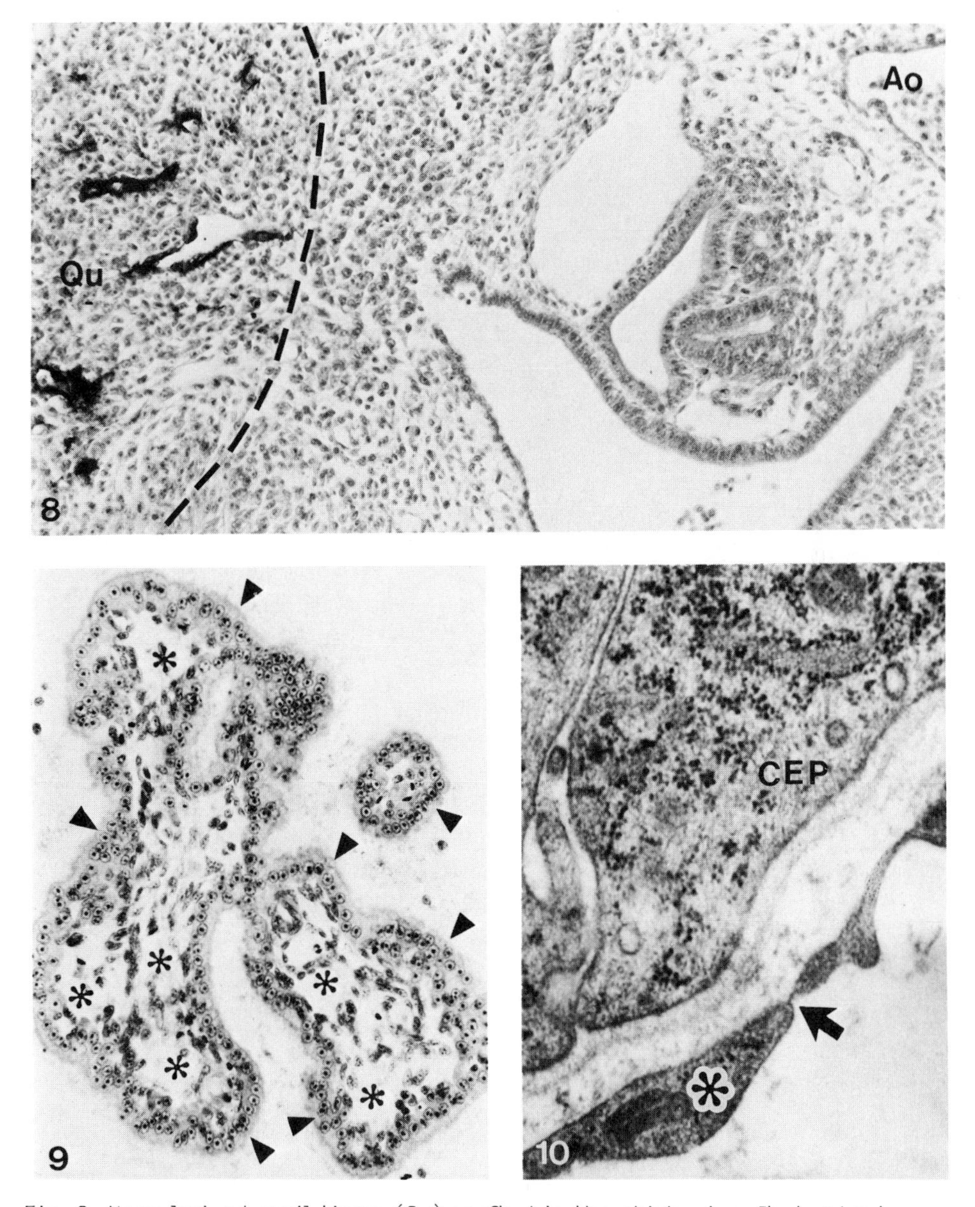


Fig. 8. Vascularized quail tissue (Qu) grafted in the chick wing. Its border is marked by a broken line. GRGDS-peptide was microinjected twice proximal to the graft during re-incubation. Ao aorta. X 220.

Figs. 9-10. Chimeric choroid plexus consisting of choroidal epithelium (CEP) of quail origin (arrowheads) and vascularized chicken connective tissue (asterisks in Fig. 9) which originates from the body wall. Note the fenestrum (arrow) in the endothelium (asterisk in Fig 10). Fig. 9. X 370. Fig. 10. X 54,000.

buds. Microinjections of the synthetic peptide GRGDS, containing the RGD-sequence, into the region proximal to the graft, stops the migration of endothelial cells (Fig 8). These observations indicate directed migratory activity of the endothelial cells. Centripetal migration favours the arterial vessels and occurs within the endothelium.

Organ-specific differentiation of endothelial cells

Differentiation of endothelial cells was studied in experimentally produced combinations of prospective telencephalic choroidal epithelium from the quail, with vessel-containing connective tissue from the body wall of the chick (Fig 9). We have found that the choroidal epithelium induces the differentiation of organ-specific fenestrated capillaries (Fig 10), which are highly permeable to intravenously injected horseradish peroxidase. The induction of capillaries typical of the choroid plexus by the epithelial cells reveals the plasticity of angioblastic cells.

## DISCUSSION

Parts of the intra-embryonic mesoderm which do not contain MB-1 or QH-1 positive precursor cells were found to give rise to endothelial cells. This finding is in line with those of Reagan (1915), Beaupain et al. (1979), Dieterlen-Lièvre (1975) and Wilms et al. (1991), who have shown that the intra-embryonic vascular system develops independently from the pre-established vessels of the yolk sac. Furthermore, we have found that the angiogenic capacity of early avian mesoderm is not restricted to specific areas.

It is a surprising observation that blastodiscs of unincubated quail eggs, as well as quail blastodiscs which have previously been treated with Cytochalasin B, are able to give rise to endothelial cells, when grafted into the wing mesenchyme (Christ et al., 1991). From these results it can be concluded that "instructive" signals from the endoderm are not necessary to switch on the differentiation of endothelial cells. This is in line with the suggestion of Holtzer (1968) that a relatively autonomous expression of the genetic program leads to cell diversification in the absence of specific environmental factors. Although it cannot be excluded that an epithelio-mesenchymal transformation occurs inside the grafts, which might be controlled by factors such as TGFß1 (Sanders and Prasad, 1991), regularly arranged germ layers could not be seen to develop from the grafts. So far as a controlling influence from the endoderm on vascularization is concerned, we obviously have to distinguish between differentiation of endothelial cells and vessel pattern formation. According to Flamme (1989), the extraembryonic vascular pattern is controlled by the endodermal pattern.

It is well known that cell migration is an important event in angiogenesis (for review see Wagner, 1980). Our grafting experiments have provided evidence that migration of endothelial cells or angioblasts even occurs over long distances in vessels already formed (Christ et al., 1990; Wilms et al., 1991). From the periphery, cells migrate within the endothelium of arteries towards the aorta. Centrifugal migration of endothelial cells was predominantly found within the endothelium of veins and capillary plexuses. The significance of this directed migration is not yet understood. It possibly represents an additional way of providing vessels with endothelial cells. The migration process has been found to depend on cell-fibronectin interaction, since it can be stopped by blocking the fibronectin receptors with the RGD sequence (for review see Ruoslahti, 1988).

The development of combinations of choroidal epithelium and vascularized connective tissue of the body wall shows that the differentiation of endothelial cells occurs in an organ-specific manner under the influence of the already determined epithelium (Wilting and Christ, 1989). This indicates the importance of ependymo-glial elements for organ-specific differentiation of vessels in various parts of the brain. This is in line with the results of Stewart and Wiley (1981), as well as those of Janzer and Raff (1987), who have shown that astrocytes induce the formation of the blood-brain barrier.

## CONCLUDING REMARKS

Early avian mesoderm, and even the blastodiscs of unincubated eggs, possess angiogenic capacity. The commitment of endothelial cells to their lineage does not depend on gastrulation. Besides commitment of cells, vascular development includes proliferation, directed migration, tube formation and organ-specific differentiation. The molecular mechanisms of these events remain to be investigated.

Acknowledgements

This work was supported by the Deutsche Forschungsgemeinschaft (Ch 44/9-1) and by Hoechst AG

## REFERENCES

Beaupain, D., Martin, C. and Dieterlen-Lièvre, F. 1979, Are developmental hemoglobin changes related to the origin of stem cells and site of erythropoiesis?, Blood, 53:212-225

Christ, B., Poelmann, R.E., Mentink, M.M.T. and Gittenberger-de Groot, A.C. 1990 Vascular endothelial cells migrate centripetally within embryonic arteries. Anat.Embryol., 181:333-339

Christ, B., Grim, M., Wilting, J., Kirschhofer von, K. and Wachtler, F. 1991, Differentation of endothelial cells in avian embryos does not depend on gastrulation, Acta Histochem., 91:in press

Dieterlen-Lièvre, F. 1975, On the origin of hemopoietic stem cells in the avian embryo. An experimental approach. J.Embryol.Exp.Morphol., 33:607-619

Dieterlen-Lièvre, F. 1984, Emergence of intra-embryonic blood stem cells in avian chimeras by means of monoclonal antibodies, Develop.Comp. Immunol., 3:75-80

Flamme, I. 1989, Is extraembryonic angiogenesis in the chick embryo controlled by the endoderm? A morphological study, Anat.Embryol., 180:259-272

Hirakow, R. and Hiruma, T. 1981, Scanning electron microscopic study on the development of primitive blood vessels in chick embryos at the early somite stage, Anat.Embryol., 163:299-306

Hirakow, R. and Hiruma, T. 1983, TEM-studies on development and canalization of the dorsal aorta in the chick embryo, Anat.Embryol., 166:307-315

Holtzer, H. 1968, Induction of chondrogenesis: A concept in quest of mechanisms. in: "Epithelial-Mesenchymal Interactions," R. Fleischmajer, ed., Williams & Wilkins, Baltimore, Maryland, pp.152-164

Janzer, R.C. and Raff, M.C. 1987, Astrocytes induce blood-brain barrier properties in endothelial cells, Nature, 325:253-257

MacLean-Fletcher, S. and Pollard, T.D. 1980, Mechanism of action of Cytochalasin B on actin, Cell, 20:329-341

Miura, Y. and Wilt, F.H. 1969, Tissue interaction and the formation of the first erythroblasts of the chick embryo, Devl.Biol., 19:201-211

Miura, Y. and Wilt, F.H. 1970, The formation of blood-islands in dissociated-reaggregated chick embryo yolk sac cells, Expl.Cell Res., 59:217-226

Noden, D.M. 1989, Embryonic origins and assembly of blood vessels, Amer.Rev. Respir.Dis., 140:1097-1103

Pardanaud, L., Yassine, F. and Dieterlen-Lièvre, F. 1989, Relationship between vasculogenesis, angiogenesis and haemopoiesis during avian ontogeny, Development, 105:473-485

Pardanaud, L., Altmann, C., Kitos, P., Dieterlen-Lièvre, F. and Buck, C.A. 1987 Vasculogenesis in the early quail blastodisc as studied with a monoclonal antibody recognizing endothelial cells, Development, 100:339-349

Péault, BM., Thiery, JP., Le Douarin, NM. 1983, Surface marker for hemopoietic and endothelial cell lineages in quail that is defined by a monoclonal antibody, Proc.Natl.Acad Sci., 80:2976-2980

Reagan, F.P. 1915, Vascularization phenomena in fragments of embryonic bodies completely isolated from yolk-sac blastoderm, Anat.Rec. 9:329-341

Risau, W. and Lemmon, V. 1988, Changes in the vascular extracellular matrix during embryonic vasculogenesis and angiogenesis, Devl.Biol., 125:441 450

Romanoff, AL. 1960, The hematopoietic, vascular and lymphatic systems. In: "The avian embryo: structural and functional development," AL.Romanoff ed., Macmillan, New York, pp.571-663

Ruoslahti, E., 1988, Fibronectin and its receptors, Ann.Rev.Biochem., 57:375-413

Sanders, E.J. and Prasad, S. 1991, Possible roles for TGFß1 in the gastrulating chick embryo, J.Cell Science, 99:617-626

Stewart, P.A. and Wiley, M.J. 1981, Developing nervous tissue induces formation of blood-brain barrier characteristics in invading endothelial cells: A study using quail-chick transplantation chimera, Devl.Biol., 84:183-192

Tannenbaum, S.F., 1978, Cytochalasins: Biochemical and biological aspects. Amsterdam: North Holland

Wagner, R.C. 1980, Endothelial cell embryology and growth, Adv.Microcirc., 9:45-75, Karger, Basel

Wilms, P., Christ, B., Wilting, J. and Wachtler, F. 1991, Distribution and migration of angiogenic cells from grafted avascular intraembryonic mesoderm, Anat.Embryol., 183:371-377

Wilting, J. and Christ, B. 1989, An experimental and ultrastructural study on the development of the avian choroid plexus, Cell Tissue Res., 255:487-494

# VASCULAR GROWTH IN THE EXTRAEMBRYONIC MESODERM OF AVIAN EMBRYOS *

Ingo Flamme[1], Marius Messerli[2], Werner Risau[3], Monika Jacob[1], and Heinz Jürgen Jacob[1]

[1] Institut für Anatomie, Abteilung für Anatomie und Embryologie, Ruhr-Universität Bochum, D-4630 Bochum, FRG
[2] Institut für Zellbiologie, Eidgenössische Technische Hochschule Zürich, CH-8093 Zürich, Switzerland
[3] Max-Planck-Institut für Psychiatrie Abteilung Neurochemie, D-8033 Martinsried, FRG

## INTRODUCTION

The precursor of each definitive blood vessel is a capillary. During embryogenesis, the pattern of adult blood vessels is preformed by patterning at the capillary level. Under pathological conditions such as inflammation, collateralization in chronic ischemia, wound healing and tumor growth, capillary vessels precede the definitive newly formed blood vessels. Therefore, the mechanisms that are involved in the formation of capillary blood vessels are crucial for the development of a vascular system. Suitable models for studying capillary formation in vivo are rare, because in most newly vascularized tissues observation and interpretation of capillary growth are hampered by a complex spatial structure.

In the avian embryo, however, abundant capillary growth takes place virtually in two dimensions when the extraembryonic mesoderm migrates out over the yolk to form the yolk sac. The vascularized part of the yolk sac - i.e. the part containing extraembryonic mesoderm - is called area vasculosa. The area vasculosa consists of three layers representing the extraembryonic parts of the three germ layers. The ectoderm that provides mechanical strength for the yolk sac, and mesoderm and endoderm together form an extraembryonic organ of complex functions: While the endoderm takes up the nutrients stored within the yolk (Bellairs and New, 1962; Litke and Low, 1975) and produces the main serum proteins of the young embryo (Young et al., 1980; Young and Klein, 1983; Darragh and Zalik, 1988), the overlying vascularized mesoderm brings nutrients into the circulation and is the site of gas exchange and the main source of hematopoiesis in the early embryo. During early development the extraembryonic mesoderm originally being a solid layer is gradually split by the extraembryonic coelom into a somatic and splanchnic mesodermal layer (Kessel and Fabian, 1985). The vascular plexus in which hematopoiesis takes place is confined to the splanchnic mesoderm lining the endoderm (Gonzalez-Crussi, 1971; Kessel and Fabian, 1985; Flamme, 1989) (Fig.1).

* Dedicated to Prof. Dr. med. Klaus V. Hinrichsen on the Occasion of his 65th Birthday.

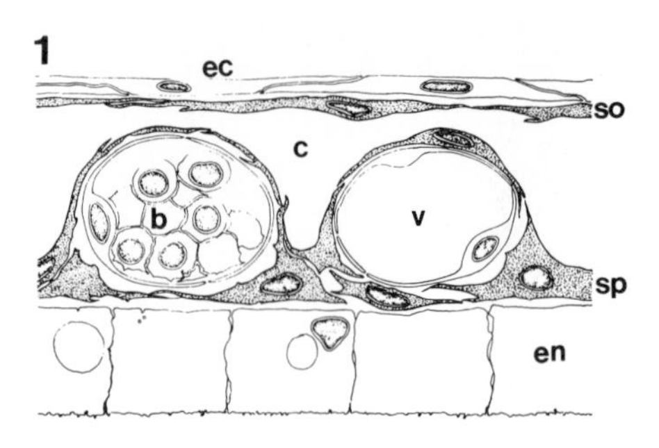


Fig. 1 Schematic drawing of a cross section through the area vasculosa of a chick embryo at the early 2nd day of incubation. ec: ectoderm, sp: splanchnopleura, so: somatopleura, c: extraembryonic coelom, v: vascular lumen, b: blood island, en: endoderm.

Several morphological studies on the emergence of blood vessels in the extraembryonic mesoderm have provided evidence that the primitive capillary plexus is formed at the beginning of the 2nd day by joining of single angioblastic cells to capillary cords. This was demonstrated by SEM studies (Lanot, 1980; Hirakow and Hiruma, 1981) and by immunohistochemical studies using the QH-1 antibody that specifically reacted with quail endothelial cells (Pardanaud et al., 1987). This mechanism is called vasculogenesis, which is defined as the formation of capillary blood vessels by fusion of angioblasts in situ (Risau, 1991). Little is known about the mechanisms of further capillary growth, in particular when the area vasculosa expands very rapidly during the following days.

Two other mechanisms of neovascularization are currently under investigation: In particular in the embryo itself vascular growth by sprouting was observed, which is called angiogenesis (Poole and Coffin, 1989). The third mechanism is termed non-sprouting-vascular growth or intussusceptional vascular growth. It has been described for the ramification of the alveolar capillary plexus and the capillary plexus of the chorioallantoic membrane (Burri and Tarek, 1990; Patan and Burri, 1991): In this mechanism the number of capillaries is augmented when endothelial cells and external cells grow into the capillary lumen and bridge it like a pillar, thus creating a new intervascular space within a pre-existing vessel.

The capacity of endothelial cells to form capillary vessels is an intrinsic property of this cell type. This is the result from in-vitro-studies in which spontaneous reorganization of dissociated capillary endothelial cells into capillary like structures was obtained (Folkman and Haudenschild, 1980; Montesano et al., 1983). The formation of an organ specific vascular pattern, however, is influenced by regulatory factors of the local environment such as extracellular matrix proteins or diffusible protein growth factors (for review see Risau, 1991). In the area vasculosa we found distinct differences in the vascular pattern dependend on the endodermal organization (Flamme, 1989). In the present article we will review our studies on extraembryonic vascular growth and mesoderm-endoderm interactions during vascular pattern formation.

VASCULAR PATTERN FORMATION IN THE AREA VASCULOSA OF BIRD EMBRYOS IS PROBABLY REGULATED BY THE ENDODERM

One of the earliest patterns formed during avian embryogenesis prior to gastrulation is the differentiation of the blastodisc into the area pellucida and opaca. These areas become distinguishable after the innermost layer of cells in the center of the blastodisc has subsided into the yolk to form the floor of the subgerminal cavity (Kochav et al. 1980; Andries et al., 1983). The remaining outer ring of cells contains large amounts of yolk and is called area opaca due to its optical density. The yolk-containing cells separate from the epiblast and form the area opaca endoderm that together with the advancing epiblast spreads out over the yolk. In the area pellucida meanwhile the primary hypoblast is formed; and this hypoblast in turn is substituted by the definitive hypoblast - or area pellucida endoderm - which derives from the primitive streak (Vakaet, 1962; Sanders et al., 1978; Bellairs, 1986). The early mesoderm destined for extraembryonic regions migrates outward from the primitive streak and passes the borderline between area opaca and area pellucida. The endodermata of area opaca and area pellucida differ greatly in their morphology. While the area opaca endoderm is a columnar epithelium containing large amounts of yolk vacuoles, the area pellucida endoderm is a flattened epithelium that contains only few lipid vacuoles (Mobbs and McMillan, 1979) (Fig 2a,b).

While the area opaca endoderm is known to synthesize the main serum proteins of the young embryo, virtually nothing is known about the biochemical properties of the area pellucida endoderm. Contact of the area opaca endoderm with the expanding mesoderm causes the cessation of cellular proliferation within the endoderm (Bennet, 1973; Flamme, 1989) and gives rise to morphological changes as well as to the expression of certain enzymes by the endoderm (Bennet, 1973; Mobbs and McMillan, 1979). Wilt (1964) and Miura and Wilt (1969) found the area opaca endoderm at the primitive streak stage to promote the formation of blood islands and to produce an important stimulus for the differentiation of endothelial cells lining the blood islands. Kessel and Fabian (1987) described a differential regulation of the erythropoiesis in the early blastodisc by the endodermata of area opaca and area pellucida.

The first assembly of blood vessels in the extraembryonic mesoderm results in a uniform lacunary vascular pattern in both regions, area opaca and pellucida. Injections of Indian ink in the vascular system of chick and quail

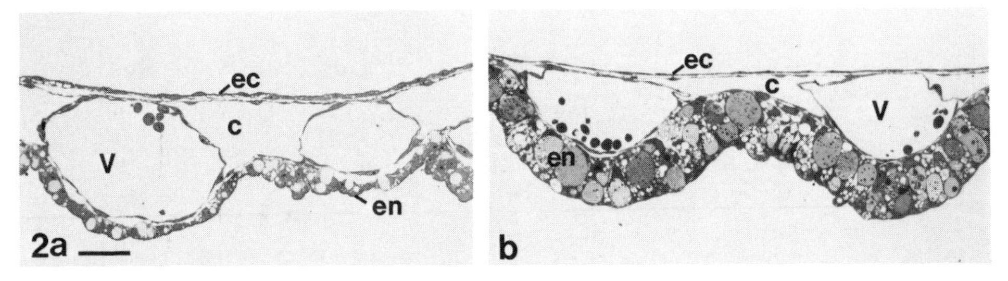


Fig. 2 Cross sections through the area pellucida (a) and the area opaca vasculosa of a two day old chick embryo. The area opaca endoderm consists of high prismatic cells which include large amounts of yolk vacuoles (b), the area pellucida endoderm consists of flat epithelial cells (a). v: vascular lumen, ec: ectoderm with somatopleura, c: extraembryonic coelom, en: endoderm. Scale bar: 50 µm.

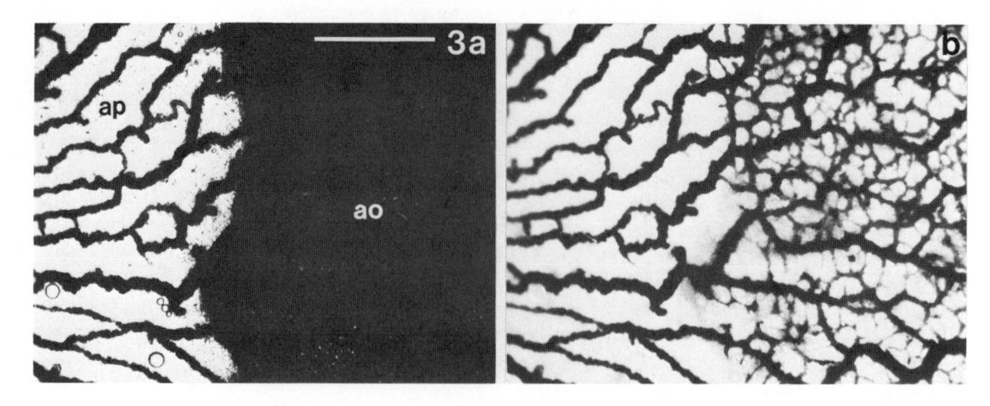


Fig. 3 (a) View of the transitional zone between area pellucida (ap) and the area opaca (ao) under transmitted light in a three day old chick embryo. The vascular system was contrasted by injection of Indian ink. (b) The same region as in (a) under direct light. The borderline between the different vascular patterns coincides with the borderline between the endodermata of area opaca and area pellucida. Scale bar: 0.5 mm.

embryos at the second and third day of incubation revealed that the extraembryonic vascular pattern underwent fundamental changes (Flamme, 1989). Whereas in the area opaca the blood vessels of the primitive plexus exhibited abundant ramification from stage 14 (Hamburger and Hamilton, 1951(HH)) onward, this was not observed in the area pellucida (Fig. 3a,b).

In this region only the diameter of intervascular spaces gradually increases. The borderline between the vascular patterns is distinct and exactly follows the borderline between area opaca and pellucida endoderm. This observation led to the conjecture that the formation of the extraembryonic vascular pattern is influenced by the endoderm. Two possible modes of this influence are conceivable: (1) The area opaca endoderm produces a strong activity that induces ramification of the vascular plexus or, (2) the area pellucida endoderm exerts inhibitory influence on the overlying mesodermal vasculature. These putative morphogenetic signals might be transmitted by soluble factors and/or by direct cell-cell contact. Mato et al. (1964) found a close association of blood islands with the area opaca endoderm when examining young blastodiscs under TEM. However, we were not able to find any direct cell-cell contact of endoderm to mesoderm in thin sections of two and three day old area vasculosa. The mesodermal layer was always found to be separated from the endoderm by a basal lamina.

To test our hypothesis of differential endodermal activities, we used two approaches:

(1) experiments in ovo in which endoderm of the area opaca was transplanted into the coelomic cavity of 3 day old chick embryos, and onto the chorioallantoic membrane. In addition, an attempt was made to substitute area opaca and area pellucida endoderm for each other in embryos cultured in vitro. In chorioallantoic assays a nonspecific fibrous reaction was observed three days after exposure to the endoderm. The other experiments have not yielded any result as yet.

(2) experiments in vitro in which the area opaca endoderm was either co-

cultured with varying cell types or in which supernatants of the endoderm were tested on endothelial cells and 3T3 fibroblasts. For in vitro culture, the area opaca endoderm was removed from the area vitellina - i.e. that part of the area opaca which lies outside the area vasculosa containing the mesoderm. Only cells taken from this region are able to attach in vitro (Young and Klein, 1983). Endoderm was peeled off the overlying ectoderm and small pieces were transferred to the culture dish containing DMEM supplemented with 10 % newborn calf serum. A short time after explantation the endoderm spread over the culture dish. Figure 4 shows typical cells of area opaca endoderm grown in vitro for two days. The synthesis of secreted proteins in culture was examined. Metabolic labelling was performed with $^{35}$S-Methionin (50 µCi/ml medium) for a period of 24 hrs. Proteins released into the culture supernatants were separated by polyacrylamide gel electrophoresis and gels were processed for fluorography on X-ray films. The fluorograms showed distinct labelled proteins. The proteins corresponded to the main serum proteins of the young avian embryo which were previously shown to be produced by the area opaca endoderm (Young et al., 1980; Young and Klein, 1983; Darragh and Zalik, 1988) (Fig. 5). Thus, we concluded that endodermal cells of the area opaca exhibited quite similar metabolical properties in vitro as in ovo.

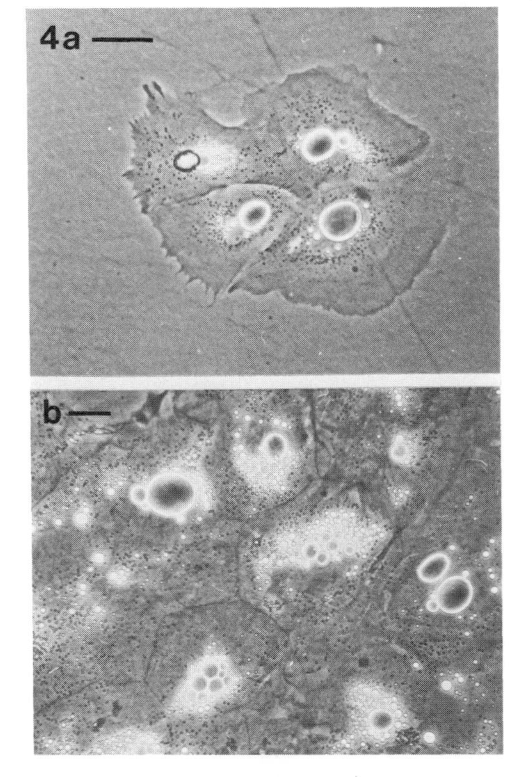


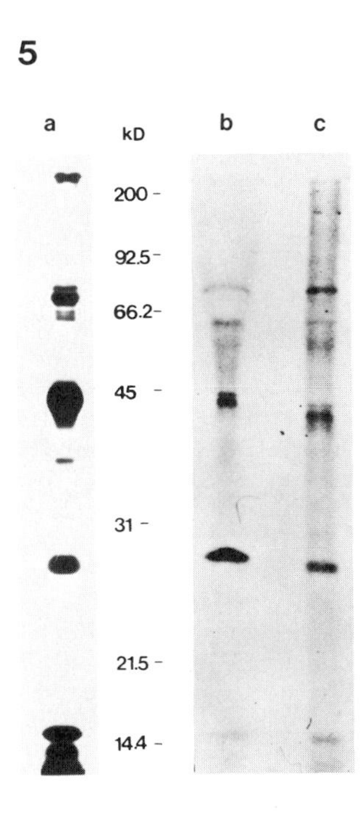


Fig.4 Area opaca endoderm maintained for two days in culture. (a) A group of cells. Scale bar: 100 µm. (b) Monolayer consisting of polygonal epithelial cells. Scale bar: 20 µm.

Fig.5 SDS-polyacrylamide gel electrophoresis of embryonic serum (day 8) after silver staining (lane A) and fluorogram of electrophoretically separated supernatants from endodermal cell cultures after metabolical labeling with $^{35}$S-Methionin for 24 hrs. Cells were grown in vitro for 2 days (lane B) and 4 days (lane C).( m.w. markers: myosin, phosphorylase B, bovine serum albumin, ovalbumin, carbonic anhydrase, soybean trypsin inhibitor, lysozyme)

In serum free supernatants of the endodermal cells tested in the 3T3-cell and endothelial cell proliferation assay no mitogenic activity could be detected. During 7 days of coculture with adult capillary endothelial cells from the bovine adrenal gland the area opaca endoderm did not exert any morphogenetic effect on these cells. These experiments were performed as transfilter experiments using Cellagen™ membranes (ICN Biochemicals, Cleveland, OH, USA) and as mixed cultures of both cell types. Cocultures of area opaca endoderm with dissociated area pellucida from primitive streak stage quail embryos had no promoting influence on the appearance of early endothelial cells as it was demonstrated by use of the QH-1 antibody (Pardanaud et al., 1987). Altogether, we conclude that the area opaca endoderm exerted no mitogenic effect on endothelial cells and fibroblasts and, furthermore, that angiogenic or angiomorphogenetic activities were not yet found to be associated with these cells. Therefore, a possible antiangiogenic activity of the area pellucida endoderm may be involved in the formation of the extraembryonic vascular pattern. This possibility is currently under investigation.

## EXTRAEMBRYONIC BLOOD VESSELS GROW BY SPROUTING AND NON-SPROUTING VASCULAR GROWTH

New blood vessels of the area opaca may form by association of single migrating angioblasts or by the formation of vascular sprouts. The identification of vascular sprouts by the technique of Indian ink injection depends on patent lumina and is complicated by the fact that obstructed lumina would give the impression of sprout formation. A marker specific for endothelial cells already integrated in blood vessels as well as for migrating angioblasts would be helpful to distinguish between the different conditions. Such a marker is the QH-1 antigen that is expressed by angioblasts and endothelium of the quail, but also by germ cells and myelomonocytic cells (Pardanaud et al., 1987). Using the QH-1 antibody the first assembly of blood vessels in the area opaca was demonstrated to result from vasculogenesis, i.e. the formation of the primitive vascular plexus of the area vasculosa by coalescence of endothelial cells of neighbouring blood islands with each other and with single migrating angioblasts (Pardanaud et al. 1987). Therefore, the resulting vascular pattern is a lacunary plexus the junctions of which are determined by the position of the blood islands. These observations were made in embryos younger than stage 14 (HH). At this stage, however - as described above - abundant ramification of the primitive vascular plexus of the area opaca commences while the area pellucida retains virtually the primitive vascular arrangement. Whereas in the area opaca numerous spike-like lumen sprouts were found to be visible after injection with Indian ink (Fig. 3a,b), in the area pellucida these structures were only rarely seen.

To elucidate the mechanisms of the rapid vascular growth of the area opaca, we subjected whole mounts of the area vasculosa of quail embryos of stages 14 -17 (HH) to immunohistochemistry using the QH-1 antibody according to the method of Poole and Coffin (1989). Briefly, after explantation from the yolk, the endoderm was removed from the area opaca to allow the access of the antibody to the mesoderm. Specimens were fixed in 4% formaldehyde for at least 12 hrs., permeabilized in 100% methanol and incubated in 2% bovine serum albumin to inhibit unspecific binding. QH-1 antibody was applied at a dilution of 1:200 for at least 12 hrs at 4 °C. Bound antibody was detected by a second rhodamine-conjugated goat-anti-mouse IgG. Specimens were evaluated under a fluorescence microscope or under a confocal laser scanning microscope (CLSM) (Zeiss Axiophot fitted with a Biorad scanner). Optical sections were made at 0.3 µm intervals and processed on a Silicon Graphics image processing system.

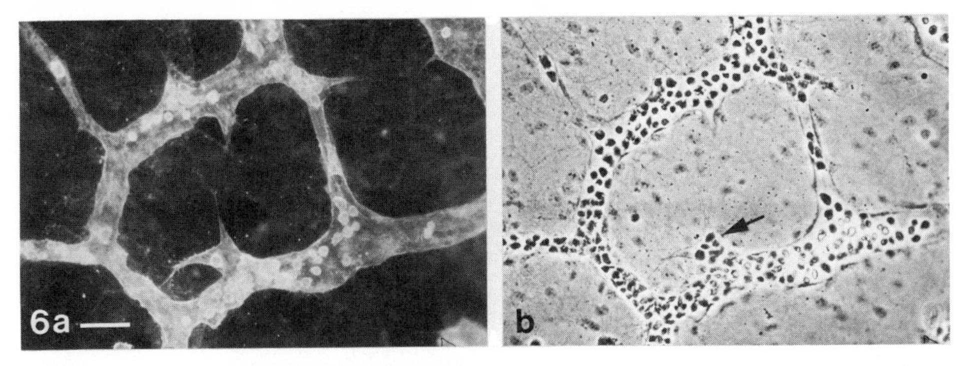


Fig. 6 (a) Indirect QH-1 immunofluorescence of a whole mount preparation from the area opaca vasculosa of a quail embryo at the 2nd day of incubation. Note the vascular sprouts. (b) Phase contrast view of (a), arrow: The lumen of the sprout is already colonized by blood cells. Scale bar: 50 µm.

The capillary network of the area opaca revealed positive labelling with the QH-1 antibody. Within the capillary loops vascular sprouts were found to extend from the pre-existing capillaries like branches of a tree (Fig. 6). These sprouts possess a broad basis which is tapered at their tip extending into a fine filopodium. In many cases also a swelling at the tip can be observed. Vascular sprouts are very numerous in the area opaca, but infrequent in the area pellucida. An important question was whether the sprouts seen in the QH-1 immunofluorescence corresponded to the lumen sprouts seen after ink injection. In fact, in some sprouts blood cells can be seen, strongly indicating that these sprouts have a lumen (Fig. 6). Furthermore, we labelled ink injected specimens with the QH-1 antibody and found that in many sprouts lumina were already present, but most lumen sprouts contrasted by ink were seen in vascular cords (not shown). In addition to QH-1 labelled sprouts, free QH-1 positive cells were found also scattered over the intervascular spaces exhibiting a morphology greatly different from that of sprout forming endothelial cells. The free cells have a polycyclic shape due to a rim of rounded lamellopodia. In most cases they are not in contact with the capillaries and vascular sprouts nor in contact with each other. With confocal microscopy single cells of this type were found in process of migration through the walls of blood vessels. Optical sections through these cells showed the QH-1 antigen concentrated at or near the cell surface in contrast to endothelial cells in which positive staining was distributed over the whole cytoplasm (Fig. 7a). Finally, by scanning electron microscopy the free QH-1 positive cells were identified as cells with the typical shape of macrophages (Fig. 7b).

We were interested in the sequence of events that occur from the formation of a sprout to the complete capillary vessel. Using high resolution optics and CLSM, we found that one or more very fine filopodia were extended from the tips of the capillary sprouts. These filopodia were two or three times as long as the sprout itself. Fine filopodia of the same type were also seen at the border of the pre-existing blood vessels (Fig. 8a). Such filopodia may be regarded as the precursors of vascular sprouts, that probably form when an endothelial cell of a capillary wall extends its cytoplasm into the direction of a filopodium. In contrast to, for instance,the situation during tumor angiogenesis (Ausprunk and Folkman, 1977), endothelial cells of the area vasculosa never seem to

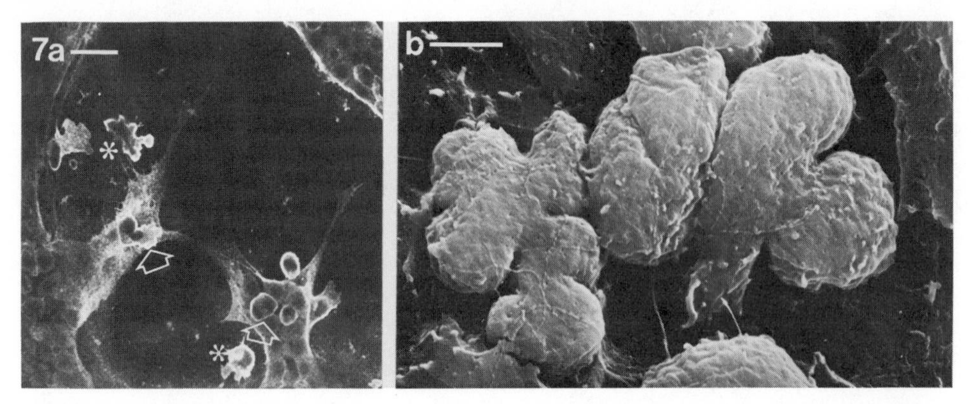


Fig. 7(a) Horizontal confocal microscopic section of 0.3 µm through a vascular sprout in the quail area opaca vasculosa after staining with the QH-1 antibody. Next to the sprouts polycyclic-shaped, free QH-1 positive cells are found (asterisks). Similar cells are seen within the lumen of the sprouts (open arrow). QH-1 fluorescence is almost restricted to the surface of these free cells in contrast to the endothelial cells which exhibit fluorescence distributed over the whole cytoplasm. Scale bar: 20 µm. (b) Scanning electron microscopic view of a group of cells with the shape of the QH-1 positive cells as seen in (a). These cells are probably phagocytic cells. Scale bar: 5 µm.

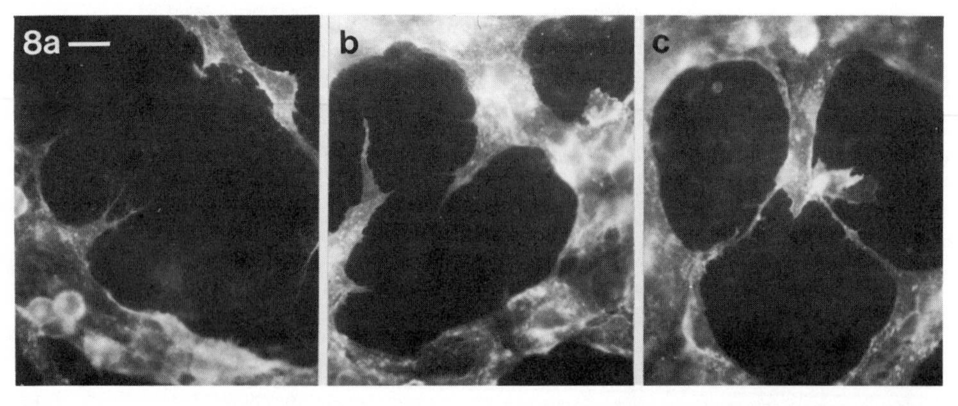


Fig. 8(a-c) Putative sequence of the de novo formation of capillary blood vessel in the area opaca of the quail embryo. (a) Filopodia are extended from opposite vascular sprouts. (b) Filopodia contact each other. (c) A stable endothelial bridge over an intervascular space is formed. Scale bar: 20 µm.

leave the association with the capillary endothelial lining and migrate as single cells.The next step that probably takes place during vascular formation in the area opaca is that filopodia of opposite vascular sprouts contact each other and form a stable connection between the sprouts to bridge an intervascular space (Fig. 8b). Such connections may transform into a solid vascular strand which consecutively forms a lumen (Fig. 8c).

Capillary sprouting is probably due to migration of cells at the tip of the sprouts since these cells have the characteristic shape of migrating cells. Whether cellular proliferation is also involved in sprouting was examined by studying the cytokinetics with the Bromodeoxyuridine-anti-Bromodeoxyuridine-technique.

Bromodeoxyuridine (BrdU) is a thymidine analogue which is incorporated into the DNA during the S-phase. Incorporated BrdU can be detected with the aid of a monoclonal anti-BrdU-antibody; S-phase nuclei are selectively stained (Gratzner 1982). BrdU was applied to explanted quail blasodiscs at concentration of 400 µM for 30 min. After fixation specimens were stained for QH-1 essentially as described above and for BrdU using a monoclonal anti-BrdU-antibody that was labelled with Fluoresceine (kindly provided by Dr. Henkes, Boehringer, Mannheim, FRG).

BrdU-positive nuclei were found randomly distributed over the area vasculosa. No accumulation of BrdU-positive nuclei, i.e. proliferative activity, could be detected in the capillary sprouts. Sprouts as well as the other elements of the vascular tree bore positive nuclei at identical frequency (Fig. 9).

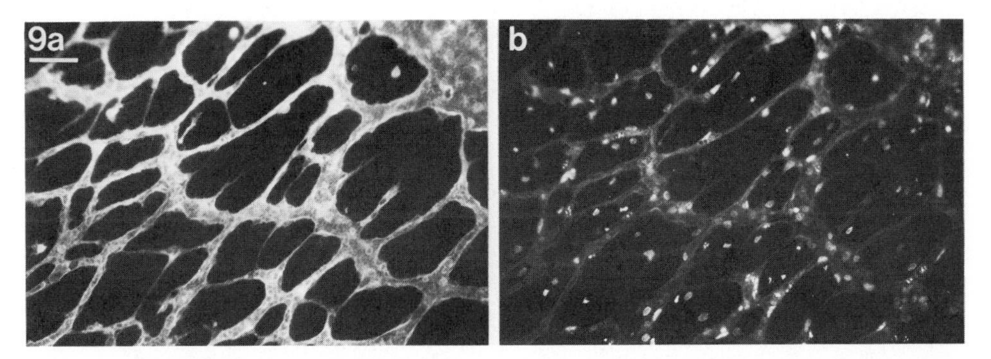


Fig. 9 Cytokinetics of blood vessels in the area opaca as studied with the BrdU-anti-BrdU method. Whole mounts treated with 400 µM BrdU were double stained with QH-1 antibody (a) and primary FITC-conjugated anti-BrdU antibody (b). Scale bar: 50 µm.

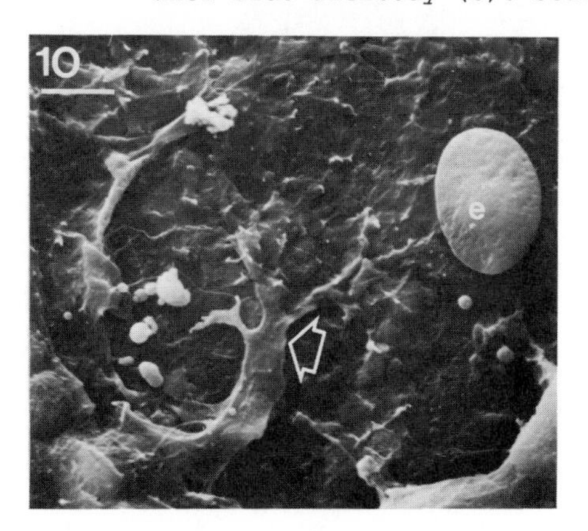


Fig. 10 SEM view of a vascular sprout (open arrow) in the area opaca after removal of the endoderm. The sprout lies embedded within the splanchnic mesodermal layer. e: erythrocyte outside the vascular bed. Scale bar: 10 µm.

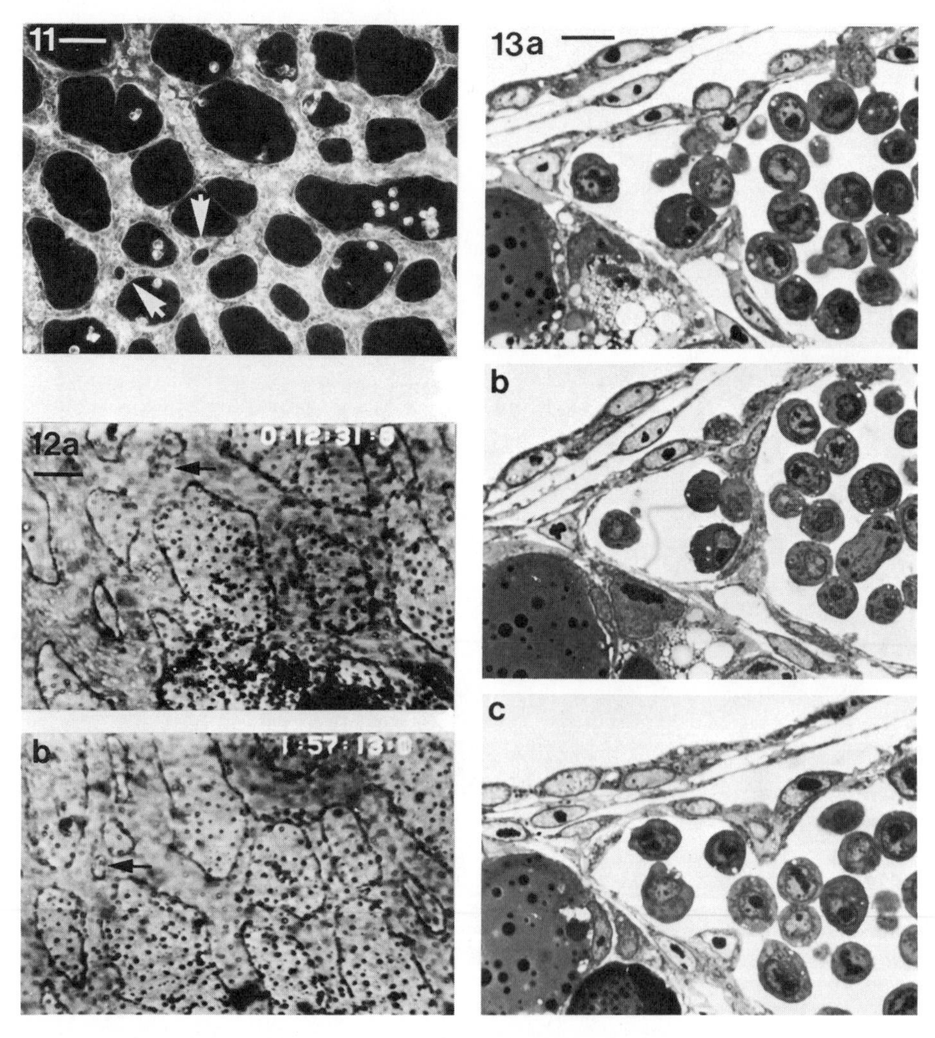


Fig. 11 QH-1 immunofluorescence view of the vascular network of the area opaca of a two day old quail embryo. Note the small intravascular mesenchymal islands (arrows). Scale bar: 50 µm.

Fig. 12 Appearance of a small intravascular mesenchymal island seen under video monitoring of a region of a chicken area pellucida (arrow). Shifting of the observed area is due to the expansion of the blastoderm. Scale bar: 100 µm.

Fig. 13 (a-c) Serial semithin transverse sections through the area opaca of a three day old chick embryo showing an intravascular pillar that corresponds with intravascular islands as shown in figures 11 and 12. Scale bar: 10 µm.

Thus, it can be concluded that sprouting involves endothelial cell migration and proliferation. Remarkably, endothelial cells at the tip of the sprout migrate in the context with other endothelial cells. Furthermore, these cells migrate within a thin mesodermal layer, i.e. the splanchnic mesoderm (Fig. 1 and 10), therefore sprouting is to be regarded an invasive process.

Moreover, a conspicuous feature of the vascular pattern after staining with QH-1 was found in the form of very small intervascular spaces. In most cases these spaces virtually resembled mesenchymal islands within a blood vessel (Fig. 11). The development of such intervascular spaces could hardly be explained as the result of vascular sprouting, since vascular sprouts were always seen to traverse wide intervascular spaces. Identical intravascular islands were also observed in the area pellucida. As this region is transparent we could record the development of such islands by in vivo-observations under transmitted light using video equipment (Fig. 12): Within blood vessels the gradual emergence of islands was observed which separated the blood stream. The diameter of these islands increased rapidly. In serially sectioned specimens the intravascular islands were identified as intravascular pillars consisting of an endothelial cell coat and a central mesenchymal core (Fig. 13a-c). Further increase of the diameter of such islands leads to real intervascular spaces. In this way one blood vessel is divided into two and the perfused area is enlarged. This type of non-sprouting vascular growth was recently termed intusscusceptional vascular growth and described as crucial pathway in the development of the capillary plexus of the lung and the chorioallantoic membrane (Burri and Tarek, 1990; Patan and Burri, 1991). Whether this type of growth also plays a relevant role in the vascular growth of the area vasculosa remains to be determined.

## CONCLUDING REMARKS

In summary, our results strongly suggest that the development of the vascular pattern in the area vasculosa is spatially regulated by the organization of the endoderm: In the mesoderm overlying the area opaca endoderm, blood vessels abundantly ramify from the 2nd day of development onward, whereas in the area pellucida ramification does not occur. We hypothesized that the endodermata of the area opaca and the area pellucida exert angiogenic or antiangiogenic activity, respectively, either independently, or in combination. So far, however, the area opaca endoderm was not found to produce any mitogenic or angiogenic activity. Therefore, the examination of the area pellucida endoderm for antiangiogenic activity may provide insights into the regulatory mechanisms. Our results on tissue cultures from early embryonic material are in contrast to the findings of Wilt (1964) and Miura and Wilt (1969) who showed that the area opaca endoderm promotes endothelial cell differentiation at early stages. However, these influences on endothelial cell differentiation may be different from those regulating pattern formation. Further investigations concerning early vascular development are required.

The mechanism by which the dense vascular network of the area opaca is formed when ramification has commenced was elucidated by examination of whole mount preparations of area vasculosa stained with the QH-1 antibody: Blood vessels grow by sprouting and - probably to a minor extent - by non sprouting (intussuceptional) vascular growth. Vascular sprouts can be found to possess already a lumen. From our studies on cytokinetics it is evident that cellular proliferation plays a subordinate role in sprouting. Sprouting is most likely the result of active cellular migration. Fine filopodia at the tips of opposite sprouts are the first that contact to bridge an intervascular space. Stable connections between such filopodia may be the precursors of solid endothelial strands which in turn form a lumen. Whether the lumen is intracellular as described for angiogenesis in vitro (Folkman and Haudenschild, 1980) or extracellular - that is by joining of several endothelial cells as described for the embryonic aorta (Hirakow and Hiruma, 1983) - has not been investigated yet. Further approaches to this problem may be expected from improved applications of confocal laser scanning microscopy. Moreover, the question of what subpopulation of endothelial cells constitutes the tips of vascular sprouts and what specific signals give rise to their migration, are major challenges of endothelial cell biology.

ACKNOWLEDGEMENTS

We thank Dr. Henkes (Boehringer, Mannheim, FRG) for kindly providing the anti-BrdU antibody. The QH-1 antibody was obtained from the Developmental studies Hybridoma Bank maintained by the Department of Pharmacology and Molecular sciences, Johns Hopkins University School of medicine, Baltimore, MD, and the Department of Biology, University of Iowa, Iowa City, IA, under contract N01-HD-6-2915 from the NICHD. The study was supported by a grant (Ja 556/1-1) from the Deutsche Forschungsgemeinschaft. We thank Ms. E. Koniezny, Mrs. M. Köhn and Mr. H. Hake for excellent technical assistance.

REFERENCES

Ausprunk, D. H., and Folkman, J., 1977, Migration and Proliferation of endothelial cells in preformed and newly formed blood vessels during tumor angiogenesis. Microvasc. Res., 14:53-65

Andries, L., Vakaet, L., Vanroelen, C., 1983, The subgerminal yolk surface and its relationship with the inner germ wall edge of the stages X to XIV chick and quail embryo. Anat. Embryol., 166:453-462

Bellairs, R., 1986, The primitive streak. Anat. Embryol., 174:1-14

Bellairs, R., and New D. A. T., 1962, Phagocytosis in the chick blastoderm. Exp. Cell. Res., 26:275-279

Bennett, N., 1973, Study of yolk-sac endoderm organogenesis in the chick using a specific enzyme (cysteine lyase) as a marker of cell differentiation. J. Embryol. exp. Morph., 29:159-174

Burri, P. H., and Tarek, M. R., 1990, A novel mechanism of capillary growth in the rat pulmonary microcirculation. Anat. Rec., 228:35-45

Darragh, E. A., and Zalik, S. E., 1988, Synthesis of serum proteins by cultured aggregates from endodermal cells of the area opaca of the primitive streak chick embryo. Roux' Arch. Dev. Biol., 197:92-100

Flamme, I., 1989, Is extraembryonic angiogenesis in the chick embryo controlled by the endoderm? A morphological study. Anat. Embryol., 180: 259-272

Folkman, J., and Haudenschild, C., 1980, Angiogenesis in vitro. Nature, 288:551-556

Gonzalez-Crussi, F., 1971, Vasculogenesis in the chick embryo. An ultrastructural study. Am. J. Anat., 130:441-460

Gratzner, H. G., 1982, Monoclonal antibody to 5-Bromo and 5-Jodo-deoxyuridine: A new reagent for detection of DNA replication. Science, 218:474-475

Hamburger, V., and Hamilton, H. L., 1951, A series of normal stages in the development of the chick embryo. J. Morphol. 88:49-92

Hirakow, R., and Hiruma, T., 1981, Scanning electron microscopic study on the development of primitive blood vessels in chick embryos at early somite-stage. Anat. Embryol., 163:299-306

Hirakow, R., and Hiruma, T., 1983, TEM-Studies on development and canalization of the dorsal aorta in the chick embryo. Anat. Embryol., 166:307-315

Kessel, J., and Fabian, B., 1985, Graded morphogenetic patterns during the development of the extraembryonic blood system and coelom of the chick blastoderm: A scanning electron microscope and light microscope study. Am. J. Anat., 173:99-112

Kessel, J., and Fabian, B., 1987, Inhibitory and stimulatory influences on mesodermal erythropoiesis in the early chick blastoderm. Development, 101:45-49

Kochav, S., Ginsburg, M., Eyal-Giladi, H., 1980, From cleavage to primitive streak formation: a complementary normal table and a new look at the first stages of the development of the chick. II. Microscopic anatomy and cell population dynamics. Dev. Biol., 79:296-308

Lanot, R., 1980, Formation of the early vascular network in chick embryo: Microscopical aspects. Arch. Biol. (Bruxelles), 91:423-438

Litke, L. L., and Low, F. N., 1975, Scanning electron microscopy of yolk absorption in early chick embryos. Am. J. Anat., 142:527-531

Mato, M., Aikawa, E., Kishi, K., 1964, Some observations on interstice between mesoderm and endoderm in the area vasculosa of chick blastoderm. Exp. Cell. Res., 35:426-428

Miura, Y., and Wilt, F. H., 1969, Tissue interaction and the formation of the first erythroblasts of the chick embryo. Dev. Biol., 19:201-211

Mobbs, I. G., and McMillan, D. B., 1979, Structure of the endodermal epithelium of the chick yolk sac during early stages of development. Am. J. Anat., 155:287-310

Montesano, R., Orci, L., Vassalli, P., 1983 In vitro rapid organization of endothelial cells into capillary-like networks is promoted by collagen matrices. J.Cell Biol., 97:1648-1652

Pardanaud, L., Altmann, C., Kitos, P., Dieterlen-Lievre, F., Buck, C. A., 1987, Vasculogenesis in the early quail blastodisc as studied with a monoclonal antibody recognizing endothelial cells. Development, 100:339-349

Patan, S., and Burri, P. H., 1991, Some microvascular systems may grow by intussusception and not by capillary sprouting. Verh. Anat. Ges., 86:228

Poole, T. J., and Coffin, J. D., 1989, Vasculogenesis and angiogenesis: two distinct morphogenetic mechanisms establish embryonic vascular pattern. J. Exp. Zool., 251: 224-231.

Risau, W., 1991, Vasculogenesis, angiogenesis and endothelial cell differentiation during embryonic development. In: The development of the vascular system. Issues biomed., 14:58-68, Feinberg, E.M., Sherer, G.K., Auerbach, R. (eds.)

Sanders, E. J., Bellairs, R., Portch, P. A., 1978, In vivo and in vitro studies on the hypoblast and definitive endoblast of avian embryos. J. Embryol. exp. Morph., 46:187-205

Vakaet, L., 1962, Some new data concerning the formation of the definitive endoblast in the chick embryo. J. Embryol. exp. Morph., 10:38-57

Wilt, F. H., 1964, Erythropoiesis in the chick embryo: the role of endoderm. Science, 147:1588-1590

Young, M. F., Minghetti, P. P., Klein, N. W., 1980, Yolk sac endoderm: exclusive site of serum protein synthesis in the early chick embryo. Dev. Biol., 75:239-245

Young, M. F., and Klein, N. W., 1983, Synthesis of serum proteins by cultures of chick embryo yolk sac endodermal cells. Dev. Biol., 100:50-58

INDEX